# Advances in Biolinguistics

Biolinguistics is a highly interdisciplinary field that seeks the rapprochement between linguistics and biology. Linking theoretical linguistics, theoretical biology, genetics, neuroscience and cognitive psychology, this book offers a collection of chapters situating the enterprise conceptually, highlighting both the promises and challenges of the field, and chapters focusing on the challenges and prospects of taking interdisciplinarity seriously. It provides concrete illustrations of some of the cutting-edge research in biolinguistics and piques the interest of undergraduate students looking for a field to major in and inspires graduate students on possible research directions. It is also meant to show specialists in adjacent fields how a particular strand of theoretical linguistics relates to their concerns, and in so doing, the book intends to foster collaboration across disciplines.

**Koji Fujita** is Professor at the Graduate School of Human and Environmental Studies, Kyoto University.

**Cedric Boeckx** is an ICREA Research Professor at the Institutió Catalana de Recerca i Estudis Avançats and is also associated with the University of Barcelona.

# Advances in Biolinguistics
## The human language faculty and its biological basis

Edited by Koji Fujita
and Cedric Boeckx

LONDON AND NEW YORK

First published 2016
by Routledge
2 Park Square, Milton Park, Abingdon, Oxon OX14 4RN

and by Routledge
711 Third Avenue, New York, NY 10017

*Routledge is an imprint of the Taylor & Francis Group,
an informa business*

© 2016 Koji Fujita and Cedric Boeckx

The right of Koji Fujita and Cedric Boeckx to be identified as
the authors of the editorial material, and of the authors for their
individual chapters, has been asserted in accordance with sections
77 and 78 of the Copyright, Designs and Patents Act 1988.

All rights reserved. No part of this book may be reprinted or
reproduced or utilised in any form or by any electronic,
mechanical, or other means, now known or hereafter invented,
including photocopying and recording, or in any information
storage or retrieval system, without permission in writing from
the publishers.

*Trademark notice*: Product or corporate names may be trademarks
or registered trademarks, and are used only for identification and
explanation without intent to infringe.

*British Library Cataloguing in Publication Data*
A catalogue record for this book is available from the British Library

*Library of Congress Cataloging-in-Publication Data*
Advances in biolinguistics : the human language faculty and its
 biological basis / Edited by Koji Fujita and Cedric A. Boeckx.
  pages cm
 Includes bibliographical references and index.
 1. Biolinguistics—Research. I. Fujita, Koji, 1958– editor.
II. Boeckx, Cedric, editor.
 P132.A383 2016
 401—dc23
 2015033055

ISBN: 978-1-138-89172-2 (hbk)
ISBN: 978-1-315-70952-9 (ebk)

Typeset in Galliard
by Apex CoVantage, LLC

# Contents

*Notes on contributors*     viii

1. Introduction: the biolinguistic program: a new beginning     1
   KOJI FUJITA AND CEDRIC BOECKX

## PART I
## Computational issues     7

2. Feature-equilibria in syntax     9
   HIROKI NARITA AND NAOKI FUKUI

3. On the primitive operations of syntax     29
   TAKAOMI KATO, HIROKI NARITA, HIRONOBU KASAI, MIHOKO ZUSHI AND NAOKI FUKUI

4. Case and predicate-argument relations     46
   MIHOKO ZUSHI

## PART II
## Development, processing and variations     67

5. Structure dependence in child English: new evidence     69
   KOJI SUGISAKI

6. Make a good prediction or get ready for a locality penalty: maybe it's coming late     83
   HAJIME ONO, KENTARO NAKATANI AND NORIAKI YUSA

7  Some things to learn from the intersection between
   language and working memory                                    103
   GONZALO CASTILLO

8  Eliminating parameters from the narrow
   syntax: rule ordering variation by third-factor
   underspecification                                             128
   MIKI OBATA AND SAMUEL EPSTEIN

## PART III
## Conceptual and methodological foundations            139

9  On certain fallacies in evolutionary linguistics
   and how one can eliminate them                                 141
   KOJI FUJITA

10 Biological pluralism in service of biolinguistics             153
   PEDRO TIAGO MARTINS, EVELINA LEIVADA, ANTONIO
   BENÍTEZ-BURRACO AND CEDRIC BOECKX

11 On the current status of biolinguistics as a
   biological science                                             170
   MASANOBU UEDA

## PART IV
## Evolutionary considerations                          187

12 Proposing the hypothesis of an earlier emergence
   of the human language faculty                                  189
   MASAYUKI IKE-UCHI

13 Two aspects of syntactic evolution                            198
   MICHIO HOSAKA

## PART V
## Topics in neurobiology                               215

14 Syntax in the brain                                            217
   NORIAKI YUSA

15 The central role of the thalamus in language
   and cognition                                                     230
   CONSTANTINA THEOFANOPOULOU AND CEDRIC BOECKX

16 A biolinguistic approach to language disorders:
   towards a paradigm shift in clinical linguistics                  256
   ANTONIO BENÍTEZ-BURRACO

   Index                                                             273

# Contributors

**Antonio Benítez-Burraco** is Professor in the Department of Spanish Philology and its Didactics, University of Huelva.

**Cedric Boeckx** is ICREA Research Professor at the Catalan Institute for Research and Advanced Studies (ICREA) and at the department of General Linguistics, University of Barcelona.

**Gonzalo Castillo** is a graduate student in the Department of General Linguistics, University of Barcelona.

**Samuel Epstein** is the Marilyn J. Shatz Collegiate Professor of Linguistics and an Arthur F. Thurnau Professor at the University of Michigan, Ann Arbor.

**Koji Fujita** is Professor of Linguistics at the Graduate School of Human and Environmental Studies, Kyoto University.

**Naoki Fukui** is Professor of Linguistics at the Graduate School of Languages and Linguistics, Sophia University.

**Michio Hosaka** is Professor of English Linguistics at the Department of English Language and Literature, the College of Humanities and Sciences, Nihon University.

**Masayuki Ike-uchi** is Professor of Linguistics at the Faculty of Liberal Arts, Tsuda College.

**Hironobu Kasai** is Professor at the Center for Fundamental Education, the University of Kitakyushu.

**Takaomi Kato** is Associate Professor of Linguistics at the Graduate School of Languages and Linguistics, Sophia University.

**Evelina Leivada** is a member of the Cyprus Acquisition Team in Nicosia, Cyprus.

**Pedro Tiago Martins** is a graduate student in the Department of General Linguistics, University of Barcelona.

**Kentaro Nakatani** is Professor of Linguistics in the Department of English Literature and Language, Konan University.

**Hiroki Narita** is Assistant Professor of Linguistics at the College of Industrial Technology, Nihon University.

**Miki Obata** is Associate Professor in the Department of Science, Tokyo University of Science.

**Hajime Ono** is Associate Professor of Linguistics in the Faculty of Liberal Arts, Tsuda College.

**Koji Sugisaki** is Professor of Linguistics at the College of Liberal Arts and Sciences, Mie University.

**Constantina Theofanopoulou** is a graduate student in the Department of General Linguistics, University of Barcelona.

**Masanobu Ueda** is Professor of Linguistics at the Graduate School of International Media, Communication, and Tourism Studies, Hokkaido University.

**Noriaki Yusa** is Professor of Linguistics in the Department of English, Miyagi Gakuin Women's University.

**Mihoko Zushi** is Professor of Linguistics at Kanagawa University, Yokohama.

# 1 Introduction
## The biolinguistic program: a new beginning

*Koji Fujita and Cedric Boeckx*

This collection of papers arose from the coalition of two independently working biolinguistic research groups, one Kyoto-Tokyo-based (Biolinguistics Project, Japan) and one Barcelona-based (Biolinguistics Initiative Barcelona), led by one of the coeditors, respectively. The chapters that follow represent some of the ongoing research which members of these groups have been devotedly engaged in. This brief introductory chapter offers some general background considerations with cursory reference to each contribution. For the reader's convenience, the following chapters are organized into five parts under different titles, but this does not imply that each section is detached from the others in any significant sense. On the contrary, the reader may easily find that all the chapters are so closely intertwined in their purposes and claims that this volume is in fact an inseparable and indivisible whole.

The term *biolinguistics* came into everyday use fairly recently, but biological approaches to human language are probably as old as science itself. Aristotle was among the first to compare human language with other animal communication systems, especially birdsong. He observed that humans and birds have similar vocal organs and vocalization capacities, but that only humans can use them to express and convey cognitive and propositional statements, as distinct from emotional and affective content. To use some contemporary terms, by noting both the evolutionary continuity and the discontinuity between human and bird communications, Aristotle arguably foresaw the progress of modern biolinguistics, where studies of birdsong enjoy a particularly important role as a key to understanding human language evolution.

Fortunately for today's biolinguists like us, Aristotle's comparative approach did not address one crucial gap between human language and animal communication – the presence vs. absence of a recursive computational system. The importance of this gap for understanding human language has been stressed by Chomsky's generative grammar over and over again, but it was only during the resurgence of biolinguistic concerns in the twenty-first century that the true meaning this gap carries came to be properly comprehended by linguists and biologists alike.

Today we cannot discuss language or its biological foundations without referring to the seminal joint article by Marc Hauser, Noam Chomsky and Tecumseh

Fitch, where they proposed the well-known distinction between FLN (those components of the language faculty which are unique to humans and human language) and FLB (every component of it, including FLN). They suggested recursion as the only candidate for FLN, but the problem was that they did so in a not very explicit way so that many unfruitful discussions or pointless criticisms followed as a result. We believe that we can safely equate what they meant by recursion with unbounded Merge, as proposed in the minimalist program of generative grammar. We believe this all the more not because unbounded Merge is a genuine part of FLN but because it offers a good opportunity to reexamine and seriously doubt the legitimacy of the FLN/FLB dichotomy.

Granted that a syntactic computational system is a uniquely human function, it is highly unlikely that this evolutionary novelty arose from nowhere, whether by mutation or by natural selection. Every biological trait has a precursor, often in an apparently unrelated domain with remote functions, and its current species- and/or domain-specificity is an end result of the Darwinian process of descent with modification. Merge serves as an ideal point of entry for a biologically/evolutionarily natural understanding of human syntax just because it is such a simple and elementary operation that one could easily find its analogues/homologues in other domains of both human and nonhuman cognitive behaviors, including tool making and tool using, in particular.

To pursue this exaptationist scenario and show that uniquely human syntactic computation indeed evolved from a not uniquely human, not specifically linguistic function, thereby establishing its biological nature, it is of supreme importance that studies of syntax be carried out with a keen interest both in securing the empirical coverage of the syntactic theory and in reducing the invoked syntactic machineries to even simpler operations, to the level where a direct comparison between syntax and other cognitive faculties makes good sense beyond a metaphor.

In this respect, the three chapters collected in Part I, despite their purely syntactic nature, are all important contributions to biolinguistics. **Hiroki Narita and Naoki Fukui** (Chapter 2) introduce the notion of feature-equilibrium to capture some interesting properties of syntactic computation, while **Takaomi Kato, Hiroki Narita, Hironobu Kasai, Mihoko Zushi and Naoki Fukui** (Chapter 3) propose to decompose Merge further into two more basic operations which they call 0-Search and 0-Merge. These two studies are significant attempts to show that simple and presumably not very language-specific principles and operations are often better at explaining ostensively linguistic phenomena, which further boosts our interdisciplinary inquiry into the biological nature of syntax. **Mihoko Zushi**'s work (Chapter 4) corroborates Kato et al.'s proposal by showing that Case valuation, a representative aspect of uniquely human (morpho)syntax which was once explained in terms of highly domain-specific analytical apparatus, now directly follows from a single computational operation.

Equally important are studies of language development and language processing, each of which is discussed neatly in Part II by **Koji Sugisaki** (Chapter 5) and **Hajime Ono, Kentaro Nakatani and Noriaki Yusa** (Chapter 6), respectively. Sugisaki addresses the issue of structure dependence, arguably the most prominent feature of human language in the biological world. Surveying English-speaking children's production of *yes/no* questions, he concludes that they are genetically predisposed to conform to structure dependence. Ono et al. argue, based on experimental data, that sentence processing is influenced by the two factors of locality effects and expectations working in a mutually exclusive way.

Processing is largely a matter of working memory, in addition to specifically linguistic knowledge, and language evolution too depends on the evolution of working memory in the brain to such an extent that we cannot discuss language evolution without considering working memory. **Gonzalo Castillo**'s contribution (Chapter 7) is highly instructive in this respect. After presenting an explicit and detailed description of working memory, Castillo explores the connection of this generic capacity to specifically linguistic unbounded Merge. This kind of connection, between what is and what is not language-specific, provides another important key to understanding how the uniquely human language faculty may have evolved through descent with modification.

In the past generative grammar, the concept of parameters was very useful to derive the vast superficial diversity observed among the world's languages, as well as to solve the logical problem of language acquisition. Language grows in children, as it was once claimed, largely as a process of internal selection (parameter setting), not by instruction from the environment, and different parametric values lead to synchronic, diachronic and developmental variations. Unfortunately, our updated understanding of biology and genetics does not support the view that these strong analytical tools belong to Universal Grammar (UG), to the extent that it is a biologically real object.

The overwhelming question is then how we can capture linguistic diversity without recourse to parameters, particularly because minimalism requires radical minimization (maximal underspecification) of UG (basically, it's Merge-only). **Miki Obata and Samuel Epstein** (Chapter 8) tackle this issue and argue that parametric variations are just a reflex of language-independent physical law (known as the "third factor" of language design) working on syntactic computation. Obviously, parameter-free universal syntax is a desideratum not only for the internal consistency of generative syntactic theory but for the overall progress in biolinguistics.

Biolinguistics, just like biological sciences in general, is not only an empirical science but it also requires a high level of conceptual and methodological considerations. In Part III, both **Koji Fujita** (Chapter 9) and **Pedro Tiago Martins, Evelina Leivada, Antonio Benítez-Burraco and Cedric Boeckx** (Chapter 10) stress the importance of a pluralistic attitude towards biolinguistics, though not necessarily for the same reasons. Fujita focuses on evolutionary issues and argues that, since language is not a monolithic object but rather a modular system consisting of several independent faculties, studies of language origins and evolution should avoid the fallacy of a single origin, the false belief that language as a whole must have evolved from one preexisting capacity. Other equally harmful fallacies in evolutionary studies of language are also discussed, and Fujita explains why the Merge-only hypothesis of the minimalist program, contrary to what one might expect, promises to remove these fallacies.

Martins et al. place their discussion in a broader context and propose to bring biolinguistics into much closer contact with modern biology. They point out that generative grammar in the past was based on some serious misconceptions about biology and evolution and show how one can remedy this situation and render biolinguistics truly biological in nature. Interestingly, both Fujita and Martins et al. argue that the FLN/FLB distinction can no longer be maintained.

*Masanobu Ueda* (Chapter 11) attempts to place the biolinguistic program in the context of the philosophy and history of natural sciences and critically evaluates its current status as a biological science. In particular, contrary to what is sometimes claimed by other practitioners of generative grammar, Ueda finds some serious mismatches between Tinbergen's four questions and the goals and proposals of biolinguistics today.

Part IV provides discussions more directly associated with evolutionary questions. *Masayuki Ike-uchi* (Chapter 12) casts doubt on the popular belief that the Merge-based human language first appeared around 60–80 kya in H. sapiens and argues that its emergence took place around 130–150 kya. This conclusion is based on recent discoveries in archaeology, paleoanthropology and genetics. Researchers' views divide between gradual/incremental vs. rapid/saltational evolution of language, the latter of which is obviously in conformity with the minimalist view of language design. Ike-uchi's observation may help resolve the tension by suggesting that the emergence of UG or human language may not have been very recent, an important antidote to the often not very productive conflict between generativists and anti-generativists.

*Michio Hosaka*'s contribution (Chapter 13) has a similar effect of bridging the gap between the two opposing camps. The original function of language has been a hot issue; some support the communication-first theory while others favor the thought-first theory. While Hosaka agrees with other generativists that language first evolved as an instrument of thought, he argues that the evolution of syntax was adaptive for communicative purposes, too. The distinction between external Merge and internal Merge corresponds to the difference between these two adaptive functions, with external Merge serving thought and both external and internal Merge (Move) serving communication. Hosaka supports the view that the evolution of syntax was somewhat gradual, from external to internal Merge, and that communication is as important a factor in understanding language evolution, and in this respect he adopts a pluralist position, much like Fujita and Martins et al.

Language is firmly based on our neurology, and biolinguistic studies hardly make sense if one fails to connect theoretical proposals about the mechanisms of language to their neuronal implementation in the brain, which in fact has proven very difficult to achieve. In Part V, *Noriaki Yusa* (Chapter 14) focuses on the role that Broca's area plays in processing the hierarchical, as opposed to sequential, structure of human language. Structure dependence is one biologically unique property of language, and it is mandatory that neuroscience explicate the neuronal mechanism of it. Whether Broca's area is the locus for this purpose, and, if so, then which subdomain of it is, remains one prominent target of inquiry. Yusa demonstrates the involvement of this region by gathering evidence from neurological studies of second language acquisition. He also suggests that within Broca's area, BA 45 may be the locus of domain-specific, syntactic Merge, whereas BA 44 may serve domain-general Merge, thereby supporting the evolutionary scenario from action to syntax in the brain.

**Constantina Theofanopoulou and Cedric Boeckx** (Chapter 15), by sharply departing from the classical cortico-centrism, highlight and examine the central role played by the thalamus to connect and regulate different regions inside the globular brain unique to H. sapiens. The suggested cortico-thalamus-cortical circuits have implications not only for language but for human cognition at large. The expansion of focus from cortical to subcortical structures should drive biolinguistics in a new, and more productive, direction.

**Antonio Benítez-Burraco** (Chapter 16) explores the possibility of restructuring clinical linguistics by bringing it into a closer relation with biolinguistic concerns. Language disorders have played a privileged role in biological studies of language, both as a window to the neurological underpinnings of language and as a clue to the supposed protolanguage. Benítez-Burraco stresses the need of a paradigm shift in studies of language disorders, by changing the focus from adult phenotypes to the dynamic process of development, much in line with what is going on in the evo-devo approach in biology and biolinguistics. His discussion offers an opportunity for us to thoroughly rethink the role of genes in language and language disorders and to move towards a biologically more natural understanding of language evolution and language development, of how they may interact with each other.

What all of these contributors and their chapters have in common, though they are dedicated to a variety of topics, is the humble realization that we are still so far from what biolinguistics should be like. We believe that biolinguistics needs to be an integral part of biological science that goes way beyond today's theoretical linguistics. We hope this volume will provide a strong driving force to reboot the biolinguistic program for the next generation.

# Part I
# Computational issues

# 2 Feature-equilibria in syntax*

*Hiroki Narita and Naoki Fukui*

## 1 Introduction

Linguistic expressions exhibit a variety of formal and interpretive properties at the SEM(antic) interface with the C(onceptual)-I(ntentional) system, depending on how they are structurally organized. In particular, it is known that for each syntactic object (SO) Σ, a certain designated lexical item (LI) – called the *head* of Σ – plays a central role in determining the formal/interpretive properties of Σ. For instance, major properties of a verbal phrase *read a book* are determined by its head verb *read*, yielding its event-predicatehood and θ-roles, among others. In contrast, nominal phrases like *a book* show properties different from verbal phrases, such as being referential, indefinite, third person, etc., each provided by their head noun (or perhaps determiner). This traditional observation is commonly referred to as the *endocentricity (headedness) of phrase structure*:

(1) *Endocentricity/headedness of phrase structure*:
Interpretive properties of an SO Σ are determined largely by the features of a unique head LI within Σ.

The proper analysis of endocentricity is one of the most important research topics in generative linguistics. Essentially since Chomsky's (1970, 1981) introduction of "X-bar theory," it has been predominantly stipulated that endocentricity is characterized by the notion of "projection." Projection can be understood as a relation in which a certain LI transmits its lexical features to nodes that "dominate" (i.e., contain) it – a very strong form of (total) feature-percolation. An X-bar-theoretic characterization of endocentricity typically consists of the following assumptions.

(2) *Universal projection*:
All SOs are not "bare" but always associated with nonterminal symbols (or "labels") such as NP, C', etc., namely "projections" of some LIs (features of a head LI, perhaps accompanied by bar-level indices).

(3) *Projection = endocentricity*:
Projection is *the* device to encode endocentricity.

(4) *Universal endocentricity:*
As a result of (2)-(3), every SO is endocentric, leaving no room for non-endocentric structures.

(2)-(4) have been very widely presumed throughout the history of generative linguistics (see Fukui and Narita 2014 for an overview). However, it should be noted that these hypotheses are no longer necessary – nor are they innocuous – in the contemporary theory of "Merge" ("bare phrase structure") launched by Chomsky (1995 *et seq.*). Merge is a simple set-formation operation that takes $n$ (typically two) SOs, $\Sigma_1, \ldots, \Sigma_n$, as its input and generates an unordered set $\{\Sigma_1, \ldots, \Sigma_n\}$ (but see Kato et al. 2016 for a further analysis of this operation):

(5) $\text{Merge}(\Sigma_1, \ldots, \Sigma_n) = \{\Sigma_1, \ldots, \Sigma_n\}$

According to bare phrase structure theory, all linguistic expressions reduce to set-theoretic objects generated via Merge. Structural descriptions provided by Merge-based syntax are therefore quite different from those of earlier phrase structure grammars (PSGs) or various versions of X-bar theory. Most notably, no notions of projection or labels are represented in the output of Merge. The stipulation of universal projection (2) therefore loses its ground, which is in significant contrast to the automatic and exceptionless generation of projection/labels in X-bar syntax. Then, if we were to somehow incorporate (2) into bare phrase structure, some additional stipulation would have to be provided. To take a familiar example, Chomsky (1995) proposes to encode the notion of label into the definition of Merge as in (6) (modulo a binarity restriction, $n = 2$).

(6) $\text{Merge}(\alpha, \beta) = \{\gamma, \{\alpha, \beta\}\}$, where the label $\gamma$ = the label of $\alpha$ or $\beta$.

In contrast, Chomsky (2004, 2008, 2013) argues for the simpler version of Merge in (5), while hypothesizing that endocentricity/projection is determined by an independent algorithm of "labeling." See also Fukui (2011), who attempts to decompose the two components of (6), with Merge as in (5) serving for hierarchical organization of constituents and an operation called "Embed" providing a label in a way similar to (6). However, whether we complicate the definition of Merge as in (6) or propose a labeling algorithm independent of Merge, those approaches count as departures from the simple Merge system formulated in (5). Thus, as a natural move, linguists of minimalist persuasions have begun to cast serious doubts on the notion of labels/projection (Collins 2002, Chomsky 2007, 2013, Narita 2014). For example, Chomsky (2007:23) remarks that "reference to labels (as in defining c-command beyond minimal search) is a departure from S[trong]M[inimalist]T[hesis], hence to be adopted only if forced by empirical evidence, enriching UG." Then, we should ask, do we have to keep the hypothesis in (2) even in the bare phrase structure

*Feature-equilibria in syntax* 11

framework? Or, if we may eliminate (2), then should we also reconsider the grounds of (3) and (4), the effects of which have been predominantly attributed to (2)? (See also Fukui 2011 and Fukui and Narita 2014.)

The purpose of this chapter is to pursue the simpler theory of Merge in (5), and argue that not only endocentric structures but also "exocentric" (label-free and non-endocentric) structures are possible in human language. In Section 2, we will first introduce Fukui's (2011) generalization concerning the distribution between endocentric (or "asymmetric") and exocentric ("symmetric") structures and explore some of its consequences. In Section 3, we will put forward a theory of "symmetry-driven" syntax that can naturally account for Fukui's (2011) generalization, which we will also refine. We will further discuss various consequences of the proposal in Section 4 and Section 5. Section 6 concludes the chapter.

## 2 Fukui's (2011) generalization

In what follows, we will explore the distribution of (non-)endocentricity in human language without recourse to the stipulations in (2)-(4). Let us start with cases whose syntactic and/or interpretive properties exhibit typical endocentricity. (7a-e) exemplify cases of the form {H, XP}, where H, an LI, typically serves as the head.[1]

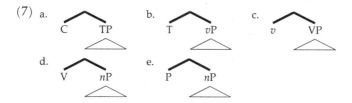

(7) a. C TP  b. T *v*P  c. *v* VP
   d. V *n*P  e. P *n*P

These are traditionally called "head-complement configurations." The structurally prominent LI H functions as a prime factor for characterizing the semantic interpretation of these structures at SEM: C determines the force of the clause (declarative, interrogative, etc.) in (7a); T feeds tense and modal properties in (7b); V characterizes lexical and aspectual properties of the denoted event and assigns a Theme/Patient θ-role to its complement, etc. Generally speaking, {H, XP} structures receive what has been traditionally called "lexical-conceptual aspects of semantics" or "d-structure interpretations," including selectional restrictions, predicate-argument structure, θ-marking, etc., in which the notion of endocentricity/headedness plays a major role.

Interestingly, such endocentric {H, XP} structures are typically generated by External Merge (EM). As pointed out by Chomsky (2004), Merge has two modes of application: EM takes two SOs that are independent of (external to) each other, as in (7a-e), while Internal Merge (IM) takes two SOs that are not independent, i.e., one is a term of the other.

(8) Merge(α, β) counts as *Internal Merge* (*IM*) if one of α, β is a term of the other. Otherwise, Merge(α, β) counts as *External Merge* (*EM*). (Chomsky 2004 *et seq.*)

For some reason, endocentric {H, XP} structures like (7) are typically generated by EM.

Then, how about SOs generated by IM? Take a representative case of IM in (9), where the subject *n*P in (9a) A-moves into the "Spec-T" position, leaving a copy of *n*P behind, as in (9b).[2] In this structure, the subject *n*P and the head T of YP undergo φ-feature-agreement, valuing the unvalued φ-features of T (in what follows, we will adopt the familiar notation where [uF] stands for a formal feature F whose value is unspecified, and [vF] stands for a valued, and hence inherently interpretable, F).

(9) a.

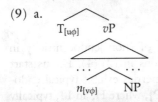

b.

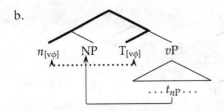

Notice that there is no obvious sense in which such an SO is "endocentric," so long as we stay away from projection-based stipulations: The output of IM is of the form {XP, YP}, where both of the constituents are phrasal, and there is no single LI that structurally stands out as the head. Further, the structure feeds various semantic effects at SEM, such as the subject-predicate relation and "aboutness," but none of these interpretations is obviously "T-like" or "*n*-like." More generally, the semantics of internally merged SOs, traditionally referred to as "s-structure interpretations," involves certain "discourse-related" properties that cannot be readily attributed to any single LI or its lexical features, hence not endocentric. The same point can be made for cases of A'-movement like (10), where a *wh*-element in (10a) moves into Spec-C, accompanying some sort of Q(uestion)-feature-agreement (see Cable 2010, Narita 2014), as in (10b). In (10) too, we cannot attribute the whole semantics of the *wh*-question to any single LI: its interrogative force can be said to be C's, while the operator-variable relations and the WH-quantification are primarily due to *which*. Again, the semantics is not endocentric and highly discourse-related.

(10) a.

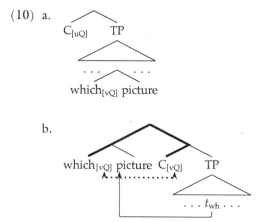

In pursuing projection-free syntax that dispenses with (2)-(4), Fukui (2011) proposes to characterize the overall tendency of syntactic computation from a novel point of view: Merge, so long as it is unconstrained and applies freely (see Chomsky 2008), should be able to generate all sorts of SOs with mixed forms and properties. However, what we observe is that the classes of SOs generated via Merge are by no means random, and they exhibit certain systematic patterns. Fukui (2011) points out that EM mostly serves for creating endocentric {H, XP} structures, which he proposes to characterize as "asymmetric," while IM almost always generates structurally "symmetric" {XP, YP} structures. It is further suggested in his analysis that {XP, YP} structures are fundamentally stable, while asymmetric {H, XP} structures are unbalanced and hence unstable, and that once free applications of EM create asymmetric/unbalanced {H, XP} structures, they should get stabilized by being mapped to some symmetric {XP, YP} structures. Thus, syntactic computation exhibits a systematic transitional pattern from EM-based asymmetric {H, XP} to IM-based symmetric {XP, YP}.[3]

(11) Fukui's (2011) *Generalization*:
"Asymmetric"/"endocentric" SOs of the form {H, XP} are unstable, and they must be mapped to "symmetric" structures of the form {XP, YP}.

Fukui's (2011) generalization in (11) pertains to an important intuition that syntactic computation is essentially driven by a need for a certain type of structural "symmetry."

Pushing Fukui's idea further, we would like to point out that the generalization in (11) can be extended to cover cases of *head-movement* ($X^0$-*movement*) as well. Although whether head-movement is syntactic or not has been contested (see, e.g., Chomsky 2001, Boeckx and Stjepanović 2001), the common assumption is that it is a form of syntactic movement that adjoins an LI $X^0$ to another LI $Y^0$ that c-commands the initial occurrence of $X^0$. According to the traditional description, head-movement of $X^0$ effectively replaces $Y^0$ in an SO with the so-called "$Y^{0max}$" category that consists of $X^0$ and a segment of $Y^0$, as shown in (12), but still behaves as $Y^0$ as a whole.

(12)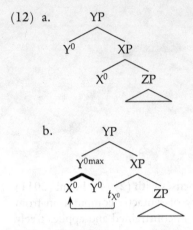

However, if Chomsky (2007, 2008) is right in claiming that every application of Merge satisfies the *No-tampering Condition* (NTC) (13), then $X^0$-$Y^0$-merger should not be able to modify the internal composition of $Y^0$.

(13) *No-tampering Condition* (*NTC*) (Chomsky 2008:138; see also Narita 2014):
Merge of X and Y leaves the two SOs unchanged.

Under the NTC, instances of $X^0$-$Y^0$-merger should not be able to replace $Y^0$ in (12a) with "$Y^{0max}$" = $\{X^0, Y^0\}$. Rather, it is predicted that such an application of Merge just generates another set $\{X^0, Y^0\}$, without tampering with the SO in (14a)/(14a'). If this operation really exists as an instance of Merge, and if we keep to the definition of Merge as a simple set-formation operation, then what it yields is two SOs, (i) $\{Y^0, \{X^0, ZP\}\}$ and (ii) $\{X^0, Y^0\}$ as shown in (14b').

(14) a. [tree: $Y^0$, $X^0$, ZP] ⟶ b. [tree]

a'. $\{Y^0, \{X^0, ZP\}\}$ ⟶ b'. (i) $\{Y^0, \{X^0, ZP\}\}$
(ii) $\{X^0, Y^0\}$

Notice that the occurrence of $Y^0$ is obviously not a term of $X^0$ (and *vice versa*). Then, such instances of Merge should count as EM under Chomsky's definition (8). However, head-movement is clearly a movement operation just like IM, and it yields copies of $X^0$. The informal tree notation in (14b) may look as though the operation counts as a kind of "sideways remerge," yielding "multidominance" structures as in (14b). No matter how we grasp the intuition behind (14), however, the point is that the replacement of $Y^0$ in an SO with a $Y^{0max}$ category is beyond the generative power of Merge. Traditional examples of head-movement, such as $T^0$-to-$C^0$ and $V^0$-to-$v^0$, should be reanalyzed along these lines of reasoning.[4]

Now, note that head-movement as depicted in (14) yields another kind of "symmetric," balanced branching, in this case of the form $\{X^0, Y^0\}$. It differs from core cases of IM only in the "size" of the relevant constituents ($\{X^0, Y^0\}$ vs. $\{XP, YP\}$), so it is reasonable to suppose that this movement operation also satisfies Fukui's (2011) generalization in (11), serving to form a symmetric structure.

In this manner, Fukui's (2011) generalization in (11) can unify the class of possible movement operations, XP-movement/IM and head-movement, under the category of symmetry-formation. Notice that in earlier theories of X-bar-theoretic syntax, it is plainly stipulated that XP-movement must target some "Spec" position (the sister of an X'-phrase), while $X^0$-movement must target another $X^0$ position. This restriction does not rest on principled grounds, and, what is worse, its formulation makes heavy recourse to projection-based notions like "Spec" and "$X^{0(max)}$," hence it has lost its basis in Merge-based syntax without projection. However, (11) naturally provides a simpler, projection-free characterization of the two positions, a highly desirable result.

## 3 Dynamic Symmetrization Condition

In the preceding section, we introduced Fukui's (2011) generalization in (11) and argued that it can naturally unify the target positions of XP-movement/IM and head-movement, i.e., $\{XP, YP\}$ and $\{X^0, Y^0\}$, respectively. Fukui's notion of symmetry is simply defined in terms of the LI vs. non-LI/phrase distinction, but we will see in this section that this is insufficient and should be modified by referring to the distribution of formal features within relevant SOs

Under Fukui's (2011) proposal, all SOs of the form $\{XP, YP\}$ are regarded as symmetric and stable. However, there are ample cases in which XP moves out of $\{XP, YP\}$, breaking the "symmetry" in Fukui's sense. For example, consider (15a), in which the external argument $n$P, once externally merged with $\{v, VP\}$, is required to move out of this SO into Spec-T (the effect of the so-called "Extended Projection Principle," EPP). Another example is the *wh*-phrase at the edge of (15b), which does not match with [–WH] C and hence is required to undergo successive cyclic movement out of this SO.

These examples clearly show that not every {XP, YP} structure is stable, contra Fukui's claim.

(15)

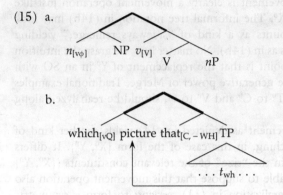

a. $n_{[v\phi]}$ NP $v_{[V]}$ / V nP

b. which$_{[vQ]}$ picture that$_{[C, -WH]}$ TP / ... $t_{wh}$ ...

What differentiates these "unstable" {XP, YP} structures from "stable" ones like (9b) and (10b)? The answer lies in the existence of Agree(ment), which can be easily seen by comparing (10b) and (15b) (see also Fukui 2011 for a different approach). The stable structure in (10b) is minimally different from the unstable SO in (15b), in that only the former accompanies Q-feature-agreement. Thus, the relevant {XP, YP} structure becomes stable and "symmetric" if there exists Q-feature-agreement between C and *wh* (as in (10b)), while it becomes unstable if no Q-feature-agreement is involved (as in (15b)). The observation that {XP, YP} without agreement is unstable can be directly extended to cases like (15a) as well. By the standard assumption, $n$ and $v$ are as distinct in featural contents as they can be, and indeed, no agreement relation is established between $n(P)$ and $v(P)$. Thus, it is tempting to hypothesize that the relevant sort of feature-agreement is a necessary condition for making XP-YP structures symmetric and stable.

To maintain the preliminary generalization in (11) in its essence while taking feature-agreement into consideration, we would like to propose the following concept of *feature-equilibrium* in (16) and advocate a new characterization of symmetry (17) in terms of this new concept.

(16) *Feature-equilibrium*:
For a formal feature F, an SO {α, β} is in an *F-equilibrium*. $\equiv_{def.}$ α and β share a matching formal feature F that is equally prominent in α and β.

(17) *Featural symmetry*:
Given a formal feature F in an SO {α, β}, the SO is *symmetric* with respect to F (or *F-symmetric*) if it is in an F-equilibrium. Otherwise, the SO {α, β} is (F-)asymmetric with respect to F.

According to the proposed definition, {XP, YP} counts as symmetric when it forms an F-equilibrium with respect to some formal feature F. Familiar XP-YP structures in (9b) and (10b) are classified as symmetric in this sense, because they form a ϕ-equilibrium and a Q-equilibrium, respectively. In contrast, cases like (15a,b) do not count as symmetric under the definition in (17), given the lack of matching features shared by XP and YP. In this approach, then, the {XP, YP} structures in (15) cease to be exceptional in the face of Fukui's (2011) generalization in (11).

Note that the notion of featural symmetry is defined in (17) relative to some feature F present in {α, β}. If a relevant feature does not exist in the SO, F-(a)symmetry is not even defined and the notion simply becomes irrelevant. Therefore, the XP-YP structure in (9b) does count as ϕ-symmetric, but it is neither symmetric nor asymmetric with respect to, say, [Q], given the lack of [Q] in the SO.

Recall that we still want to classify cases of {H, XP} structures like (7) as asymmetric SOs, typical of EM. How does our new concept of featural symmetry achieve this result? It is not clear whether H and XP share any matching feature in these cases. Observationally, there are various sorts of dependencies between H and the head of XP in these structures: C assigns finiteness to T in (7a); T determines tense and aspectual properties of *v* in (7b); *v/n/a/p* assign their categorial features to V/N/A/P in SOs like (7c) (which may originally be uncategorized root LIs); V and P select/subcategorize *n* in (7d-e); and so on. A variety of linguistic theories hypothesize abstract features and mechanisms of feature-checking to capture these dependencies. Quite independently of the nature of these dependencies and any analyses we may give them, it is clear that SOs of the form {H, XP} are "unbalanced" with respect to the size of their two constituents, or the "depth" of features embedded in H and XP: H is an LI, which is the smallest possible syntactic element, immediately presenting its features, but XP is a phrasal composite whose featural content can be determined only by further inspection of its LI terms. Building on this observation (see also Chomsky 2013), we define the notion of feature-equilibrium in (16) in such a way as to require that the matching feature F must be "equally prominent" in α and β. Here we characterize *structural prominence* inductively as in (18).[5]

(18) Structural Prominence (see Kato et al. 2016; see also Ohta et al. 2013 for a related notion of "Degree of Merger"):
Suppose that Depth(α) = $m$ ($m \geq 0$) is the order of depth – the inverse relation of prominence – associated with an SO α, with lower prominence indicated by a higher value of depth. Then, we can say:

   a. Depth(α) = 0 if there is no SO β such that α ∈ β (i.e., α is a root SO dominated by no other SO).
   b. If Depth(α) = $m$, then, Depth(β) = $m + 1$ for any β such that β ∈ α (i.e., β is a daughter of α).

(18a) says that every root SO within a given workspace is of the highest order of prominence (0), and (18b) says that the prominence of an SO decreases as it gets more deeply embedded into a larger SO. To extend this inductive definition to formal features, let us assume that each LI can be characterized as a set of features and that, therefore, any feature F of an LI H counts as a daughter of H. Then, it follows that no feature F of H and XP can be equally prominent in {H, XP}, because, given Depth(H) = Depth(XP), it is necessarily the case that Depth(F of H) < Depth(F of some other LI H' within XP). Therefore, {H, XP} can never count as symmetric according to the definition in (18).

In contrast, (9b) and (10b) can still successfully count as symmetric, given the definition of structural prominence in (18). Furthermore, SOs created by head-movement also count as symmetric. To take a familiar example, consider a prototypical case of V-to-$v$ head-movement in (19).

(19) a. [tree: $v_{[V]}$ dominating $V_{[uCat]}$ and $n$P]    b. [tree: $v_{[V]}$ dominating $V_{[V]}$ and $n$P]

a'. {$v$, {V, $n$P}}    b'. (i) {$v$, {V, $n$P}}
                              (ii) {$v$, V}

V-to-$v$ head-movement can be regarded as an instance of root-incorporation, driven by the need to categorize the root (the same can be said of N-to-$n$ and A-to-$a$ movement as well). We may describe this situation by assuming that root categories are associated with an unvalued Cat(egorial)-feature that is in need of matching with a neighboring categorizer. Then, root-incorporation can also be seen as driven by the need for Cat-feature symmetry. (19) illustrates the situation with V-to-$v$ IM, where V/root's [uCat] enters into a Cat-equilibrium with $v$'s [V-Cat] (a categorial-feature valued as [V(erb)]). Further, if we follow Chomsky (2007, 2008) and Richards (2007) in assuming that T's tense feature is dependent on C, then we might also say that T-to-C incorporation is just another instance of movement driven by the need for feature-equilibrium. We assume that T has an unvalued T(ense)-feature [uT] that undergoes matching with an interpretable counterpart in C. Thanks to this T-feature-matching, T-to-C movement results in a symmetric structure {T, C}, just like root-incorporation.

(20) a. [tree: $C_{[vT]}$ dominating $T_{[uT]}$ and $v$P]    b. [tree: $C_{[vT]}$ dominating $T_{[vT]}$ and $v$P]

a'. {C, {T, $v$P}}    b'. (i) {C, {T, $v$P}}
                              (ii) {C, T}

Thus, the formation of {T, C} can also be seen as driven by T(ense)-feature symmetry. More generally, head-movement can be regarded as fulfilling an important function of fixing the asymmetry of {X, YP} and forming a feature-equilibrium. Therefore, our notion of featural symmetry provides an important rationale for head-movement, reducing it to a special case of symmetry-driven computation.

Exploring these lines of reasoning, we put forward a revised version of Fukui's (2011) generalization, where the notion of symmetry/asymmetry is redefined in terms of feature-equilibrium.

(21) SOs asymmetric with respect to some formal feature F are unstable, and they must be mapped to SOs symmetric with respect to F.

(21) correctly differentiates stable vs. unstable {XP, YP} structures, while successfully incorporating head-movement structures into the category of symmetric structures.

Without further stipulations, Merge can freely combine SOs, generating both asymmetric and symmetric structures, but we observed that movement operations typically result in symmetric structures: Subject-raising via IM yields a ɸ-equilibrium and *wh*-movement via IM yields a Q-equilibrium, whereas head-movement ($X^0$-$Y^0$-remerge) yields a Cat/T-equilibrium. In contrast, EM, the first application of Merge for each LI and SO, typically yields F-asymmetric structures, as we saw above. This pattern seems to be systematic, and we would like to know why this is almost always the case. To capture this state of affairs, we propose that syntactic derivation is driven by the need for featural symmetry. This idea can be stated as in (22):

(22) *Dynamic Symmetrization Condition (DSC)*:
Each formal feature F must form an F-equilibrium in the course of the derivation.

LIs are associated with their own feature contents. EM freely introduces LIs and their features (including formal features) into a syntactic derivation, but the DSC (22) essentially states that once a formal feature F is introduced, the subsequent derivation must guarantee that F moves to a position where it can form an F-equilibrium. The DSC (22) thus articulates the view that linguistic computation is irreversibly directed toward symmetrization, and this idea provides a natural ordering of operations in (23) below, further refining Fukui's (2011) generalization (cf. (11), (21)).

(23) For any formal feature F, an application of Merge that creates an F-asymmetric SO (i.e., an SO that is asymmetric with respect to F) entails a later application of Merge that yields an F-equilibrium.

This completes our rationalization of Fukui's hypothesis that syntax is driven by a need for symmetry.

Notice that asymmetric and symmetric structures, after their generation by unconstrained Merge, are utilized differently by CI: As we saw in Section 2, the former yield various "endocentric," d-structure semantics such as predicate-argument structure and selection, whereas the latter yield "exocentric," s-structure semantics such as topic-focus, operator-scope, theme-rheme, and so on. Given our symmetry-driven syntax, we may make better sense of why this holds at SEM: Merge-based SOs are "bare" and free from asymmetric representations such as left-to-right ordering (X precedes Y) and projection (X projects over Y), which were intrinsic to PSG-based/X-bar-theoretic syntax. SOs are no longer universally endocentric, so we can reduce the notion of endocentricity to LIs' unbalanced/asymmetric contributions to semantic interpretations at SEM, typical of externally merged asymmetric SOs. In contrast, symmetrically organized SOs show no such property, and hence their s-structure interpretations are not dependent on any single LI (i.e., they are non-endocentric), so they exhibit various discourse-related properties.

We can incorporate this observation into our generalization in (23) and sum up the whole picture of symmetry-driven syntax as in (24).

(24) a. F-asymmetry
 ⇒ introduced by EM
 ⇒ formed before (b)
 ⇒ exhibits endocentricity
 ⇒ contributes to lexical, d-structure interpretation (predicate-argument structure, selection, etc.)

b. F-symmetry
 ⇒ achieved by IM, head-movement, and agreement
 ⇒ formed after (a)
 ⇒ exhibits no endocentricity (exocentric)
 ⇒ contributes to discourse-related, s-structure interpretation (quantificational, topic-focus, etc.)

We specifically hypothesized that the DSC is the prime factor behind this overall tendency in linguistic computation. Note again that this result can be achieved only if we eliminate universal projection (2) and universal endocentricity (4), which lends further support to the truly "bare" and "symmetric" formulation of Merge in (5).

## 4 Equilibrium Intactness Condition

In the rest of this chapter, we will discuss some further consequences of symmetry-driven syntax. In this section, we will propose another condition on F-symmetry, the "Equilibrium Intactness Condition" (EIC), in an attempt to capture the stable nature of F-symmetry.

As we saw above, the need for F-symmetry (the DSC) serves to trigger various "symmetrizing" operations in syntactic derivation, such as IM, head-movement, and feature agreement. Notice that the F-symmetric structures created by these operations seem to be "stable" (as noted by Fukui 2011), and resist further computations. For example, once a φ-symmetric SO Σ = {α, β} in (25a) is constructed, then neither Σ nor its constituents may be subjected to further movement. (25b-d) show that α and β cannot undergo further φ-symmetry-related operations. (25e) further shows that the φ-symmetric SO Σ itself is also inaccessible to further φ-related computation.

(25) a. _____ seems that [$_\Sigma$ [$_\alpha$ John]$_i$ [$_\beta$ is believed $t_i$ to be a great linguist]].
b. *[$_\alpha$ John]$_i$ seems that [$_\Sigma$ $t_i$ [$_\beta$ is believed $t_i$ to be a great linguist]].
c. *[$_\beta$ is believed $t_i$ to be a great linguist]$_j$ seems that [$_\Sigma$ [$_\alpha$ John]$_i$ $t_j$].
d. *There seems that [$_\Sigma$ [$_\alpha$ John]$_i$ [$_\beta$ is believed $t_i$ to be a great linguist]].
e. *[$_\Sigma$ [$_\alpha$ John]$_i$ [$_\beta$ is believed $t_i$ to be a great linguist]] seem(s) (that) $t$.
f. It seems that [$_\Sigma$ [$_\alpha$ John]$_i$ [$_\beta$ is believed $t_i$ to be a great linguist]].

The same sort of observation can be made regarding Q-equilibrium in (10). For example, (26b-d) show that *wh*-movement of constituents in a Q-equilibrium results in much stronger deviance than the "weak island" effect in (26e).

(26) a. (Guess) _____ C he wonders [$_\Sigma$ [$_\alpha$ which boy]$_i$ [$_\beta$ C $t_i$ read which book]].
b. *Guess [$_\alpha$ which boy]$_i$ C he wonders [$_\Sigma$ $t_i$ [$_\beta$ C $t_i$ read which book]].
c. *Guess [$_\beta$ C $t_i$ read which book]$_j$ C he wonders [$_\Sigma$ [$_\alpha$ which boy]$_i$ $t_j$].
d. *Guess [$_\Sigma$ [$_\alpha$ which boy]$_i$ [$_\beta$ C $t_i$ read which book]] C he wonders $t$.
e. ??Guess [which book]$_j$ C he wonders [$_\Sigma$ [$_\alpha$ which boy]$_i$ [$_\beta$ C $t_i$ read $t_j$]].

These data suggest that constituents of an F-equilibrium cannot enter into further F-related operations. We will refer to this restriction as the *Equilibrium Intactness Condition* (*EIC*).

(27) *Equilibrium Intactness Condition* (*EIC*):
Constituents of an F-equilibrium are stable and invisible for further F-related computation.

Note that examples like (25) and (26) have been attributed to the traditional idea that Move/IM is a "costly" operation (unlike EM) and can apply only when it yields otherwise impossible feature-checking (such as checking of abstract Case-features). As pointed out by Chomsky (2007, 2008), this "last resort" conception of Move runs directly afoul of the free and unconstrained conception of Merge, hence this stipulation should be eliminated. Our proposal is that this elimination can in fact be naturally achieved while maintaining unconstrained Merge, if the results of its free application are subject to general – possibly third-factor – conditions such as the DSC (22) and the EIC (27) on formal features.[6,7]

In the next section, we will further argue that the DSC and the EIC provide room for certain parametric variation concerning the distribution of argument *n*Ps across languages, focusing on English and Japanese as representative examples.

## 5 Further consequences for comparative syntax

Japanese is a language that exhibits no active ϕ-features for nominal declensions and subject-verb agreement. Thus, regardless of the semantic person, number and gender of the subject $n$P, the verb form always stays the same, as exemplified in (28).

(28) Watasi-ga/anata-ga/gakusei-ga maitosi ronbun-o kak-u.
 I-NOM/you-NOM/student-NOM every.year paper-ACC write-PRES
 'I/you/a student/students write(s) a paper/papers every year.'

This traditional observation has been carried over to the generative literature at least since Kuroda (1965). Adopting the Principles-and-Parameters approach, Fukui (1986/1995, 1988, 2006) and Kuroda (1988, 1992) independently put forward the hypothesis that the lack of obligatory ϕ-feature-agreement in Japanese yields a highly intricate array of facts about this language that are unobservable in languages like English. In this section, we would like to maintain that our theory of symmetry-driven syntax may provide an interesting set of links between certain peculiar facts about Japanese, incorporating some of the major insights behind Fukui and Kuroda's macro-parametric accounts.

As a starting point, let us adopt (29), a particular formulation of the aforementioned observation put forward by Fukui (1986/1995, 1988, 2006):

(29) Hypothesis (Fukui 1986/1995, 1988, 2006):
 Japanese is a language that lacks active formal ϕ-features in its Lexicon.

An immediate consequence of this premise is that there is no notion of ϕ-feature symmetry/asymmetry in Japanese. Then, the language is also expected to lack ϕ-symmetry-forming operations, such as ϕ-driven A-movement. This is exactly what happens in Japanese, which is independently known as a language that shows no evidence for obligatory A-movement: see Fukui (1986/1995), Kuroda (1988), Ishii (1997), Kato (2006), and Narita (2014) among others for the view that Japanese subjects can (at least optionally) stay *in situ*.

(30) Japanese lacks obligatory A-movement.

A related consequence is that the edge of T remains unused/vacant in Japanese-type languages. We maintain that this position can be optionally utilized by a "major subject," an $n$P that does not participate in predicate-argument structure of the main verb and receives a topic-like interpretation.

(31) The edge of T in Japanese can be optionally utilized by a major subject, yielding multiple subject constructions.

Examples of the major subject construction in Japanese are provided in (32):

(32) Japanese:
    a. Taro-ga      musuko-ga    nyuugaku.siken-ni    sippaisita.
       Taro-NOM     son-NOM      entrance.exam-DAT   failed
       'As for Taro, his son failed the entrance examination.'
    b. Aki-ga        sanma-ga     umai.
       autumn-NOM    saury-NOM    tasty
       'As for autumn, saury is tasty.'

The merger of such an extra *n*P into Spec-T (or any other positions) is unproblematic in Japanese, because such a merger is free from any φ-symmetry requirements. Thus, it is predicted that any number of *n*Ps can be merged into this domain. Indeed, more than one major subject can be freely merged to a sentential structure in Japanese, as shown in (33) (see Kuno 1973, Fukui 1986/1995, 1988, 2006).

(33) Japanese: Sentence with more than one major subject (Kuno 1973)
      bunmeikoku-ga      dansei-ga  heikinzyumyoo-ga    mizikai.
      civilized.countries-NOM   male-NOM   average.lifespan-NOM   is.short
      'It is civilized countries that men, their average lifespan is short in.'

Compare the situation with English, where each merger of an *n*P with [φ] always results in a φ-asymmetric SO, necessitating later applications of symmetry-forming movement and agreement. Since positions entering into φ-equilibrium are limited in syntax (Spec-T, Spec-V, etc., where the "matching" LIs, T, V, etc., can offer unvalued φ-features for agreement), no free merger of a major subject *n*P is possible in this language:

(34) a. *Taro, his son failed the entrance examination.
     b. *Autumn, saury is tasty.
     c. *Civilized countries, male, the average lifespan is short (with the intended meaning 'it is civilized countries that men, their average lifespan is short in.')

Still another consequence of Japanese φ-free syntax is that *n*Ps do not form any φ-equilibrium with their Merge-mates. Thus, they never stabilize or become invisible via the EIC (27). Therefore, not having any "magnetic" power around, *n*Ps can freely undergo scrambling (optional dislocation).

(35) Japanese *n*Ps can freely undergo scrambling.

Japanese is a textbook example of a scrambling language, in which *n*Ps can freely move around as exemplified in (36). This peculiarity now follows as a consequence of its lack of φ-features.

(36) Japanese:
   a. John-ga     Mary-ni     sono hon-o     watasita
       John-NOM    Mary-DAT    that book-ACC    handed
       'John handed that book to Mary.'
   b. Mary-ni      John-ga      sono hon-o     watasita
   c. sono hon-o    John-ga      Mary-ni       watasita
   d. sono hon-o    Mary-ni      John-ga      watasita
   e. Mary-ni      sono hon-o    John-ga      watasita

In contrast to Japanese, English (among many other languages) does not exhibit optional scrambling, which follows from the fact that such $n$Ps have φ-features and they are "frozen" by the EIC in positions forming a φ-equilibrium, as depicted in (37).

(37) English-type: $n$Ps are "frozen" in positions forming a φ-equilibrium:

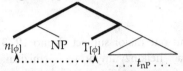

In contrast, there is no φ-equilibrium in Japanese-type languages. Thus $n$Ps can freely move around without getting frozen via the EIC.

We saw in this section that a single parametric statement, namely that Japanese lacks active φ-features, yields a number of intricate consequences in Japanese syntax (see Narita and Fukui 2012, 2014 for further discussion). No brute-force stipulation on the constituents of UG is necessary to account for this "macro-parametric" variation, simplifying linguistic theory to a considerable extent.

## 6 Conclusion

In this chapter, we argued for and extended Fukui's (2011) hypothesis that linguistic computation is essentially driven by symmetry, deriving the overall directionality from F-asymmetry to F-symmetry in (24) (repeated here as (38)).

(38) a. F-asymmetry
       ⇒ introduced by EM
       ⇒ formed before (b)
       ⇒ exhibits endocentricity
       ⇒ contributes to lexical, "d-structure" interpretation (predicate-argument structure, selection, etc.)

b. F-symmetry
    ⇒ achieved by IM, head-movement, and agreement
    ⇒ formed after (a)
    ⇒ exhibits no endocentricity (exocentric)
    ⇒ contributes to discourse-related, "s-structure" interpretation (quantificational, topic-focus, etc.)

The discussion was based on the fundamental hypothesis that Merge-based syntax is essentially "symmetric," free from any asymmetric representations like left-to-right ordering or projection, and further driven by the need for feature-equilibria. Note that this is in significant contrast to the still-dominant hypothesis that "asymmetry" of some sort is the norm for syntax. The traditional stipulations of universal endocentricity/projection (2)-(4) are examples of such an asymmetry-oriented approach, and so are various versions of Kayne's (1994 et seq.) "antisymmetry" approach (see also Moro 2000), all of which clearly fall short of capturing the fundamental transitional directionality from F-asymmetry to F-symmetry in (24). Although a number of important implications of the proposal remain to be explored and examined carefully, we hope that the analysis presented in this chapter provides a promising insight into the nature of linguistic computation and the syntax-semantic interface of human language.

## Notes

* Part of this research is supported by the Japan Science and Technology Agency (CREST) and by the Japan Society for the Promotion of Science (Grant-in-Aid for Scientific Research, Scientific Research (A) (General) #23242025, and Challenging Exploratory Research #25580095).
1 Here and in what follows, we will freely use notations like XP, YP to refer to phrasal constituents. However, it should be noted that we are not granting any ontological status to nonterminal label-symbols such as XP, or to any label-based structural notions like Spec or complement. That is to say, no notion of labeling or projection is implied in our informal usage of these terms. Thus, notations like TP, $v$P, and so on, refer to nothing more than phrasal constituents containing T, $v$, etc.
2 We may say that nominals are headed by D(eterminer) (Fukui and Speas 1986, Abney 1987, and Fukui 1986/1995, 1988) or by the categorizer head $n$ into which D incorporates (Chomsky 2007). We refrain from making specific assumptions regarding the exact nature of the nominal head, for want of better understanding of nominal-internal syntax.
3 This systematic transitional pattern cannot be captured by, e.g., Moro's (2000) approach, which holds that movement is instead a "symmetry-breaking" (i.e., asymmetrizing) phenomenon serving for linearization (or labeling, see also Chomsky 2008). We will provide various arguments bearing on Moro's and others' hypotheses that asymmetrization is the norm of syntactic computation.
4 When we discuss the $X^0$-$Y^0$-merger, it is not accurate to provide an asymmetric characterization like "$X^0$ moves to $Y^0$." In the symmetric $\{X^0, Y^0\}$ structure, the two LIs are equally prominent, and neither is a "target" or "adjunct" of the other.
    In the traditional description, the $X^0$-$Y^0$-merger has been assumed to feed "upward head-movement" at the PHON(ological) interface, ensuring that $\{X^0, Y^0\}$ is pronounced at the higher position of $Y^0$. However, since $X^0$ is no less prominent than $Y^0$ in $\{X^0, Y^0\}$, there is no reason to exclude the possibility that $\{X^0, Y^0\}$ instead gets pronounced at $X^0$, resulting in a "downward head-movement" form at PHON. For example, Pollock (1989) argues that the $V^0$/$v^0$-movement in Romance languages gets pronounced at $T^0$, whereas its counterpart in English yields $T^0$'s downward attachment to $V^0$/$v^0$. Under our analysis, we may suppose that one and the same SO $\{v^0, T^0\}$ may underlie these two different phonetic forms. We will leave further exploration of this approach for future research.

5 Narita and Fukui (2012, 2014) alternatively propose the notion of "feature prominence" in (i), which defines the order of prominence inductively "from bottom-up," in contrast to the "top-down" induction in (18b).

(i) *Feature prominence*:

Suppose that $\text{Prom}(F)(\alpha) = n$ ($n \geq 0$) is the order of prominence associated with a feature F with respect to an SO $\alpha$, with, let us say, lower prominence indicated by higher number, 0 being the highest order of prominence. Then, we can say:

a. $\text{Prom}(F)(H) = 0$, if F is a feature of an LI H.
b. If $\text{Prom}(F)(\alpha) = n$, then $\text{Prom}(F)(\{\alpha, \beta\}) = n + 1$ for any SO $\beta$.

(i) serves just as well for the purposes of this chapter as the notion of structural prominence in (18), though the latter may have somewhat broader applications; see Kato et al. (2016).

6 We maintain that the EIC also plays an important role in the determination of featural (a)symmetry. Recall that featural symmetry is defined relative to the presence of some formal feature F in $\{\alpha, \beta\}$, thus $\{n\text{P}, \text{T'}\}$ in (19) counts as $\phi$-symmetric thanks to the presence of [$\phi$], but not as, say, Q-symmetric or Q-asymmetric given the lack of [Q] in the SO. However, one may still wonder whether the same holds if the subject $n$P contains Q-features that are more deeply embedded: for example, [*the question of what John bought yesterday*] or [*the boy who John met last week*]. Obviously, the desirable result is that these occurrences of [Q] within those embedded clauses are irrelevant to the featural (a)symmetry defined for $n$Ps containing them. We argue that this desideratum can be achieved once we allow [Q] and other features in these lower domains (probably demarcated by notions like "phase"; see Chomsky 2000 *et seq.*, Narita 2014) to become invisible due to the EIC, after forming an equilibrium on their own. More generally, we maintain that the calculation of F-(a)symmetry also counts as "F-related computation" in the definition of the EIC (26).

7 Narita and Fukui (2012, 2014) point out that the interplay of the DSC and the EIC can provide a full-fledged account of A-movement in terms of $\phi$-(a)symmetry, thereby nullifying the recourse to the technical concept of Case-checking. They argue that the notion of abstract Case can be thereby eliminated altogether from the theory of syntax.

# References

Abney, Steven Paul. 1987. The English noun phrase in its sentential aspect. Doctoral Dissertation, MIT.

Boeckx, Cedric, and Sandra Stjepanović. 2001. Heading toward PF. *Linguistic Inquiry* 32:345–355.

Cable, Seth. 2010. *The grammar of Q: Q-particles, wh-movement, and pied-piping*. Oxford: Oxford University Press.

Chomsky, Noam. 1970. Remarks on nominalization. In *Readings in English transformational grammar*, ed. Roderick A. Jacobs and Peter S. Rosenbaum, 184–221. Waltham, MA: Ginn.

Chomsky, Noam. 1995. *The minimalist program*. Cambridge, MA: MIT Press.

Chomsky, Noam. 2000. Minimalist inquiries: The framework. In *Step by step: Essays on minimalist syntax in honor of Howard Lasnik*, ed. Roger Martin, David Michaels, and Juan Uriagereka, 89–155. Cambridge, MA: MIT Press.

Chomsky, Noam. 2001. Derivation by phase. In *Ken Hale: A life in language*, ed. Michael Kenstowicz, 1–52. Cambridge, MA: MIT Press.
Chomsky, Noam. 2004. Beyond explanatory adequacy. In *Structures and beyond: The cartography of syntactic structures*, ed. Adriana Belletti, volume 3, 104–131. New York: Oxford University Press.
Chomsky, Noam. 2007. Approaching UG from below. In *Interfaces + recursion = language? Chomsky's minimalism and the view from semantics*, ed. U. Sauerland and H.-M. Gärtner, 1–29. Berlin and New York: Mouton de Gruyter.
Chomsky, Noam. 2008. On phases. In *Foundational issues in linguistic theory*, ed. Robert Freidin, Carlos Otero, and Maria Luisa Zubizarreta, 133–166. Cambridge, MA: MIT Press.
Chomsky, Noam. 2013. Problems of projection. *Lingua* 130:33–49.
Collins, Chris. 2002. Eliminating labels. In *Derivation and explanation in the minimalist program*, ed. Samuel David Epstein and T. Daniel Seely, 42–64. Oxford: Blackwell.
Fukui, Naoki. 1986/1995. A theory of category projection and its applications. Doctoral Dissertation, MIT. Published in 1995 with revisions as *Theory of projection in syntax*, Kurosio Publishers and CSLI publications.
Fukui, Naoki. 1988. Deriving the differences between English and Japanese: A case study in parametric syntax. *English Linguistics* 5:249–270.
Fukui, Naoki. 2006. *Theoretical comparative syntax: Studies in macroparameters*. London/New York: Routledge.
Fukui, Naoki. 2011. Merge and bare phrase structure. In *The Oxford handbook of linguistic minimalism*, ed. Cedric Boeckx, 73–95. Oxford: Oxford University Press.
Fukui, Naoki. 2012. *Linguistics as a natural science* (new and expanded edition). Chikuma Gakugei Bunko (Math & Science). Tokyo: Chikuma Syoboo.
Fukui, Naoki, and Hiroki Narita. 2014. Merge, labeling, and projection. In *Routledge handbook of syntax*, ed. Andrew Carnie, Dan Siddiqi, and Yosuke Sato, 3–23. London/New York: Routledge.
Fukui, Naoki, and Margaret Speas. 1986. Specifiers and projection. *MIT Working Papers in Linguistics* 8:128–172. Reprinted in Fukui (2006).
Ishii, Toru. 1997. An asymmetry in the composition of phrase structure and its consequences. Doctoral Dissertation, University of California, Irvine.
Kato, Takaomi. 2006. Symmetries in coordination. Doctoral Dissertation, Harvard University.
Kato, Takaomi, Hiroki Narita, Hironobu Kasai, Mihoko Zushi, and Naoki Fukui. 2016. On the primitive operations of syntax. In *Advances in biolinguistics: The human language faculty and its biological basis*, ed. Koji Fujita and Cedric Boeckx, 29–45. London/New York: Routledge.
Kayne, Richard S. 1994. *Antisymmetry of syntax*. Cambridge, MA: MIT Press.
Kuno, Susumu. 1973. *The structure of the Japanese language*. Cambridge, MA: MIT Press.
Kuroda, S.-Y. 1965. Generative grammatical studies in the Japanese language. Doctoral Dissertation, MIT.
Kuroda, S.-Y. 1988. Whether we agree or not: A comparative syntax of English and Japanese. In *Papers from the second international workshop on Japanese syntax*, ed. W. J. Poser, 103–143. Stanford, CA: CSLI Publications. Reprinted in Kuroda (1992).

Kuroda, S.-Y. 1992. *Japanese syntax and semantics: Collected papers.* Dordrecht, Netherlands: Kluwer.

Moro, Andrea. 2000. *Dynamic antisymmetry.* Cambridge, MA: MIT Press.

Narita, Hiroki, and Naoki Fukui. 2012. Merge and (a)symmetry. ms. Waseda Institute for Advanced Study and Sophia University. Paper presented by the second author at the Kyoto Conference on Biolinguistics, Kyoto University, March 12, 2012.

Narita, Hiroki, and Naoki Fukui. 2014. Toji kozo no naishinsei to taishosei ni tsuite [On the notions of endocentricity and symmetry in syntactic structures]. In *Gengo no Sekkei, Hattatsu, Shinka: Seibutsugengogaku Tankyu* [*The design, development and evolution of language: Explorations in biolinguistics*], ed. Koji Fujita, Naoki Fukui, Noriaki Yusa, and Masayuki Ikeuchi, 37–65. Tokyo: Kaitakusha.

Ohta, Shinri, Naoki Fukui, and Kuniyoshi L. Sakai. 2013. Syntactic computation in the human brain: The degree of merger as a key factor. *PLoS ONE* 8(2):e56230. doi:10.1371/journal.pone.0056230.

Pollock, Jean-Yves. 1989. Verb movement, Universal Grammar and the structure of IP. *Linguistic Inquiry* 20:365–424.

Richards, Marc D. 2007. On feature inheritance: An argument from the phase impenetrability condition. *Linguistic Inquiry* 38:563–572.

# 3 On the primitive operations of syntax*

*Takaomi Kato, Hiroki Narita, Hironobu Kasai, Mihoko Zushi and Naoki Fukui*

## 1 Introduction

Human language in its essence is a system that generates an infinite set of hierarchically structured syntactic objects (SOs) which in turn can be used to fulfill various cognitive needs, such as to relate "meaning" and "sound" (the "Conceptual-Intentional" (CI) and "Sensorimotor" (SM) systems, respectively). It has been widely assumed in the literature that the simplest characterization of the relevant computation builds upon the notion of *Merge*, a combinatory operation that takes $n$ (often two) SOs (lexical items or composite phrases) and constructs an unordered set of these objects: Merge($\Sigma_1, \ldots, \Sigma_n$) = {$\Sigma_1, \ldots, \Sigma_n$}. We can safely assume that iterative application of Merge is sufficient to yield an infinite set of hierarchical SOs (Chomsky 2004 *et seq.*). Thus, it has been asked in various manners whether we may ultimately conclude that Merge is all we need to stipulate in the theory of human language (cf. Boeckx 2009, Berwick 2011, Fujita 2013). In so doing, the literature seems to share the conclusion that Merge counts as a "virtual conceptual necessity," and hence may be the most elementary and the most basic operation of syntax.

The purpose of this chapter is to challenge this established assumption. We will put forward the hypothesis that Merge is not the most elementary syntactic operation, and that it should rather be understood as a composite of two primitive operations, i.e., (i) the selection of $n$ elements, $\alpha_1, \ldots, \alpha_n$, from a designated domain of computation, and (ii) the formation of an unordered set of these $n$ elements, {$\alpha_1, \ldots, \alpha_n$}. We will argue that the proposed analysis is not only required in terms of descriptive preciseness, but also favorable on several empirical grounds. Among other things, we will argue that the relevant operations (i) and (ii) can be further generalized to capture a number of linguistic "relations" that Merge alone as usually construed cannot capture, including Agree(ment), chains formed by Internal Merge (IM), binding, and also headedness (or labeling; cf. Chomsky 2005, 2007, 2008).

This chapter is organized as follows. In Section 2, we will review the notion of *Search* discussed by Kato et al. (2014), which was proposed as a way to unify the otherwise disparate relation-formation operations, including Agree, chain-formation, and binding. We will show that the relevant notion of Search can be decomposed into two more primitive operations: the first operation, which we will call *0-Search* ($Search_0$), is the selection of *n* (typically two) elements, corresponding to what are generally called the "probe" and the "goal"; then, the second operation *0-Merge* ($Merge_0$) is the formation of a linguistic relation between X and Y, which we propose to characterize as {X, Y}. Therefore, our hypothesis is that Search = $Merge_0(Search_0(\Sigma))$ = $Merge_0 \circ Search_0(\Sigma)$ for some SO $\Sigma$. In Section 3, we will propose the decomposition of the standard notion of "Merge" discussed above and argue that (i) and (ii) are exactly the same operations as $Search_0$ and $Merge_0$, respectively. Therefore, Merge reduces to another instance of $Merge_0 \circ Search_0$. In Sections 4 and 5, we will discuss various consequences and challenges of the proposed unification of Search and Merge as $Merge_0 \circ Search_0$. Section 6 will conclude the chapter.

## 2 Decomposing Search

Merge has been taken to be an indispensable basic operation of the syntax of human language, since it has the fundamental function of constructing an infinite set of hierarchically structured SOs, without which human language simply cannot have the property of discrete infinity that it exhibits. Although it is further suggested in the literature (cf. Boeckx 2009, Berwick 2011, Fujita 2013, etc.) that Merge is the only operation that exists in human language, the actual situation in practice is such that various other miscellaneous operations have been postulated to describe a variety of linguistic phenomena. This is partly because Merge as usually construed does not suffice to capture the "relations" that are established among (portions of) SOs, feeding interpretations at the CI and SM interfaces (SEM and PHON, respectively). Such relations include Agree(ment), chains formed by IM and binding.

Kato et al. (2014) attempt to unify the operations proposed to capture these relations under a general search operation, which is referred to as *Search* and characterized as in (1).

(1) Let α be an element which initiates Search and β be the c-command domain of α. Then, Search is an operation which searches through β for a feature or a complex of features identical to the one contained in α and establishes a relation between those (complexes of) features. (Kato et al. 2014:204)

Search functions, as it were, as a generalized probe-goal mechanism and covers the cases of Agree, chain-formation and binding.

Let us briefly see how Search applies in the course of a derivation, taking a case of *wh*-movement as an example.[1] Consider the derivation of (2a) below.

(2) a. What did John buy?
  b. [C [T [John [what *v* [buy what]]]]]
      [Q]      [Q]
      └─────────────┘ *Search*
  c. [what C [T [John [what *v* [buy what]]]]]
      └──────────────────┘ *Search*

At the derivational stage illustrated in (2b), C initiates Search, so that an agreement relation is established between the Q feature ([Q]) of C (which Kato et al. take to be unvalued) and the Q feature of *what* (which they take to be valued) at the edge of *v*P.[2] After *what* is internally merged to the edge of CP from the edge of *v*P as in (2c), it also initiates Search from the higher position and a chain is created between the two copies of the *wh*-phrase. Agree and chain-formation are thus unified under a single operation Search.[3]

Kato et al. (2014) present Search as one of the basic operations of syntax (or a version of the basic operation of syntax, Merge[4]). It is implicitly assumed there that Search (as well as Merge) is a primitive operation, which cannot be decomposed any further. As can be seen in (1), however, Search clearly consists of two parts: one that picks out two elements (that is, a feature or feature complex which the Search initiator contains and a feature or feature complex identical to it contained in the c-command domain of the Search initiator) and one that establishes a relation between them. Thus, it seems quite plausible to regard Search as being decomposable, or a composite operation.

In this chapter, we propose that Search is a composition of two more primitive operations, which we call *0-Search* ($Search_0$) and *0-Merge* ($Merge_0$) and formulate as in (3) and (4), respectively (the formulation of $Search_0$ will be slightly changed in the next section).

(3) $Search_0$:
Search$_0$ is an operation such that given an SO $\Sigma$, it picks out $n$ elements contained in $\Sigma$ (i.e. $Search_0(\Sigma) = \alpha_1, \ldots, \alpha_n$, where $\alpha_1, \ldots, \alpha_n$ are elements contained in $\Sigma$).[5,6]

(4) $Merge_0$:
Merge$_0$ is an operation such that given $n$ objects, it forms the set of these objects (i.e. $Merge_0(\alpha_1, \ldots, \alpha_n) = \{\alpha_1, \ldots, \alpha_n\}$).

We claim that what is called Search in Kato et al. (2014) is $Merge_0(Search_0(\Sigma))$ (henceforth, $M_0 \circ S_0(\Sigma)$), and that when the output of $M_0 \circ S_0(\Sigma)$ is a two-membered set {X, Y}, it is interpreted as "there is a relation between X and Y" at SEM and/or PHON.[7]

Let us see how agreement, chain-formation and binding, which Kato et al. (2014) unify under Search, are recaptured under $M_0 \circ S_0(\Sigma)$.[8] First, consider $\phi$-agreement between T and the subject *n*P.[9] Suppose that the derivation has reached the stage where T and *v*P have been merged to form {T, *v*P}. First, Search$_0$ applies to this SO and picks out the (unvalued) $\phi$-features ([$\phi$]) of T and the (valued) $\phi$-features of the subject *n*P, located at the edge of *v*P. Then, Merge$_0$ applies to these sets of features and forms a set containing them. Thus, $M_0 \circ S_0(\Sigma)$ takes {T, *v*P} as an input and yields {[$\phi$] (of T), [$\phi$] (of the subject *n*P)} as the output, as shown in (5).[10,11]

(5) $M_0 \circ S_0(\{T, vP\}) = \{[\phi]$ (of T), $[\phi]$ (of the subject *n*P)$\}$

We propose that the output of this application of $M_0 \circ S_0(\Sigma)$ will be interpreted as "there is a $\phi$-agreement relation between T and the subject $n$P" at PHON.[12]

Next, let us see how $M_0 \circ S_0$ works for chain-formation, taking (6a) as an example. Consider the derivational stage where as a result of IM, a copy of *what* has been created at the edge of $v$P, as shown in (6b).

(6) a. What did John buy?
   b. [$_{v\text{P}}$ what [John $v$ [$_{\text{VP}}$ buy what]]]

First, Search$_0$ applies to the entire SO at hand (i.e., $v$P) and picks out the two copies of *what*. Then Merge$_0$ applies to these copies and forms the set of them. Thus, $M_0 \circ S_0(\Sigma)$ works as shown in (7) here.

(7) $M_0 \circ S_0(v\text{P}) = \{$*what* (at the edge of $v$P), *what* (within VP)$\}$

The set on the right-hand side of (7) will be interpreted as a chain of *what* at PHON and SEM.

Anaphor binding as in (8a) also falls under $M_0 \circ S_0(\Sigma)$.

(8) a. John loves himself.
   b. [$_{v\text{P}}$ John [$_{\text{VP}}$ love himself]]

Departing from Kato et al. (2014), we assume that binding is a relation established between two $n$Ps, rather than agreement in some specific feature (see note 3). At the derivational stage given in (8b), Search$_0$ first applies to $v$P and picks out *himself* and *John*. Then Merge$_0$ applies to these $n$Ps and forms the set of these objects, as shown in (9).

(9) $M_0 \circ S_0(v\text{P}) = \{$*John*, *himself*$\}$

This output of $M_0 \circ S_0(\Sigma)$ will be interpreted as "there is a binding relation between *John* and *himself*" at SEM (in order for *John* to be interpreted as a "binder" and *himself* as a "bindee," SEM also needs to access information on the configurational relation between the two $n$Ps, namely that *John* is structurally higher than (or "c-commands") *himself*).

Unlike the Search operation proposed by Kato et al. (2014), which establishes a relation only between identical elements (i.e., identical features or feature complexes), $M_0 \circ S_0(\Sigma)$ can form a pair of non-identical elements as well. We claim that the formation of a pair of non-identical elements occurs in the case of labeling.[13] More specifically, assuming that what has to be done in labeling is to form a(n "is headed by") relation between an SO and its label, we claim that in labeling, $M_0 \circ S_0(\Sigma)$ works in such a way that Search$_0$ first picks out the SO $\Sigma$ itself and its label and then Merge$_0$ forms the set of these objects, as shown in (10) (where λ = the label of Σ).[14,15] This set is read as "Σ is headed by λ" at the interfaces.

(10)  $M_0 \circ S_0(\Sigma) = \{\Sigma, \lambda\}$

For example, in the labeling of the SO {*love, himself*}, $Search_0$ applies to this SO and picks out the SO itself and *love*, and $Merge_0$ applies to these objects and forms the set of them, as in (11). As a result of this application of $M_0 \circ S_0(\Sigma)$, *love* will be interpreted as the label of the relevant constituent.

(11)  $M_0 \circ S_0(\{love, himself\}) = \{\{love, himself\}, love\}$

If the discussion so far is on the right track, labeling is unified with other operations such as Agree, binding and chain-formation, and we now have a very general unifying mechanism. This is one of the most notable results obtained here. Chomsky has suggested in his various writings that (minimal) search is involved in labeling (see, for example, Chomsky 2005, 2007, 2008). Although it is safe to bet that some search mechanism is also involved in operations like Agree (see, for example, Chomsky 2015b), it is not clear how to unify these operations with labeling under the standard probe-goal system. This is because under the latter system, the two items to be related (that is, the probe and the goal) must be located in such a way that one c-commands the other, but in the case of labeling, an SO and its label are not in such a positional relationship.[16] In contrast, under the current proposal, in which the traditional notion of probe/goal is abandoned, we can unify labeling and other operations such as Agree in a very natural way under a single operation $M_0 \circ S_0(\Sigma)$ (see note 7 for relevant discussion).

## 3 Decomposing Merge

According to Chomsky, Merge is an operation that takes *n* objects and combines them by forming the set of these objects (cf. Chomsky 2005, 2007, 2008, among others). Given these two functions of Merge, it can be also regarded as a composite operation. Recall from the discussion in the preceding section that Search is decomposed into two more primitive operations $Search_0$ and $Merge_0$. The former picks out *n* objects and the latter applies to these objects and forms the set of them. In this section, we propose that Merge is also an instance of the composition of $Search_0$ and $Merge_0$, $M_0 \circ S_0$.[17] In the preceding section, we formulated $Search_0$ in such a way that it applies to SOs. Below, we argue that this formulation of $Search_0$ should be modified, thereby allowing $M_0 \circ S_0$ to cover the cases of Merge, particularly the cases of External Merge (EM).

Let us begin our discussion with IM. Under our proposal, IM involves an application of $M_0 \circ S_0$ as shown below:

(12)  Internal Merge (IM)
      $M_0 \circ S_0(\Sigma) = \{\alpha, \Sigma\}$ (where α is contained in Σ)[18]

Given an SO $\Sigma$, $Search_0$ picks out $\Sigma$ and an element $\alpha$ contained in it. Then $Merge_0$ applies to these objects and forms a set $\{\alpha, \Sigma\}$. Let us consider the following example:

(13) a. What did John buy?
 b. [$_{vP}$ John [$v$ [$_{VP}$ buy what]]]

The derivation of (13a) has reached the stage shown in (13b), where *John* has been merged with $\{v, \{buy, what\}\}$ to form $\{John, \{v, \{buy, what\}\}\}$. $Search_0$ applies to this SO and picks out *what* and the whole SO. $Merge_0$ applies to these objects and forms the set of them. In short, $M_0°S_0$ works as in (14).

(14) $M_0°S_0(\{John, \{v, \{buy, what\}\}\})$ = $\{what, \{John, \{v, \{buy, what\}\}\}\}$

The output of this application of $M_0°S_0$ becomes the input to another application of $M_0°S_0$ for chain-formation (see Section 2 for details).[19]

Now, let us turn to EM. Consider first the cases where both of the targets of EM are lexical items (LIs).[20] Suppose that *the* and *book* undergo EM to form *n*P. So far, we have assumed that the input of $Search_0$ is an SO, as formulated in (3). However, since *the* and *book* are LIs, they are contained in the Lexicon, not in any SO. This would prevent $Search_0$ from picking out these objects. Thus, it is necessary to modify the formulation of $Search_0$ so that $Search_0$ can also pick out LIs contained in the Lexicon. A possible reformulation of this operation to obtain the desired effect is the following, where "linguistic object" is used as a cover term for both SOs and the Lexicon.

(15) $Search_0$ (revised):
 $Search_0$ is an operation such that given a linguistic object $\Lambda$, it picks out *n* elements contained in $\Lambda$.

Given the revised version of $Search_0$ in (15), $M_0°S_0$ can take the Lexicon as an input and pick out two lexical items $LI_i$, $LI_j$ contained in it and form the set of them, as shown in (16).

(16) $M_0°S_0(Lexicon)$ =$\{LI_i, LI_j\}$

Thus, $M_0°S_0$ can now construct SOs like $\{the, book\}$, where both of the members are LIs.

However, the formulation given in (15) fails to accommodate the cases where at least one of the targets of EM is not an LI. For example, let us consider the derivational step where *read* undergoes EM with *the book*, as shown in (17).

(17) *read* ← EM → $\{the, book\}$

In order to achieve the effect of this EM, $Search_0$ in $M_0°S_0$ must be able to pick out an object contained in the Lexicon (*read*) and an object not contained in the Lexicon ({*the*, *book*}). Under the formulation in (15), however, Search$_0$ fails to do this: if it takes the Lexicon as an input, it cannot pick out {*the*, *book*}, while if it takes this SO as an input, it cannot pick out *read*. Thus, we propose to revise the formulation of Search$_0$ further in the following way:

(18) Search$_0$ (final version)
Search$_0$ is an operation that picks out *n* elements contained in the workspace (WS).

We assume that WS is the set consisting of SOs already constructed and LIs in the Lexicon, that is, WS = {$\Sigma_1, \ldots, \Sigma_n$} ∪ Lexicon = {$\Sigma_1, \ldots, \Sigma_n, LI_1, \ldots, LI_m$} (note that the Lexicon is the set of LIs, which we tentatively take to be finite). Reformulated as in (18), Search$_0$ can now pick out an LI and a non-LI SO $\Sigma$, since both of them are contained in WS. Thus, $M_0 \circ S_0$(WS) can work as shown below.

(19) $M_0 \circ S_0$(WS) = {$\Sigma$, LI}

We argue that this is what happens in (17): Search$_0$ applies to WS = {{*the*, *book*}, $LI_1, \ldots$, *read*, $\ldots, LI_m$} and picks out *read* and {*the*, *book*}; Merge$_0$ applies to these objects and forms a set {*read*, {*the*, *book*}}.

Search$_0$ as formulated in (18) can also pick out two non-LI SOs $\Sigma_i, \Sigma_j$ contained in WS. In that case $M_0 \circ S_0$(WS) = {$\Sigma_i, \Sigma_j$}. This takes place, for example, in the EM of {*the*, *woman*} and {*v*, {*criticize*, *John*}} to form *v*P:

(20) {*the*, *woman*} ← EM → {*v*, {*criticize*, *John*}}

Search$_0$ applies to WS = {{*the*, *woman*}, {*v*, {*criticize*, *John*}}, $LI_1, \ldots, LI_m$} and picks out {*the*, *woman*} and {*v*,{*criticize*, *John*}}; Merge$_0$ applies to these objects and forms a set {{*the*, *woman*}, {*v*,{*criticize*, *John*}}}.

To sum up the discussion in this section and the previous one: we have proposed that the Search operation of Kato et al. (2014) ought to be regarded as a composition of the two more primitive operations, Search$_0$ and Merge$_0$, which we notate as $M_0 \circ S_0$. Search$_0$ picks out *n* objects and Merge$_0$ forms the set of these objects. Under the proposed mechanism, labeling is naturally unified with other operations such as Agree, binding and chain-formation. We have also shown that Merge can also be regarded as an instance of $M_0 \circ S_0$ by modifying the formulation of Search$_0$ so that $M_0 \circ S_0$ takes WS as an input. If the discussion so far is on the right track, all the syntactic operations are unified under $M_0 \circ S_0$(WS), a welcome result from a minimalist perspective.

## 4 Refining the notion of occurrence

In the preceding discussion, we established that Merge and Search, as well as labeling (Chomsky 2005, 2007, 2008), can be uniformly characterized as M∘S(WS). We specifically hypothesized that Merge$_0$ always yields an unordered set in every case, regardless of whether the result is chain-formation, Agree, labeling or binding. As a concrete example, consider again the $v$P structure in (21), and specifically the instance of $M_0°S_0$ that serves to generate a chain with two copies of *what* in (22).[21]

(21)   $v$P = {*what*, {*John*, {*v*, {*buy*, *what*}}}}

(22)   $M_0°S_0$(WS) = {*what* (at the edge of $v$P), *what* (within VP)}

In (22), the two copies of *what* are identical and indistinguishable from each other, assuming that there are no indices or other representational tricks that differentiate the two tokens (cf. the Inclusiveness Condition of Chomsky 1995 et seq.). Then, readers may wonder if chain-notations like (22) cause a problem from the perspective of set theory. The reason is that the output of (22) should be equivalent to that of (23) in terms of extension.

(23)   $M_0°S_0$(WS) = {*what*}

(23) can be achieved when $Search_0$($v$P) picks out only one instance of *what* within $v$P. The two sets in (22) and (23) contain exactly the same (in fact unique) element, and hence count as identical, if we follow the basic principle of set theory that a set is determined uniquely by its members (the Zermelo-Fraenkel axiom of extensionality). This is problematic because the conception that (22) = (23) fails to capture the chain relation between the two occurrences of *what* in (21).

One way to circumvent this problem is to adopt Chomsky's (2001) idea, namely that each occurrence of an SO is in fact defined in terms of its "mother" node SO. According to this hypothesis, the chain corresponding to (22) is represented as in (24).

(24)   $M_0°S_0$(WS) = {{*what*, {*John*, {*v*, {*buy*, *what*}}}}, {*buy*, *what*}}

This approach nicely resolves the problem of extensional equivalence, since the non-distinguishable copies of *what* in (22) are successfully replaced with their "mother" SOs, which are distinguishable from each other.

Moreover, the mother-based definition of chains can also offer a unified analysis of the notion of "feature-chain" (that is, the agreement relation established between features of LIs). Recall that we defined the output of $M_0°S_0$ in service of $\phi$-agreement informally as in (25):

(25)   $M_0°S_0$(WS) = {[$\phi$] (of T), [$\phi$] (of the subject $n$P)}

If a valued feature and its unvalued counterpart count as identical to each other (cf. Chomsky's (2000:124) remark, "We therefore understand "feature-identity" [. . .] to be identity of choice of feature, not of value."), then the two sets of $\phi$-features may yield another case of extensional equivalence as shown in (26).

(26)   $M_0°S_0$(WS) = {[$\phi$], [$\phi$]} = {[$\phi$]}

Here, our mother-based solution to chains can be applied to the cases of feature-chains as well. Suppose that feature-chains are defined in terms of the "mothers" of relevant feature-occurrences. Note that LIs have been traditionally understood as bundles (i.e., sets) of features. Thus, each LI counts as the "mother" of the features it contains, if we assume that a set in the set-theoretic notation of phrase structure is regarded as the "mother" of its members. Then, we might say that what $M_0°S_0$ creates, when applying in the case of feature-chain formation, is in fact a set of two LIs, {T, $n$} in the case of (25) (assuming that $n$ is the LI that contains φ-features within a nominal). The two LIs each define an occurrence of the relevant agreement feature.

(27) $M_0°S_0$(WS) = {T, $n$}

Given these considerations, we propose that chains are uniformly characterized as *sets of the mothers of* the relevant elements. This approach not only eliminates the problem of extensional equivalence, but also has a broader empirical application, unifying chains of movement and feature-chains created by Agree(ment).[22]

## 5 Minimizing Search$_0$

We saw that $M_0°S_0$ not only derives recursive structure-generation via Merge, which constantly extends and rearranges the elements within WS, but also provides a unified account of various relation-forming computations, such as chain-formation, Agree(ment), binding and labeling. In this section, we will provide arguments that $M_0°S_0$ in service of relation-formation obeys certain locality conditions.

To begin, let us discuss the case of $M_0°S_0$ in service of labeling, which takes the following form.

(28) $M_0°S_0$(WS) = {Σ, λ}

It has been proposed in various forms that labeling obeys a strong locality condition, or what Chomsky (2012, 2013, 2015a) calls a "minimal search" requirement (see also Narita 2014). For example, consider the case of labeling in (29), which yields (30a).

(29) *v*P = {*v*, {*read*, {*n*, *books*}}} ∈ WS

(30) $M_0°S_0$(WS) = a. {*v*P, *v*}
　　　　　　　　　 b. *{*v*P, *read*}
　　　　　　　　　 c. *{*v*P, *n*}

Whenever labeling targets an SO Σ within WS, it must select the highest possible element λ within Σ as the label of Σ. Thus, only (30a) among (30a-c) counts as a legitimate instance of labeling.

What seems to be at work here is the requirement that the two outputs of $\text{Search}_0$, $\Sigma$ and $\lambda$, be minimally distant. In order to formulate this requirement, let us introduce the notion of "Structural Prominence" as a measure for structural distance:

(31) Structural Prominence (see Narita and Fukui 2016; see also Ohta et al. 2013 for a related notion of "Degree of Merger"):[23]
Suppose that $\text{Depth}(\alpha) = m$ ($m \geq 0$) is the order of depth – the inverse relation of prominence – associated with an SO $\alpha$, with lower prominence indicated by a higher value of depth. Then, we can say:

   a. $\text{Depth}(\alpha) = 0$ if there is no SO $\beta$ such that $\alpha \in \beta$ (i.e., $\alpha$ is a root SO dominated by no other SO).
   b. If $\text{Depth}(\alpha) = m$, then $\text{Depth}(\beta) = m + 1$ for any $\beta$ such that $\beta \in \alpha$ (i.e., $\beta$ is a daughter of $\alpha$).

(31a) states that "undominated" SOs are the most prominent in WS. (31b) states that the value of $\text{Depth}(\alpha)$ increases as $\alpha$ gets more deeply embedded and less prominent. In (29), for example, $\text{Depth}(v\text{P}) = 0$, $\text{Depth}(v) = 1$, $\text{Depth}(read) = 2$, $\text{Depth}(n) = 3$, etc. As shown in (30), $v$ but not *read* or $n$ is a legitimate output of $\text{Search}_0$ in service of labeling, given that $v$ is the most prominent lexical element that can define a label for some SO.

Capitalizing on (31), we can formulate a minimality condition on $M_0 \circ S_0$ as in (32), which basically serves to minimize the distance between the two outputs of $\text{Search}_0$.

(32) Minimality Condition on $M_0 \circ S_0$:
For any linguistic relation R, $M_0 \circ S_0(\text{WS})$ may generate $\{\alpha, \beta\}$ as an instance of R only if

   a. $\{\alpha, \beta\}$ meets formal restrictions on R, and
   b. There is no $\gamma$ such that $\{\alpha, \gamma\}$ also meets the formal restrictions on R, and $\text{Depth}(\gamma) < \text{Depth}(\beta)$.

(32) essentially recaptures the intuition of Rizzi's (1990) "Relativized Minimality" in the theory of $M_0 \circ S_0$. It holds that a relation between $\alpha$ and $\beta$ cannot be established if there is any intervening element $\gamma$ that can formally participate in R with $\alpha$ and is "closer" to $\alpha$ than $\beta$. Specifically for the Label(ing)-relation, we propose (33) as its formal restriction.

(33) Label:
$\{\alpha, \beta\}$ may count as an instance of Label only if

   a. $\alpha$ is a (bundle of) feature(s) (typically an LI),[24] and
   b. $\alpha$ is contained in $\beta$.

Given the formal restriction in (33), the minimality condition in (32) can explain cases like (30): $\text{Search}_0(\text{WS})$ can access any of $v\text{P}$, $v$, *read*, $n$, etc. in

(29), but $v$ is the closest term of $v$P that can define its label, and hence is the only LI that can enter into a labeling relation with $v$P via minimal Search$_0$.

Now, if we are right in claiming that every operation of syntax in fact involves $M_0 \circ S_0$, then it is naturally predicted that (32) generally applies to any instance of $M_0 \circ S_0$(WS), irrespective of which relation R it serves to establish. We will argue that this prediction is indeed borne out.

Let us consider binding, which is known to be possible only when the binder c-commands the bindee (in fact, c-command is part of the definition of binding). Thus, *John* can bind *himself* in (34a) but not in (34b), since *John* c-commands *himself* in (34a) but not in (34b).

(34) a. John$_i$ loves himself$_i$.
  b. *John$_i$'s father loves himself$_i$.

We argue that this "c-command requirement" can be seen as a natural consequence of minimality (32). To begin, let us introduce a formal condition on the Bind(ing) relation:[25]

(35) Bind:
  {α, β} may count as an instance of Bind only if α and β are headed by *n*.

In (34a), the subject *n*P *John* occupies the "Spec-T" position as shown in (36a), while this position is occupied by the *n*P that contains *John* (i.e., *John's father*) in (36b).

(36) a. {*John*, {T, {. . . *himself* . . .}}}
  b. {*n*P (= *John's father*), {T, {. . . *himself* . . .}}}

In (36b), Depth(*John*) > Depth(*John's father*). Thus, *John's father*, a possible binder, is closer to *himself* than *John*, and hence the Bind-relation in (34b) is blocked by the minimality condition in (32), coupled with the formal restriction in (35).

We can see that cases of chain-formation also satisfy (32). Building on the discussion in Section 4, we define the formal restriction on the Chain-relation as in (37).

(37) Chain (cf. Chomsky 2001):
  {α, β} may count as an instance of Chain only if there exists an element γ such that γ ∈ α and γ ∈ β.

Readers can easily confirm that the Chain-objects in (24) and (27) meet (32) and (37).

We argue that these conditions can also account for the general immobility of "trace" objects. For instance, once *John* gets stabilized in a Case-marked, subject position in (38a), neither *John* nor its trace (copy) *t* can move into another Case position, as shown in (38b).

(38) a. ___ seems that [$_{\alpha(=TP)}$ *John* will [$_{\beta(=\nu P)}$ *t* meet Mary]]
 b. *[$_\Sigma$ *John* seems that [$_{\alpha(=TP)}$ *t'* will [$_{\beta(=\nu P)}$ *t* meet Mary]]]

Since *John* already checks its Case in $\alpha$, it cannot move into another Case position, resulting in the immobility of *John* from *t'* in (38b). See Narita and Fukui (2016) for an account of the relevant fact in terms of their "Equilibrium Intactness Condition." See also Rizzi (2006) for the notion of "Subject Criterion." However, these authors have nothing to say concerning why the lower occurrence of *John* within $\beta$ (indicated by *t*) can never move, skipping over the other occurrence in *t'* in (38b).[26] We argue that our minimality condition in (32) can explain this state of affairs: $M_0 \circ S_0$(WS) cannot generate $\{\Sigma, \beta\}$, because there exists an SO $\alpha$ such that $\{\Sigma, \alpha\}$ also meets the formal restriction on Chain and Depth($\alpha$) < Depth($\beta$); hence, $\alpha$ is closer to $\Sigma$ than $\beta$.

Finally, let us turn to Merge, understood here as another instance of $M_0 \circ S_0$(WS). It combines two SOs (lexical or constructed) within WS, $\alpha$ and $\beta$, and creates a new SO $\{\alpha, \beta\}$. We take the primary function of Merge to be formation of a new sister relation. We can define Sister in terms of structural prominence discussed in (31).

(39) Sister:
 $\{\alpha, \beta\}$ may count as an instance of Sister only if Depth($\alpha$) = Depth($\beta$).

Each application of Merge($\alpha, \beta$) creates a new SO $\{\alpha, \beta\}$, and this new SO is added to WS. Since $\{\alpha, \beta\}$ is by definition not dominated by any other SO in WS so long as the NTC is satisfied, Depth($\alpha$) = Depth($\beta$) = 1 holds for every instance of Merge($\alpha, \beta$) = $M_0 \circ S_0$(WS) = $\{\alpha, \beta\}$ (see note 19).

Now we should ask if this form of $M_0 \circ S_0$ satisfies the minimality condition in (32). Consider (40) below:

(40) $\Sigma$ = {*John*, {*v*, {*buy*, *what*}}}

At first sight, IM of *what* to the edge of $\Sigma$ appears to be blocked by the existence of, for example, *v*: $\{\Sigma, v\}$ would satisfy the formal restriction in (39) and Depth(*v*) < Depth(*what*) in (40). This problem will be resolved if Depth($\gamma$) and Depth($\beta$) in (32b) are interpreted as "Depth($\gamma$) when $\{\alpha, \gamma\}$ meets the relevant formal restrictions" and "Depth($\beta$) when $\{\alpha, \beta\}$ meets the relevant formal restrictions," respectively. Under this assumption, Depth(*what*) and Depth(*v*) for the application of IM of *what* in (40) are calculated not based on (40), but based on (41a) and (41b), respectively:

(41) a. {*what*, {*John*, {*v*, {*buy*, *what*}}}}
 b. {*v*, {*John* {*v*, {*buy*, *what*}}}}

Depth($v$) in (41b) is not less than Depth($what$) in (41a) (since both of them are 1). We claim that this is why IM of $what$ satisfies the minimality condition in (32) in spite of the existence of the apparently intervening elements.[27]

The above discussion shows that the generalized minimality condition in (32) can allow unconstrained applications of Merge (i.e., Sister-formation), while it can also account for various effects of "relativized minimality" when applied to non-Sister-relations (Label, Bind, Chain, etc.). As for non-Sister-relations, it is also important to note that (32) not only sets an account of the locality of "goal" elements, but also derives the shallowness of "probe" elements in certain environments, such as the case of binding in (34a) vs. (34b).

It is clear that this unification and extension of the minimal search requirement can be achieved only if the traditional notion of probe/goal is eliminated, and various linguistic relations are unified into generalized $M_0 \circ S_0$(WS). Thus, the above discussion lends further support to our unified approach to various linguistic relations.[28]

## 6 Conclusion

In this chapter, we pointed out that Merge as traditionally construed is in fact a composite of two operations, i.e., (i) Search$_0$, the selection of $n$ elements, $\alpha_1, \ldots, \alpha_n$, from a designated domain of computation (WS), and (ii) Merge$_0$, the formation of an unordered set of the $n$ objects, $\{\alpha_1, \ldots, \alpha_n\}$. We argued that the proposed analysis of Merge as $M_0 \circ S_0$ is not only required in terms of descriptive preciseness, but also favorable in terms of breadth of empirical coverage. Specifically, we showed that $M_0 \circ S_0$ can be generalized to cases including Agree(ment), chain-formation, binding (cf. Kato et al.'s earlier notion of Search) and also labeling (cf. Chomsky 2005, 2007, 2008). In this manner, we can make a stronger and more precise sense of the recent claim that Merge is all we need to assume as a basic operation of syntax (cf. Boeckx 2009, Berwick 2011, Fujita 2013).

### Notes

* This research is supported in part by AMED-CREST.
1 Space limitations prevent us from providing a full review of how Search works. We refer the reader to Kato et al. (2014) for details.
2 Kato et al. (2014) argue that the mechanism of feature valuation must be eliminated if a theory of no-tampering syntax is seriously pursued (see Kato et al. 2014 for details). They instead put forward the following hypothesis:

 (i) When an Agree(ment) relation is established between an unvalued feature and a valued feature, the interface systems access it, so that the unvalued feature will be processed at SEM/PHON in relation to the valued feature. (Kato et al. 2014:214)

 In this chapter, we also assume that feature valuation does not exist and adopt the hypothesis in (i).
3 It is assumed in Kato et al. (2014) that binding is a sort of agreement (in φ-features or some independent referential feature [Ref]).

4  It is suggested in Kato et al (2014:Section 4) that Search could ultimately be reduced to Merge. Below we will suggest a different way of unifying the two operations.
5  Given what it actually does, a more appropriate – though cumbersome – term for this operation might be Search/Select.
6  In what follows we will not discuss cases where more than two elements are obtained as a result of $Search_0$, but leave open the possibility that such cases exist.
7  It is suggested in Kato et al. (2014:218) that the output of Search, which functions as a sort of generalized probe-goal mechanism, should be represented as an ordered pair, not a(n unordered) set, because a probe and a goal have to be distinguished. Departing from Kato et al. (2014), we claim that the output of $M_0°S_0(\Sigma)$ is represented as a set, not an ordered pair. This departure is reasonable to the extent that the traditional notion of probe/goal is to be abandoned under the current proposal. In this connection, it may be worth noting that Chomsky has recently suggested the possibility that there is no probe in the application of Agree, which should be reduced to some search procedure along with labeling (see Chomsky 2015b).

   Obviously, a set {X, Y} itself does not represent a relation R holding between X and Y. Our claim is that the interface systems somehow derive R from {X, Y} (and other information available to them). How this is done is an issue full investigation of which we leave for future research. See Section 5 for some related discussion.
8  Kato et al. (2014) hypothesize that what initiates Search are all and only elements at phase-edges, but we do not adopt a similar hypothesis for $M_0°S_0(\Sigma)$ here: we assume that $M_0°S_0(\Sigma)$ can apply at any point of the derivation in principle and its application is not forced by phase-edge elements (see note 17 below).
9  In this chapter, we use "$n$P" to refer to nominal phrases.
10  We will discuss how the outputs of $M_0°S_0(\Sigma)$ should be represented in Section 4. Until then, we will represent them in an informal way as in (5).
11  Here we tentatively assume that subject-T agreement occurs as soon as T is introduced into the derivation (see Kato et al. 2014 for the argument that the mechanism of feature inheritance should be eliminated). If it turns out that subject-T agreement occurs after C is introduced into the derivation (as suggested, for example, by Chomsky 2007, 2008, and Richards 2007), $M_0°S_0(\Sigma)$ would take CP, rather than TP, as an input for the agreement.
12  Note that the distinction between valued vs. unvalued features plays little role in $M_0°S_0(\Sigma)$. Thus, it can in principle establish agreement relations not only between a "higher" unvalued feature and a "lower" valued feature (as in traditional theories of Agree; see Chomsky 2000 *et seq.*), but also between a "higher" valued feature and a "lower" unvalued feature (see Baker 2008), and in fact between two valued features or between two unvalued features as well.
13  Kato et al. (2014) do not discuss labeling.
14  In Section 5, we will discuss how $M_0°S_0(\Sigma)$ picks out a particular element within $\Sigma$ as its label.
15  The tentative assumption here is that labeling occurs every time a new SO is created. If it turns out that labeling occurs at phase levels, as suggested in Chomsky (2013), we would claim that what $Search_0$ in $M_0°S_0(\Sigma)$ for labeling picks out is an SO contained in $\Sigma$ and its label.
16  A similar point applies to Search in Kato et al. (2014).
17  Since External Merge (EM) is also an instance of $M_0°S_0$ under our proposal, it is assumed here that $M_0°S_0$ in principle can apply at any point of the derivation (see note 8).

18 Note that IM and labeling are formally similar to each other, as shown in (10) and (12). See also Fujita (2009, 2012) for the view that labeling is a special case of IM.
19 We assume here that IM can occur not only at phase levels but also at other derivational points (contra Chomsky 2007, 2008, among others). Thus, subject-raising to "Spec-T" is assumed to occur before the phase head C is introduced into the derivation, satisfying the No-tampering Condition (NTC). See Epstein et al. (2012), Kato et al. (2014), Narita (2014) and Chomsky (2007, 2008, 2013, 2015a) for various proposals to the effect that IM occurs at phase levels while satisfying the NTC.
20 See also Chomsky (2012:23), where it is suggested that EM, as well as IM, involves some kind of search.
21 Note that all instances of $M_0 \circ S_0$ take WS as an input, insofar as the conclusion we reached in Section 3 is on the right track. Thus, the left-hand side of (22) and all other equations discussed in the previous sections is to be replaced with $M_0 \circ S_0(WS)$. It can reach elements inside $v$P when $v$P is a term of WS.
22 Chomsky's (2000) earlier set-based approach to chain is to assume that each occurrence is defined in terms of its sister (not its mother). This approach has a much earlier origin in Chomsky (1955/1975), adapted from Quine's (1940) notion of "occurrence of a variable." However, the sister-based definition of occurrence is not readily applicable to occurrences of feature-chains. Therefore, for the purpose of this chapter, we propose to adopt Chomsky's (2001) mother-based version of "occurrence."
23 Structural prominence is defined for SOs, which we take to be LIs and objects constructed from them via $M_0 \circ S_0$ in the form of Merge, which serves for Sister-relation-formation, as we will propose shortly. Crucially, we do not regard outputs of $M_0 \circ S_0$ representing non-Sister-relations as SOs.
24 Chomsky (2013) leaves room for cases in which a bundle of features smaller than an LI (say a collection of $\phi$-features) may participate in labeling. We will leave this possibility for future research, but the definition in (33a) can naturally incorporate such cases as well.
25 Obviously, (35) leaves many cases of binding failures unaccounted for. In order to achieve a full-fledged account of binding, much more than (35) should be supplied to constrain proper Bind-formation (conditions of binding theory, possibility of vehicle change, and so on), which falls beyond the scope of this short chapter.
26 Readers may wonder whether this effect can be attributed to some other factor, such as the "Phase-Impenetrability Condition" (PIC) (Chomsky 2004 *et seq*.). As pointed out by Kato et al. (2014), there is good reason to cast doubts on the current formulation of the PIC. These doubts pertain to the clear existence of long-distance (i.e., cross-phasal) dependencies. See Bošković (2007) and references cited therein for ample crosslinguistic examples of long-distance Agree(ment) that clearly violate the PIC. Moreover, binding (say, of pronouns or subject-oriented anaphors) can no doubt apply in a long-distance fashion as well, and thus binding is not constrained by the PIC, either. The theory of $M_0 \circ S_0$ eventually has to subsume such cases by assumption, and these considerations support the view that the PIC must be reconsidered.
27 Labeling of $v$P with $v$ in (29) is not blocked by the existence of, say, *read*, though it appears that Depth(*read*) in (30b) is not less than Depth($v$) in (30a). This is because structural prominence is defined only for SOs, and the sets in (30a-b) are not SOs. See note 23.
28 Note that, given $\Sigma_1, \Sigma_2 \in$ WS, $M_0 \circ S_0(WS)$ as proposed so far can merge $\Sigma_1$ and an element properly contained in $\Sigma_2$. Suppose that this type of Merge, known

as "parallel Merge" or "sideward movement" in the literature, is what should be blocked (cf. Chomsky 2015b). One way to make sense of this restriction is to assume that WS has a property such that only root SOs and LIs in the Lexicon count as its terms, and elements properly contained in root SOs are not terms of WS, hence inaccessible to $Search_0(WS)$. Suppose further that $Search_0$ so constrained can take either the entire WS or some root SO $\Sigma \in WS$ as its input. Then, specifically in order to pick out an element properly contained in $\Sigma$, $M_0°S_0$ has to take $\Sigma$ as an input, rather than WS. Therefore, neither $M_0°S_0(WS)$ nor $M_0°S_0(\Sigma)$ can achieve parallel Merge/sideward movement. We leave further exploration of the issue for future research.

# References

Baker, Mark. 2008. *The syntax of agreement and concord*. Cambridge, UK: Cambridge University Press.

Berwick, Robert. 2011. All you need is Merge: Biology, computation, and language from bottom up. In *The biolinguistic enterprise: New perspectives on the evolution and nature of the human language faculty*, ed. Anna Maria Di Sciullo and Cedric Boeckx, 461–491. Oxford: Oxford University Press.

Boeckx, Cedric. 2009. On the locus of asymmetry in UG. *Catalan Journal of Linguistics* 8:41–53.

Bošković, Željko. 2007. On the locality and motivation of Move and Agree: An even more minimal theory. *Linguistic Inquiry* 38:589–644.

Chomsky, Noam. 1955/1975. *The logical structure of linguistic theory*. Ms. Harvard University, 1955. Published in part in 1975, New York: Plenum.

Chomsky, Noam. 1995. *The minimalist program*. Cambridge, MA: MIT Press.

Chomsky, Noam. 2000. Minimalist inquiries: The framework. In *Step by step: Essays on minimalist syntax in honor of Howard Lasnik*, ed. Roger Martin, David Michaels, and Juan Uriagereka, 89–155. Cambridge, MA: MIT Press.

Chomsky, Noam. 2001. Derivation by phase. In *Ken Hale: A life in language*, ed. Michael Kenstowicz, 1–52. Cambridge, MA: MIT Press.

Chomsky, Noam. 2004. Beyond explanatory adequacy. In *Structures and beyond: The cartography of syntactic structures*, ed. Adriana Belletti, vol. 3, 104–131. Oxford: Oxford University Press.

Chomsky, Noam. 2005. Three factors in language design. *Linguistic Inquiry* 36:1–22.

Chomsky, Noam. 2007. Approaching UG from below. In *Interfaces + recursion = language? Chomsky's minimalism and the view from syntax-semantics*, ed. Uli Sauerland and Hans-Martin Gärtner, 1–29. Berlin and New York: Mouton de Gruyter.

Chomsky, Noam. 2008. On phases. In *Foundational issues in linguistic theory*, ed. Robert Freidin, Carlos Otero and Maria Luisa Zubizarreta, 133–166. Cambridge, MA: MIT Press.

Chomsky, Noam. 2012. Introduction. In *Gengokisoronshuu* [*Foundations of biolinguistics: Selected writings*], Noam Chomsky, ed. and trans. Naoki Fukui, 17–26. Tokyo: Iwanami Shoten.

Chomsky, Noam. 2013. Problems of projection. *Lingua* 130:33–49.

Chomsky, Noam 2015a. Problems of projection: Extensions. In *Structures, strategies and beyond: Studies in honour of Adriana Belletti*, ed. Elisa Di Domenico, Cornelia Hamann, and Simona Matteini, 3–16. Amsterdam/Philadelphia: John Benjamins.

Chomsky, Noam 2015b. A discussion with Naoki Fukui and Mihoko Zushi (March 4, 2014). In *The Sophia Lectures* (*Sophia Linguistica* 64), Noam Chomsky, 69–97. Tokyo: Sophia Linguistic Institute for International Communication, Sophia University.

Epstein, Samuel David, Hisatsugu Kitahara, and T. Daniel Seely. 2012. Structure building that can't be! In *Ways of structure building*, ed. Myriam Uribe-Etxebarria and Vidal Valmala, 253–270. Oxford and New York: Oxford University Press.

Fujita, Koji. 2009. A prospect for evolutionary adequacy: Merge and the evolution and development of human language. *Biolinguistics* 3:128–153.

Fujita, Koji. 2012. Toogoenzannooryoku to gengonooryoku no shinka [The evolution of the capacity for syntactic operations and linguistic competence]. In *Shinka-gengogaku no koochiku – Atarashii ningenkagaku o mezashite* [*Constructing evolutionary linguistics – Toward new human science*], ed. Koji Fujita and Kazuo Okanoya, 55–75. Tokyo: Hituzi Syobo.

Fujita, Koji. 2013. Evolutionary problems of projection. Paper read at Tokyo Workshop on Biolinguistics, Sophia University, Tokyo.

Kato, Takaomi, Masakazu Kuno, Hiroki Narita, Mihoko Zushi, and Naoki Fukui. 2014. Generalized Search and cyclic derivation by phase: A preliminary study. *Sophia Linguistica* 61:203–222.

Narita, Hiroki. 2014. *Endocentric structuring of projection-free syntax*. Amsterdam/Philadelphia: John Benjamins.

Narita, Hiroki, and Naoki Fukui. 2016. Feature-equilibria in syntax. In *Advances in biolinguistics: The human language faculty and its biological basis*, ed. Koji Fujita and Cedric Boeckx, 9–28. London/New York: Routledge.

Ohta, Shinri, Naoki Fukui, and Kuniyoshi L. Sakai. 2013. Syntactic computation in the human brain: The degree of merger as a key factor. *PLoS ONE* 8(2):e56230. doi:10.1371/journal.pone.0056230.

Quine, Willard V. O. 1940. *Mathematical logic*. Cambridge, MA: Harvard University Press.

Richards, Marc D. 2007. On feature inheritance: An argument from the phase impenetrability condition. *Linguistic Inquiry* 38:563–572.

Rizzi, Luigi. 1990. *Relativized minimality*. Cambridge, MA: MIT Press.

Rizzi, Luigi. 2006. On the form of chains: Criterial positions and ECP effects. In *Wh-movement: Moving on*, ed. Lisa Lai-Shen Cheng and Norbert Corver, 97–133. Cambridge, MA: MIT Press.

# 4 Case and predicate-argument relations*

*Mihoko Zushi*

## 1 Introduction

Linguistic items (LIs) are bundles of features. Among the features of LIs, the role of formal features in narrow syntax is very important, as formal features in their interaction with semantic features crucially characterize human language in a way that distinguishes it from other computational systems and other types of symbolic systems. What role do computational atoms like formal features play in human languages? In particular, what role do formal features, such as case and φ features, play in syntax? These are among the most contested issues in contemporary theoretical linguistics. This chapter intends to contribute to the understanding of these issues by intensively investigating the nature of formal features with a special focus on case features.

The dominant view on case in current linguistic theory is the one put forth by Chomsky (2001), according to which a case feature of nominal expressions is valued by means of Agree. On this view, case valuation takes place contingent on φ-feature agreement. Although this way of case valuation appears to work well in agreeing languages, including well-studied Indo-European languages, it faces serious empirical challenges when our investigation goes beyond such familiar languages and begins to explore empirically adequate theories of other types of languages, including Japanese, in which φ-features obviously play no role in syntax. How do case features of nominal expressions obtain their values without recourse to φ-feature agreement in non-agreeing languages? How can various peculiar case phenomena be explained? Several conditions of the application of Agree are seemingly violated in these phenomena. In order to resolve these questions, Zushi (2014a, b) proposes that case valuation in terms of external Merge, in addition to Agree, should be available in UG. This chapter presents how this proposal provides natural accounts of peculiar case phenomena in Japanese, such as multiple occurrences of identical cases and case alternation. Additionally, we attempt to reduce the two modes of case valuation into a single operation by incorporating the proposal made by Kato et al. (this volume) that the alleged basic operations of Agree and Merge are essentially the same operations.

Furthermore, this chapter addresses the question as to what role formal features play in human language. It has been generally assumed in current linguistic theory that while φ-features carry out important tasks in narrow syntax, the role of case features remains rather obscure. This chapter suggests that case features of nominal expressions, as well as φ-features of verbal elements, do in fact play an active role in determining labels of syntactic objects (SOs). Two ways of labeling are available. One way is based on case features of nominal elements, while the other way is based on φ-features of verbal elements. The former is typically employed in dependent-marking languages (to use Nichols's 1986 terms) such as Japanese, whereas the latter is utilized in head-marking languages like Mohawk. Thus, in dependent-marking languages, case serves as a key feature in building up the system of predicate-argument relations, which constitutes a core portion of the human language of thought. The role of case in this type of language, therefore, actually corresponds to the role of agreement in head-marking languages.

This chapter is organized as follows. Section 2.1 puts forth a proposal for case valuation in terms of Merge, in contradistinction to the widely assumed Agree-based case valuation. Section 2.2 discusses how the above-mentioned two modes of case valuation can be reduced to a single operation, by incorporating the ideas proposed by Kato et al. (this volume). Section 2.3 discusses the theoretical status of formal features such as case and φ features in narrow syntax. Section 3 presents some consequences of the proposed case theory for the analysis of peculiar case phenomena in Japanese. Section 3.1 deals with multiple occurrences of identical cases, and Section 3.2 discusses the phenomenon of case alternations, particularly nominative-genitive conversion. Finally, Section 4 presents the conclusion.

## 2 The case system

### 2.1 Two modes of case valuation

The standard view of case valuation in recent generative literature is to relate case valuation with φ-feature agreement (Chomsky 1995, 2001). This view faces empirical challenges from the type of language that does not exhibit φ-feature agreement. How does case valuation take place in such non-agreeing languages? How does it account for various peculiar case phenomena in those languages? In such phenomena, the conditions for Agree are seemingly broken. In order to resolve these questions, Zushi (2014a, b) revives and develops the view of case assignment on a structural basis, drawing on "classical" works such as Kuroda (1965, 1978, 1983, 1986, 1988), Fukui (1986, 1988), and Saito (1982), in a way that accommodates minimalist conceptions. Central to these classical works is the fact that the case assigned to a nominal depends on its position in the phrase structure. Therefore, Zushi reformulates the structural case assignment view in terms of external Merge, a structure-building operation in the minimalist program. Zushi claims that a case feature is valued by the following mechanisms in non-agreeing languages such as Japanese.[1]

(1) a. When a nominal is merged with a lexical head, its case feature is valued as accusative.
   b. When a nominal is merged with a phase head ($v$ or $n$), its case feature is valued as nominative or genitive.
   c. Otherwise, the case feature of a nominal is valued as dative.[2]

For example, when the nominal *hon* 'books' is merged with the verb *yomu* 'read' to form {hon, yomu}, the case feature of the nominal is valued as accusative by (1a). Likewise, when the nominal *Taroo* is merged with a syntactic object headed by $v$, its case feature is valued as nominative by (1b).[3] When a case feature of the nominal is not valued by either (1a) or (1b), (1c) features as a last resort to save the nominal that has an unvalued case feature.

While case valuation occurs in terms of external Merge as described in (1) in non-agreeing languages, Zushi argues, following Chomsky (2001) and subsequent work, that a case feature is valued in terms of Agree in languages that exhibit φ-feature agreement. Therefore, this proposal suggests two distinct modes of valuing case features – Agree-based and Merge-based – one of which is chosen for a particular language.

One question that arises here is what determines the mode of case valuation in a given language. Given that UG has the option of case valuation via Merge or Agree, why is the Merge-based case valuation not an option in agreeing languages? Similarly, why is the Agree-based case valuation not an option in non-agreeing languages? To address these questions, Zushi proposes that case valuation is conducted by a syntactic operation – either Merge or Agree – if the operation establishes a *covariance* relation R, which is defined as follows.

(2) R holds between α and β iff α and β covary with respect to a set of φ-features F.
(3) α and β covary with respect to F iff either (i) both α and β are specified with respect to F, or (ii) neither α nor β is specified with respect to F.

From this viewpoint, case valuation is a consequence of the covariance relation between α and β. The covariance relation holds if and only if either of two conditions is satisfied: both α and β bear a set of φ-features (3i) or neither one bears any φ-features (3ii). The former condition is tantamount to a standard agreement relation established by Agree between a probe and a goal, where both are specified with a set of φ-features. The gist of this proposal is that the covariance relation also holds between two elements, both of which have no φ-feature specifications, as stated in (3ii). Such a covariance relation is established by Merge when two elements with no φ-features are introduced into a structure and combined by Merge. When a covariance relation is constructed in this way, case valuation becomes available based on Merge. This makes the operation Agree unnecessary, insofar as a covariance relation (hence, a case valuation) is concerned. Introducing the notion of covariance relation into the general system of case valuation opens a novel and natural way to capture case valuation uniformly in non-agreeing languages and agreeing languages. This possibility is not readily available through the standard case system under Agree without stipulating an otherwise unmotivated (abstract) ϕ-feature agreement in non-agreeing languages.

In this proposal, ϕ-feature specifications of lexical items in a language determine how a covariance relation is established, which in turn selects a mode of case valuation in the language. Japanese is among the languages that allow the covariance relation to be built up by Merge; hence, case valuation occurs through Merge. Based on the assumption that no ϕ-features are involved in Japanese syntax, the covariance relation is established by Merge when it applies to two elements, such as a verb and an object, since neither is specified with ϕ-features.[4] Therefore, case valuation takes place at the merger of the two elements, and the case feature of the object is valued as accusative according to (1a). Similarly, the merger of a subject and *v* establishes a covariance relation, since neither has ϕ–features. Thus, the case feature of the subject is valued as nominative by (1b). In this manner, case valuation in this type of language occurs in terms of Merge that establishes a covariance relation, rendering case valuation in terms of Agree unnecessary and impossible.

In the Agree-based case system put forth by Chomsky, Agree holds between a probe with unvalued ϕ-features and a goal with valued ϕ-features when certain locality conditions are met. The assumption underlying this view is that both the probe and the goal are specified with ϕ-features, regardless of whether they are valued. As stated above, this concept is included in (3i); that is, a covariance relation is established between the probe and the goal. As a consequence, the case feature of the goal is valued under Agree in agreeing languages.

Why is case valuation in terms of Merge impossible in this type of language? In other words, why is a covariance relation not established by Merge in this type of language? Keys to the answers to these questions appear in the feature specifications of lexical items. Following the general assumption that it is *v* rather than a verb that comes with ϕ-feature specifications, the merger of a verb and an object does not establish a covariance relation in this type of language because only the object is specified with F. Therefore, a case feature of the object cannot be valued at the point of merger with a verb. When a merger of *v* occurs that combines it with VP (a verb + an object), a covariance relation is established via Agree, and thus the object receives an accusative case value.

Note that the unvalued ϕ-features of *v* are valued by this Agree relation, and then deleted in the domain of the *v*P-phase. Thus, when a subject is merged with *v*, its case feature cannot be valued at the merger of the two, since the subject and *v* do not covary with respect to F. It follows that the unvalued case feature of the subject should be valued by Agree in the next phase domain of C-T. This proposal, given in (2) and (3), together with the properties of ϕ-features in the agreeing languages, suggests that Agree is the only option for case valuation in such agreeing languages.

## 2.2 Generalized case valuation

We have seen a recent claim made by Zushi (2014a, b) that case valuation in terms of Merge should be available in addition to Agree. The fundamental theoretical assumption that underlies this proposal is that Merge and Agree are among the elementary computational operations in human language. This is the widely held view shared by most of the studies within the minimalist program. However, more recent studies claim that Agree, as well as other notions including chains (formed by internal Merge) and binding, ought to be subsumed under a more general operation of minimal Search. This operation seeks two elements that have an identical feature complex and establishes a relation between them (Kato et al. 2014). Developing this line of research further, Kato et al. (in this volume) argue that the operation of Search defined by Kato et al. (2014) is decomposed into more primitive operations, *0-Search* ($Search_0$) and *0-Merge* ($Merge_0$). The former searches certain features and selects *n* (typically two) elements from a designated domain of computation, while the latter forms a set of the selected elements. The set formed through this operation is interpreted as a certain linguistic relation at the CI interface. Kato et al. also propose that the standard notion of Merge should be decomposed into the two elementary operations, $Search_0$ and $Merge_0$, suggesting that Search and Merge are essentially the same operations of a composite of $Search_0$ and $Merge_0$ (see Kato et al. for details). This is a promising line of research that enables us to eliminate additional devices that have been stipulated in various aspects of grammar.

Now, if we follow Kato et al. (this volume) in assuming that Merge and Search (hence Agree) are the same operations, it turns out that the two modes of case valuation proposed by Zushi (2014a, b) can be consolidated into a single operation. Let us first consider how a covariance (case) relation based on Agree can be reformed. According to the proposal made by Kato et al., φ-feature agreement takes place in the following way. Suppose that T and *v*P are merged, forming an SO {T, *v*P} at some point of the derivation. When applied to this SO, $Search_0$ searches (unvalued) φ-features of the subject *n*P and (unvalued) φ-features of T, and picks out these two elements (*n* and T). $Merge_0$ applies these sets of features and forms a set of them. It is plausible to say, then that based on the set formed by $Merge_0$, a covariance relation defined in (3i) is established between the two elements. Therefore, the case feature of the subject is valued as nominative.

External Merge can also be decomposed into the two primitive operations of $Search_0$ and $Merge_0$ based on the proposal made by Kato et al. In this case, $Search_0$ applies to the workspace, a domain that the computational system can operate in, including SOs and the lexicon. Suppose that $Search_0$ picks out two elements, for example *hon* 'books' and *yomu* 'read,' from the workspace. Then, $Merge_0$ forms a set {{*hon*}, *yomu*}. A covariance relation defined in (3ii) is formed based on this set, since neither of the two elements have φ-features. Accordingly, the case feature of the nominal is valued as accusative. In this way, incorporating the proposal made by Kato et al. into Zushi's (2014a, b) system of case valuation enables us to formulate a generalized system of case valuation by reducing the two distinct operations postulated in the previous system into a single operation of the composite of $Merge_0$ and $Search_0$.

## 2.3 The role of formal features: their contribution to determining labels

Let us consider what role formal features play in narrow syntax – and on what grounds feature valuation needs to occur in the first place. It is generally assumed that φ-features of nominal expressions enter into interpretation at the CI interface and should not be eliminated at that level. On the other hand, φ-features of $v$/C-T have no role in interpretation, and should be eliminated before the CI interface (Chomsky 1995). Their role in syntax is solely to initiate the operation of Agree to establish an agreement relation (Chomsky 2001). Although the φ-features play a key role in computation, the status of case features in syntax is rather obscure. Chomsky (1995) argues that case features do not enter into interpretation and therefore they are formal features par excellence. As a theory of case and agreement develops, however, the status of case in syntactic theory is being threatened with the theory of Agree put forth by Chomsky (2001). From the viewpoint of case valuation via Agree, the case valuation is seen as a side effect of the primary operation of Agree. The role of case features is merely restricted to making a nominal expression visible for the probe-goal agreement. This and other considerations led some researchers to argue that there is no role for case features in narrow syntax, and that their contribution is limited to determining the morphological form of a particular noun at the Sensorimotor interface (Bobaljik 2008, Marantz 1991).

Contrary to such a view, this chapter suggests that case should play a more active role in syntax, following the basic insight of Keenan (1999) and Keenan and Stabler (2003) (see also Baker and Vinokurova 2010). In particular, we propose that case features of nominal elements, as well as φ-features of verbal elements, contribute to determining labels of SOs.

In order for a SO to be interpreted at CI interface, the SO must have a label. Therefore, there must be some syntactic mechanism to determine SO labels. Chomsky (2013) claims that an operation of minimal search serves to provide labels to SOs. Elaborating on this idea, Kato et al. (this volume) argue that the process involved in labeling is to establish a relation of "is headed by" between a SO and its label. They propose that the application of the two primitive operations, $Merge_0$ and $Search_0$, gives the result of forming a set consisting of these two elements, and that a semantic operation establishes the relevant relation based on the set. For example, $Search_0$ takes a SO, say *read books*, and seeks its head *read* within the domain, and $Merge_0$ forms a set {{read books}, read}. The application of the semantic operation yields the interpretation of "*read books* is headed by *read*" at the CI interface.

However, it remains unclear in this proposal (and in Chomsky's labeling algorithm) how a LI is chosen as a head from the domain.[5] What property helps to characterize a LI to be selected as a head in a way that distinguishes it from, for example, a phrase? This is not obvious in the theory of bare phrase structure where the notion of projection is eliminated. We would like to suggest that formal features, as computational atoms, play a vital role in determining a label (and hence a head) for SOs.

Consider first non-agreeing or dependent-marking languages, such as Japanese. As we presented in the above sections, external Merge forms a covariance relation, and case valuation takes place based on this relation. For example, $Search_0$ applies to the workspace and picks out *hon* 'books' and *yomu* 'read,' and $Merge_0$ forms a set of the two.

(4) $M_0 \circ S_0 (WS) = \{\{hon\}, yomu\}$

According to (1a), the case feature of *hon* is valued as accusative. Now, let us take the accusative case value on the nominal as a marker of an "argument," following the basic insight of Keenan (1999) and Keenan and Stabler (2003). This means that the other element of the SO, *yomu*, turns out to be a predicate, namely the head of the SO. In order to determine the label of the SO {{hon}, yomu}, $Search_0$ searches *yomu* which does not have a case feature, and picks out the SO and *yomu* from the domain. $Merge_0$ forms a set {{*hon, yomu*} *yomu*}. Then, *yomu* is interpreted as a head of the SO in the CI interface. In this way, a valued case feature helps $Search_0$ to correctly seek a head from the domain it applies to.[6]

On the other hand, there are languages in which case assignment/valuation does not seem to apply to nominal expressions. These are so-called head-marking languages (Baker 1996; Nichols 1986). The prominent feature of these languages is the fact that their grammatical relations are marked only by inflectional affixes, as the following example from Mohawk shows.

(5) Sak ra-nuhwe'-s ako-[a]tya'tawi.
    Sak MsS-like-HAB FsP-dress
    'Sak likes her dress'

(Baker 1996:10)

How can labels be given to SOs in these languages, given that case features of nominal expressions play no role? It is reasonable to claim that the valued φ-features of verbal elements such as $v$ and C/T take the place of valued case features in directly determining the label. That is, such features identify the verbal expression as heads. Suppose that a derivation has reached the stage where a subject has merged with TP. An agreement relation between the φ-features of the subject and the φ-features of T is established via $M_0 \circ S_0$, yielding valued φ-features of T. $Search_0$ applies to the SO {subject, TP} and searches the T, bearing valued φ-features. $Merge_0$ forms a set of the SO and T, from which the interpretation that T is a head of the SO is derived. In this way, the valued φ-features of T play a role in providing labels to SOs.

We propose that the primitive operation Search$_0$ makes use of two kinds of features to determine SO labels: (i) case features of nominal expressions and (ii) φ-features of verbal expressions. Languages employ at least one type of feature to provide labels to SOs. This claim enables us to capture the three-way typological distinction of languages proposed by Nichols (1986). In dependent-marking languages, such as Japanese, the former means of labeling based on case features is employed, whereas in head-marking languages, such as Mohawk, the latter way of labeling based on φ-features is utilized. This claim allows for the third type of language that makes use of both means to provide labels to SOs. English seems to be an example of this type. In its CP domain, the φ-features of C/T determines SO labels, while a case feature of nominal may play this role in the VP domain where no obvious agreement is found in this language. On the other hand, the system we are proposing here predicts that if neither of these two features were employed, then the result would be ruled illegitimate, since there would then be no way of labeling SOs.[7]

## 3 Consequences for peculiar case phenomena in Japanese

### 3.1 Multiple occurrences of identical case

Having laid out the theory of case valuation, let us now examine how it accounts for case peculiarities in Japanese. First, we deal with multiple occurrences of identical case. This phenomenon has challenged Agree-based case theory, since Agree (or feature checking) is generally assumed to be a biunique relation between a probe and a goal. To apply Agree/case checking to Japanese, one must modify the theory to allow more than one nominal to enter an Agree relation. Ura (2000) takes this direction. He proposes, couched in the framework of Chomsky (1995), that UG provides a [±multiple] feature with a functional head. A [+multiple] feature allows more than one element to enter into feature checking, whereas a [−multiple] feature permits only one element to involve feature checking. Thus, it is argued that in Japanese, unlike in English, T has a [+multiple] feature, which allows more than one nominal to occur in Spec of TP and enter into feature checking with it. As a result, more than one nominal receives a nominative case. Hiraiwa (2001) adopts and elaborates upon this idea within the framework of Chomsky (2001) regarding the notion of multiple Agree.

On the other hand, researchers that adopt a structure-based view of case assignment approach this phenomenon from a more general and crosslinguistic perspective. They relate this phenomenon to other syntactic peculiarities (e.g., scrambling) in Japanese, in contradistinction to English, to see what accounts uniformly for typological differences between the two types of languages. On this view, the phenomenon of multiple nominative/genitive is not explained by the case theory per se, but by a more general principle that accounts for the lack of biuniqueness effects in other domains of grammar as well.[8] Thus, Kuroda (1988) argues that Agreement is forced in English but not in Japanese, which subsequently allows multiple specifiers of a functional head. On the basis of the relativized X-bar theory, Fukui (1986, 1988) claims that functional categories in languages like English close off their projection by inducing agreement. However, they do not induce agreement in Japanese, thus enabling any element to adjoin freely to their projections in that language. Given that Japanese permits multiple specifiers on general grounds, it naturally follows that more than one nominal in the specifier positions receives an identical case under the contextual case assignment view. Fukui (2011) advances the development of this idea on the basis of Chomsky's (1995) claim that Merge applies freely without any driving force. He proposes that Japanese allows unbounded Merge to be applied at the edge of phase heads, provided that the elements merged at the edge satisfy such a licensing condition as predication to receive an appropriate interpretation. Since these approaches to the phenomenon potentially have a greater explanatory force than the Agree-based case account in grasping parametric differences between the two types of languages, we develop their basic idea.

Let us agree with Fukui (2011) that unbounded Merge ($M_0 \circ S_0$ to use Kato et al.'s terms) applies to the edge of a phase head in Japanese. We then show how the proposed case system that embodies the idea of contextual case assignment accounts for the phenomenon. First, consider multiple nominatives in stative sentences like (6a), in which both the experiencer "subject" and the theme "object" are marked nominative.

(6) a. Taroo-ga   okane-ga   hosii/hituyoo-da/aru.
    Taro-NOM   money-NOM   want/need-COP/have
    'Taro {wants/needs/has} some money'
  b.

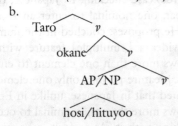

Unbounded Merge is applied to the edge of $v$P, allowing both experiencer and theme arguments to be merged with $v$, as illustrated in (6b). The question of why the theme (an apparent "object" of the verb) is merged outside VP will be discussed later.

Application of unbounded Merge at the edge of $v$P makes it possible to establish a covariance relation between *okane* 'money' and $v$, since neither of them has $\phi$-features. A covariance relation also holds between *Taro* and $v$ for the same reason. Notice that since no $\phi$-feature agreement is involved here, the property of $v$ without $\phi$-features remains intact even once case valuation is carried out. This allows the $v$ to enter into a covariance relation with *Taro*. Therefore, the case features of both experiencer and theme are valued as nominative.

*Case and predicate-argument relations* 55

This account of multiple nominatives can naturally be extended to multiple genitives, as seen in the following example of NPs.

(7) a. kinoo-no       Yamada-sensei-no       gengogaku-no     koogi
       yesterday-GEN  Yamada-professor-GEN   linguistics-GEN  lecture
       '(Lit.) yesterday's Professor Yamada's lecture on linguistics'

  b.
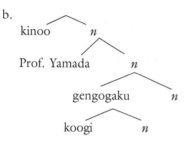

The proposed theory conducts a valuation of the genitive case in a manner parallel to the valuation of the nominative case. As noted in footnote 3, any nominal element in the nominal structure is merged with *n* in our analysis, which adopts the idea proposed by Baker (2003) and Kayne (2009). Since *n* is assumed to be a phase head (Fukui and Zushi 2008), unbounded Merge is applied to its edge, as illustrated in (7b). Each nominal element merged at the edge of *n* receives a genitive case value, and hence multiple occurrences of genitive case.[9]

It is important to note that the fact that multiple occurrences of identical case are impossible in agreeing languages, such as English, is straightforwardly derived from the proposal that case valuation in that type of language is carried out via Agree. Whether unbounded Merge is applied to the edge in English-type languages is a matter of debate. If unbounded Merge does not apply to those languages, the impossibility of multiple occurrences of identical case immediately follows. The same conclusion can be drawn even if unbounded Merge is assumed to be possible at the edge in this type of language. Suppose such languages allow more than one nominal to be merged with *v*, as in the structure (6b). Under the assumption that Agree is a biunique relation, once a case feature of one nominal can be valued by a probe under Agree, further case valuation by the same probe becomes unavailable, since its $\phi$-features are inactivated. Unless some special device of case valuation is stipulated, the unvalued case feature would remain, crashing the derivation.

Let us now return to the question regarding the structure of stative sentences. We argue that both experiencer and theme arguments are outside the domain of VP, which differs from the standard view of the structure of stative sentences. It is standardly assumed that the theme argument is an "object" that occupies a complement position of the predicate. We attribute this analysis of the structure of stative sentences to the theory of lexical categories proposed by Baker (2003) and Kayne (2009). They argue that verbs differ from both adjectives and nouns in how they take arguments when functioning as predicates. Verbs can take arguments directly, whereas adjectives and nouns must combine with a phase head. Given that most stative predicates are adjectives and nominal adjectives (Kuno 1973), it is natural to extend the Baker-Kayne theory and propose that adjectives and nominal adjectives in stative sentences must be combined with a phase head to take arguments. This claim is supported by Nishiyama's (1999) observation that overt manifestations of the head combined with adjectives and nominal adjectives appear in Japanese. Assuming that nouns and adjectives are combined with a phase head to become predicates, the structure of stative sentences is like that of (6b). The adjective or nominal adjective is first merged with a light verb $v$ to become a predicate. Then, the derived verbal complex headed by $v$ takes its arguments at its edge.[10]

### 3.2 Nominative-genitive conversion

In Japanese, nominative case is optionally converted to genitive case in certain environments. This phenomenon is known as *ga-no* (nominative-genitive) conversion (henceforth, NGC) since Harada (1971). The phenomenon also poses a problem for the Agree-based case theory, since it shows that a single NP apparently receives case values from distinct probes. The biunique relation between a probe and a goal assumed in the theory of Agree is seemingly broken in this case. This section discusses how various properties of the phenomenon can be derived by means of the proposed case theory.

### 3.2.1 Basic properties of nominative-genitive conversion

Optional conversion between nominative and genitive case is typically observed in prenominal clauses, including relative clauses like (8a) and nominal complements like (8b).

(8) a. [kinoo   Taroo-{ga/no}   kaita] kizi-ga   sinbun-ni   notta.
      yesterday Taro-{NOM/GEN} wrote article-NOM newspaper-DAT appeared
      'The article that Taro wrote yesterday appeared in the newspaper'
   b. Taroo-wa [kinoo   Hanako-{ga/no}   kita koto]-o   sira-nakat-ta.
      Taro-TOP yesterday Hanako-{NOM/GEN} came fact-ACC know-not-PAST
      'Taro did not know (about the fact) that Hanako came yesterday'

NGC has the following central properties, which have attracted considerable discussion in the literature.[11]

(9) a. It is optional.
    b. It occurs typically in prenominal clauses.
    c. No accusative-genitive conversion is possible.

d. It occurs with a predicate featuring special verbal morphology (attributive form).
e. A nominative NP with adverbial particle attached to it cannot be a target.

Numerous analyses have been proposed to account for this peculiarity in NGC. The central issue is to determine what is responsible for genitive case marking. There are two different views on this. One attributes genitive case marking to the nominal head outside the prenominal clauses on the basis of the fact that genitive case generally appears within noun phrases in Japanese. Adherents of this view have advanced various means of associating a clause-internal element with an external nominal head: the optional application of the restructuring rule (Bedel 1972); overt or covert movement of the relevant element into the domain of the external head (Miyagawa 1993, Sakai 1994); feature-checking/Agree (Ochi 2001, Maki and Uchibori 2008; Miyagawa 2012, 2013); or scrambling of the relevant element to the clause initial position, allowing the external head to govern and case license it (Fukui and Nishigauchi 1992).

When considering (8d) rather than (8b) as a fundamental property of NGC, an alternative view argues that the genitive case marking is irrelevant to the presence of the external nominal head. Thus, Watanabe (1994, 1996) claims that genitive case is a consequence of *wh*-agreement within the attributive clauses. Hiraiwa (2001) argues that the attributive form is a realization of the Agree relation of C-T-*v*-V, and that genitive case marking is a consequence of the Agree relation.

Given that genitive case is assigned to any nominal elements within Japanese noun phases, attributing the genitive case to the external nominal head is more natural and appropriate. In addition, the latter view does not clarify why no case other than genitive emerges. Therefore, we develop the former view in our proposed theory of case valuation. However, Watanabe and Hiraiwas's insight into genitive case marking that is associated with the attributive form is also important for understanding the nature of NGC. To see the relevance of the attributive form to NGC, it is instructive, following Hiraiwa (2001), to glance at the historical development of attributive forms.

Old Japanese featured a clear distinction between attributive and conclusive forms: the former was used only in certain types of subordinate clauses and the latter only in matrix clauses. Examples of subordinate clauses in which predicates take an attributive form are given below: (10a) is an example of a relative clause, (10b) an internally headed relative clause, and (10c) a (nominal) complement.

(10) a. [tubakurame-no *motitaru*] koyasugai
 swallow-GEN have cowry
 'a cowry that the swallow has' (*Taketori-monogatari*, cited in Kinsui 1995)
 b. [Kogimi tikau *husi-taru*]-o okosi-tamahe-ba. . .
 Kogimi near lay=asleep-ACC wake-HON-as

'as [Prince Genji] woke up Kogimi [, who] lay asleep near by, . . .
(*Genji*, vol 1, p. 117. Cited in Kuroda 1992: 136)

c. [ume-no hana   ori kazasitutu morohito-no      *asobu*]-o mireba
plum-GEN blossom cut hold=up   many people-GEN play-ACC see
miyakozo omohu
capital  think
'To see many people playing with holding a branch of plum blossoms up over their heads reminds me of the things about the capital'
(*Manyoshu* 843. Cited in Kinsui 1995)

It is notable in the examples from (10) that the subordinate clauses exhibit a nominal property. This is illustrated by the internally headed relative clause (10b) and the complement clause (10c), whereby an accusative case particle is attached directly to these clauses, a clear indication of nounhood. This form of subordinate clause disappeared from Middle Japanese. Instead, a nominalizer (or a complementizer-like element) *no* came to be attached to the attributive form. For example, (10c) has to be ". . . morobito-no asobu-*no*-o mireba . . ." (to see many people playing . . .) in Modern Japanese. Kinsui (1995) relates this diachronic change to the collapse of the attributive and conclusive forms in the thirteenth century, when the former came to be used in matrix clauses for rhetorical reasons. The collapse of the two forms made it difficult for the attributive form alone to mark subordinate clauses, which might have promoted the emergence of a marker to introduce subordinate clauses. Although the form of internally headed relative clauses without a nominalizer like (10b) disappeared because of the diachronic change, the form of nominal clauses like (10c) remains in Modern Japanese as a fossilized expression. The following example illustrates that an attributive clause can be a subject to which a nominative case particle is directly attached.

(11) *nigeru*-ga     saizyoo-no saku     da
     run=away-NOM  best-GEN   strategy is
     'To run away is the best strategy' (Kuno 1976: 38)

Kuno (1976) observes that the attributive form can occur as an object with an accusative case particle being attached in Raising-to-Object constructions, as shown in (12b).

(12) a. Yamada-wa [*nigeru*-ga      saizyoo-no saku     da to] omotta.
        Yamada-TOP run=away-NOM  best-GEN   strategy is that thought
        'Yamada thought that to run away was the best strategy'
     b. Yamada-wa *nigeru*-o      [saizyoo-no saku     da to] omotta.
        Yamada-TOP run=away-ACC  best-GEN   strategy is that thought
        'Yamada thought that to run away was the best strategy' (Kuno 1976: 39)

How the accusative-marked element in (12b) is derived is not our concern here. An important point is that the fact that an accusative case particle can be attached to the attributive form reinforces the idea that the attributive form behaves like a nominal.[12]

One conclusion we draw from these observations is that the clauses containing an attributive form had both nominal and clausal properties in Classical Japanese. Its nominal properties remain to some degree in Modern Japanese.[13] Given this nature of attributive forms, Hiraiwa's (2001) idea that genitive case is associated with the attributive form becomes more plausible. However, his idea does not fully account for the alternation of nominative with genitive case. We show below how the nominal property of the attributive forms brings about case alternation from a crosslinguistic perspective. A key to solutions for this question can be found in Ott's (2011) proposal for a free relative, a construction also known as a clause with nominal properties.

### 3.2.2 Nominative-genitive conversion in the theory of phases

Ott (2011) examines free relatives in comparison to embedded interrogatives and claims that the categorial duality exhibited by free relatives can be explained in terms of head Transfer. In a nutshell, he argues that at the stage of derivation, where *what you cook* is formed, a head C is removed along with its complement from the workplace by Transfer on the assumption that no interpretable features of C remain. This makes it possible for the *wh*-phrase that remains at the edge to be the only element visible at the next $v$P phase, rendering its categorial label as DP. Therefore, a free relative is a CP up to the derivational stage, where Transfer of C occurs along its complement, but it becomes a DP after the next $v$P phase level. Thus, the dual nature of free relatives is derived. On the other hand, a head C of the embedded interrogatives cannot undergo Transfer because interpretable features are present. Consequently, it retains the clausal status throughout the derivation. In short, the possibility of the head C's Transfer, which depends on particular feature specifications, can capture the categorial duality of free relatives and the distinction between the two closely related but structurally different constructions (free relatives versus embedded interrogatives).

Let us show how the proposed theory of case gives a natural account for the NGC by adopting the basic idea propose of Ott (2011). We take Japanese prenominal clauses, including relative clauses like (13a), to be TP, following Murasugi (1991), Saito (1985), Sakai (1994), and many others. We also assume that TP can be a phase only in embedded contexts (Zushi 2005).

(13) a. [Taroo-ga kaita] (kizi)
    Taro-NOM wrote (article)
    '(the article) which Taro wrote'
  b. [ Taro$_i$ [ T [$_{vP}$ t$_i$ [$_{VP}$ kaita pro ]]]]

In the structure (13b), the subject *Taro* is assumed to be raised to the edge of T. This is possible, since T, as a phase head, allows unbounded Merge to be applied to its edge, and the NP at its edge can be interpreted in terms of "aboutness" relation (Saito 1985, Rizzi 2010). Regarding the label for the structure (13b) as a whole, there is an option in Japanese whereby category determines the label. Suppose that the subject receives a nominative case value within $v$P, and moves to the edge of T. The label of the structure (13b) is determined by T, according to the proposal of labeling presented in Section 2.3. When T determines the label, the structure retains clausal properties throughout the derivation.

Suppose that the case feature of the subject is not valued within $v$P. This is possible because no φ-agreement relation between the subject and $v$ is assumed under the proposed case theory.[14] Since T has no role in this case, it is transferred along with its complement, making it possible for the element that remains at the edge to determine the label. As a result, the structure turns out to be an NP, yielding the nominal character of the clause.

(14) $[_{nP}\ [_{NP}\ Taro_i\ [T\ [_{vP}\ t_i\ [_{VP}\ kaita\ pro]]]]\ n\ ]$

Transfer of T enables the subject *Taro* to satisfy the condition of genitive case valuation, since it becomes a sister to the external head *n*. Therefore, the case feature of the subject is valued as genitive. In this way, genitive case markings on the subjects of the prenominal clauses are derived.

Let us briefly demonstrate how the basic properties of NGC listed in (9) follow from the proposed analysis. The first property deals with the optionality of NGC (9a). In the proposed analysis, there is no optionality in case valuation per se. Alternation of nominative with genitive case is caused by the different structures determined by the choice of how the prenominal clauses are labeled. The label of the prenominal structure can be determined either by T or by the edge element. In the former instance, the subject of the clause receives a nominative case value; in the latter, a genitive case value.

The second property (9b) is that NGC occurs in prenominal clauses, but not in embedded clauses headed by an overt complementizer, as shown in (15).

(15) Taroo-ga [Hanako-{ga/*no} kenkoo-da to] itta
Taro-NOM Hanako-{NOM/GEN} is-healthy that said
'Taro said that Hanako is healthy'

Unlike prenominal clauses, clauses headed by an overt complementizer are regarded as CPs. Given the standard assumption that the element merged with C should be an operator-like element, an argument NP cannot be raised to the edge of C in the embedded CP clauses in (15). Thus, CPs of this sort cannot be nominalized in the way that TPs can be in prenominal clauses. Therefore, no NGC is possible in these clauses.

The third property (9c) is that there is no conversion of accusative to genitive, which also follows directly from our proposed case mechanism.

(16) [sono kizi-{o/*no}    kaita] kisya-ga     uttae-rare-ta.
     the article-{ACC/*GEN} wrote  journalist-NOM accuse-PASS-PAST
     'The journalist who wrote the article was accused'

Unlike nominative case valuation, accusative case valuation occurs within the $v$P phase and is not affected by the nature of the higher T/CP phase. Therefore, the object never receives a genitive case value.

The fourth property (9d) – the relevance of attributive forms to NGC – was discussed above. The fifth property (9e) reflects the fact that a nominative NP with an adverbial particle such as *saemo* 'even' cannot be a target of conversion, if we assume that such a particle requires a clausal structure throughout the derivation for licensing purposes (see Miyagawa 2013, Nakai 1980).

(17) [Taroo-saemo-{ga/*no} katta] hon
     Taro-even-{NOM/*GEN} bought book
     'a book that even Taro bought'

The nominative subject featuring *saemo* 'even' in (17) is allowed, as it occurs in the clausal structure throughout the derivation. On the other hand, the prenominal clause becomes NP at some point of derivation after Transfer of T, and the licensing condition for adverbial particles attached to the genitive subject cannot be met at that level. Therefore, the genitive subject with the adverbial particle is disallowed, exactly like the genitive element in regular noun phrases, \**Taroo-saemo-no hantai* '(even) Taro's objection.' In this manner, the central properties of NGC follow straightforwardly from the proposed case system and the theory of phases.

## 4 Conclusion

This chapter investigated what role formal features, such as case and φ features, play in narrow syntax, with a special focus on case. It was proposed that case valuation takes place based on a covariance relation between the two elements with respect to their φ-feature specifications. Depending on the feature specifications of lexical items in a given language, either Merge or Agree is employed for case valuation. In non-agreeing languages, such as Japanese, a covariance relation is established by Merge, and thus Merge is chosen as a device for case valuation, rendering Agree unnecessary. In agreeing languages, on the other hand, a covariance relation is established by Agree; therefore Agree is chosen for case valuation. It was also claimed that by incorporating the idea proposed by Kato et al. (this volume) that Agree and Merge are essentially the same operations, the two modes of case valuation can be reduced to a single operation.

Although case valuation based on covariance relation is a single system, different case arrays emerge between agreeing languages and non-agreeing languages because of the difference between the two operations. Namely, Agree is a biunique relation, while Merge is not. The presence of multiple occurrences of identical case and case alternation in Japanese, and their absence in agreeing languages, reflects this difference between the two operations.

It was also claimed that case features of nominal expressions and φ-features of verbal expressions play an important role in determining SO labels. This claim enables us to capture the three-way typological distinction of languages proposed by Nichols (1986). Dependent-marking languages, such as Japanese, make use of case features to determine labels, whereas head-marking languages like Mohawk utilize φ-features to provide labels to SOs. The third type of language, such as English, uses both features for this purpose. Thus, in dependent-marking languages, case serves as a key feature in establishing the system of predicate-argument relations that is a central component of human language of thought. Therefore, the investigation of the nature of case as well as φ-agreement, as they pertain to predicate-argument relations, continues to be one of the major topics in theoretical linguistics, and as such to make a significant contribution to understanding the human language.

## Notes

* This research is supported in part by AMED-CREST.
1 Zushi (2014a, b) considers the mechanism (1) to be an algorithm that applies in syntax. Whether the algorithm is reduced to a more basic operation deserves further investigation. The mechanism could be reformulated as a case licensing. We could argue that a nominal expression is merged with a case marker to form KP, and that the KP is case licensed when it is merged with a particular element. For example, a nominal combined with an accusative marker (AccP) is licensed when it is merged with a verb. We leave further elaboration on this line of approach for future research.
2 Space limitations do not allow us to discuss dative case. See Zushi (2014a, b) for discussion.
3 The structural environment in which genitive case valuation occurs is basically analogous to valuation of nominative case. This is made possible if we assume that a noun, unlike a verb, cannot directly take its argument and that it must be combined with a functional head, as proposed by Baker (2003) and Kayne (2009). See Zushi (2014a, b) for discussion.
4 Note that the covariance relation does not hold in modification relations in Japanese. For example, attributive adjectives cannot be merged directly with nouns in this language, but instead they constitute (reduced) relative clauses when they modify nouns. Therefore, a covariance relation is not formed between an adjective and a noun.
5 Note that the operation involving labeling defined by Kato et al. (this volume) is not ambiguous. The operation determining labels, $M_0 \circ S_0(\Sigma) = \{\Sigma, LI\}$, applies to a syntactic object ($\Sigma$) and selects it together with a LI, and forms a set of them. Subsequently, the LI is interpreted as a head of the object. The question that was addressed in the text is how one can recognize a particular expression as a LI.
6 This idea can be extended to particles other than case markers, such as the topic marker *wa*. It is reasonable to argue that *wa* identifies the nominal to which it is attached as an argument. When XP-*wa* is merged with an element in the CP phase, for example CP, *wa* makes it possible for C to be selected as a head. Thus, case markers basically determine SO labels in the *v*P domain,

and other markers, like a topic marker, provide labels to SOs in the CP domain.
7. In languages like Chinese that do not have visible case markers and agreement morphology, perhaps affixes on verbal elements, instead of agreement morphology, determine labels of SOs.
8. Lack of biuniqueness effects includes the absence of overt *wh*-movement (i.e., the absence of spec of CP), no EPP effect (i.e., the absence of spec of TP), and the presence of scrambling (more than one spec of TP). See Fukui (1986, 1988) and Kuroda (1988).
9. The general ban against multiple occurrences of the accusative case can be explained if we assume that unbounded Merge is permitted only at the edge. Alternatively, one could say that unbounded Merge is allowed within VP. However, the element within the phrase cannot be appropriately licensed. See Zushi (2014a, b) for discussion.
10. Zushi (2014a, b) extends this structural analysis to stative sentences based on stative verbs. She argues that the class of stative verbs is a complex predicate formed by the process posited for adjectives, in a way that Hale and Keyser (1993) posit for denominal verbs, such as *laugh*. See Zushi (2014a, b) for details.
11. Another property of NGC has been discussed in the literature, based on Harada's (1971) observation about an idiolectal variation among Japanese speakers with respect to NGC. Harada observes that NGC is prohibited among one group of speakers when some element intervenes between the (nominative) subject and the predicate, whereas no such intervention effects appear among the other group. Since the nature of intervention effect remains unclear, we set aside this issue (see Harada 1976, Hiraiwa 2001, and Watanabe 1994, 1996 for some discussions).
12. Similarly, a case particle can be directly attached to the attributive form of adjectives. An accusative case particle can be attached to the attributive form *yowaki* in (ia), for example, with the meaning that the one/thing that is weak.

   (i) a. *yowaki*-o tasuke, *tuyoki*-o kujiku
   weak-ACC help    strong-ACC crush
   'Fight for the oppressed against the oppressor'
   b. *furuki*-o tazunete, *atarasiki*-o siru
   old-ACC learn    new-ACC know
   '(You) discover new things by learning from the past'

13. This idea is in accord with Kuroda's (1992, 1999) proposal that internally headed relative clauses occur in the argument position, and thus are nominal in that N is their label in Classical and Modern Japanese.
14. Note that this way of case alternation is not allowed in languages like English, in which case valuation is contingent on agreement. In such languages, case valuation must be carried out in the designated domain; otherwise (unvalued) φ-features of C/T cannot be valued. However, case alternation may be possible in some special environments, such as gerunds, even in English. See Zushi (2014b) for discussion.

# References

Baker, Mark. 1996. *The polysynthesis parameter.* Oxford: Oxford University Press.
Baker, Mark. 2003. *Lexical categories: Verbs, nouns, and adjectives.* Cambridge, UK: Cambridge University Press.
Baker, Mark, and Nadya Vinokurova. 2010. Two modalities of case assignment: Case in Sakha. *Natural Language & Linguistic Theory* 28:593–642.

Bedel, George. 1972. On *no*. In *Studies in East Asian Syntax. UCLA Papers in Syntax* 3:1–20.

Bobaljik, Jonathan. 2008. Where's phi? Agreement as a post-syntactic operation. In *Phi theory: Phi features across interfaces and modules*, ed. David Adger, Daniel Harbour, and Susanna Béjar, 295–328. Oxford: Oxford University Press.

Chomsky, Noam. 1995. *The minimalist program*. Cambridge, MA: MIT Press.

Chomsky, Noam. 2001. Derivation by phase. In *Ken Hale: A life in language*, ed. Michael Kenstowicz, 1–52, Cambridge, MA: MIT Press.

Chomsky, Noam. 2013. Problems of projection. *Lingua* 130:33–49.

Fukui, Naoki. 1986. *A theory of category projection and its applications*. Doctoral dissertation, MIT. [Published as *Theory of projection in syntax*, in 1995. CSLI.]

Fukui, Naoki. 1988. Deriving the differences between English and Japanese: A case study in parametric syntax. *English Linguistics* 5:249–270.

Fukui, Naoki. 2011. Merge and bare phrase structure. In *The Oxford Handbook of Linguistic Minimalism*, ed. Cedric Boeckx, 73–95. Oxford: Oxford University Press.

Fukui, Naoki, and Taisuke Nishigauchi. 1992. Head-movement and Case-marking in Japanese. *Journal of Japanese Linguistics* 14:1–46.

Fukui, Naoki, and Mihoko Zushi. 2008. On certain differences between noun phrases and clauses. In *Essays on nominal determination*, ed. Henrik Muller and Alex Klinge, 265–286. Amsterdam: John Benjamins Publishing Company.

Hale, Kenneth, and Samuel J. Keyser. 1993. On argument structure and the lexical expression of syntactic relations. In *The view from building 20*, ed. Kenneth Hale and Samuel J. Keyser, 53–109. Cambridge, MA: MIT Press.

Harada, Shin-Ichi. 1971. *Ga-no* conversion and idiolectal variations in Japanese. *Gengo Kenkyu* (Journal of the Linguistic Society of Japan) 60:25–38.

Harada, Shin-Ichi. 1976. *Ga-No* conversion revisited: A reply to *Shibatani*. *Gengo Kenkyu* (Journal of the Linguistic Society of Japan) 70:23–38.

Hiraiwa, Ken. 2001. On nominative-genitive conversion. In *A Few from Building E39: Papers in Syntax, Semantics, and their Interface, MIT Working Papers in Linguistics* 39, ed. E. Guerzoni and O. Matushansky, 66–125. Cambridge, MA: MITWPL.

Kato, Takaomi, Masakazu Kuno, Hiroki Narita, Mihoko Zushi, and Naoki Fukui. 2014. Generalized Search and cyclic derivation by phrase: a preliminary study. *Sophia Linguistica* 61:203–222.

Kato, Takaomi, Hiroki Narita, Hironobu Kasai, Mihoko Zushi, and Naoki Fukui. 2016. On the primitive operations of syntax. In *Advances in biolinguistics: The human language faculty and its biological basis*, ed. Koji Fujita and Cedric Boeckx, 29–45. Abingdon, Oxon/New York: Routledge.

Kayne, Richard, 2009. Antisymmetry and the lexicon. In *Linguistic variation yearbook*, ed. Jeroen van Craenenbroeck, 1–32. Amsterdam: John Benjamins Publishing Company.

Keenan, Edward. 1999. Language invariants: The syntax and semantics of case marking. *Linguistics in the Morning Calm* 4: 21–39. The Linguistic Society of Korea.

Keenan, Edward, and Edward Stabler. 2003. *Bare grammar: Lectures on linguistic invariants*. Stanford: CSLI.

Kinsui, Satoshi. 1995. Iwayuru nihongo-no N'-sakujo ni tuite (On the so-called N'-deletion in Japanese). *The Proceedings of 1994 Nanzan Symposium*, 153–176.

Kuno, Susumu. 1973. *The structure of the Japanese language*. Cambridge, MA: MIT Press.
Kuno, Susumu. 1976. Subject raising. In *Syntax and Semantics 5: Japanese Generative Grammar*, ed. Masayoshi Shibatani, 17–49. New York: Academic Press.
Kuroda, S.-Y. 1965. *Generative grammatical studies in the Japanese language*. Doctoral dissertation, MIT.
Kuroda, S.-Y. 1978. Case-marking, canonical sentence patterns, and Counter-equi in Japanese. In *Problems in Japanese syntax and semantics*, ed. John Hinds and Irwin Howard, 30–51. Tokyo: Kaitakusha.
Kuroda, S.-Y. 1983. What can Japanese say about government and binding. *Proceedings of the West Coast Conference on Formal Linguistics* 2:153–164. Stanford Linguistics Associations.
Kuroda, S.-Y. 1986. Movement of noun phrases in Japanese. In *Issues in Japanese Linguistics*, ed. Takashi Imai and Mamoru Saito, 229–271. Dordrecht: Foris.
Kuroda, S.-Y. 1988. Whether we agree or not: A comparative syntax of English and Japanese. *Linguisticae Investigationes* 12:1–47.
Kuroda, S.-Y. 1992. *Japanese Syntax and Semantics*. Dordrecht: Kluwer.
Kuroda, S.-Y. 1999. Shubu naizai kankeisetu (Internally-headed relative clauses). In *Kotoba-no Kaku to Shuhen* (Language, its core and periphery), ed. S.-Y. Kuroda and Masaru Nakamura, 278–103. Tokyo: Kurosio Publishing.
Maki, Hideki, and Asako Uchibori. 2008. *Ga/no* conversion. In *Handbook of Japanese Linguistics*, ed. Shigeru Miyagawa and Mamoru Saito, 192–216. Oxford: Oxford University Press.
Marantz, Alec. 1991. Case and licensing. In *ESCOL '91: Proceedings of the Eighth Eastern States Conference on Linguistics*, 234–253. Columbus: The Ohio State University.
Miyagawa, Shigeru. 1993. LF Case-checking and minimal link condition. In *Papers on Case and Agreement, MIT Working Papers in Linguistics 19*, ed. Colin Phillips, 213–254. Cambridge, MA: MITWPL.
Miyagawa, Shigeru. 2012. *Case, Argument Structure, and Word Order*. Routledge Leading Linguists Series. New York: Routledge.
Miyagawa, Shigeru. 2013. Strong uniformity and *ga/no* conversion. *English Linguistics* 30:1–24.
Murasugi, Keiko. 1991. *Noun phrases in Japanese and English: A study in syntax, learnability, and acquisition*. Doctoral dissertation, University of Connecticut.
Nakai, Satoru. 1980. A reconsideration of *ga-no* conversion in Japanese. *Papers in Linguistics* 12:247–286.
Nichols, Johanna. 1986. Head-marking and dependent-marking grammar. *Language* 62:56–119.
Nishiyama, Kunio. 1999. Adjectives and the copulas in Japanese. *Journal of East Asian Linguistics* 8:183–222.
Ochi, Masao. 2001. Move F and *ga/no* conversion in Japanese. *Journal of East Asian Linguistics* 10:247–286.
Ott, Dennis. 2011. A note on free relative clauses in the theory of phases. *Linguistic Inquiry* 42:183–192.
Rizzi, Luigi. 2010. On some properties of criterial freezing. In *The Complementizer Phrases: Subjects and Operators*, ed. E. Phoeros Panagiotidis, 17–32. Oxford: Oxford University Press.
Saito, Mamoru. 1982. *Case marking in Japanese: A preliminary study*. Ms., MIT.

Saito, Mamoru. 1985. *Some asymmetries in Japanese and their theoretical implications.* Doctoral dissertation, MIT.

Sakai, Hiromu. 1994. Complex NP constraint and case-conversions in Japanese. In *Current topics in English and Japanese*, ed. Masaru Nakamura, 179–203. Tokyo: Hituzi Syobo.

Ura, Hiroaki. 2000. *Checking theory and grammatical functions in universal grammar.* Oxford: Oxford University Press.

Watanabe, Akira. 1994. A crosslinguistic perspective on Japanese nominative-genitive conversion and its implications for Japanese syntax. In *Current topics in English and Japanese*, ed. Masaru Nakamura, 341–369. Tokyo: Hituzi Syobo.

Watanabe, Akira. 1996. Nominative-genitive conversion and agreement in Japanese: A cross-linguistic perspective. *Journal of East Asian Linguistics* 5:373–410.

Zushi, Mihoko. 2005. Deriving the similarities between Japanese and Italian: A case study in comparative syntax. *Lingua* 115:711–752.

Zushi, Mihoko. 2014a. Kaku to heigoo (Case and merge). In *Gengo no sekkei, hattatu, sinka* (The design, development and evolution of language: Explorations in biolinguistics), ed. Koji Fujita, Naoki Fukui, Yusa Noriaki, and Masayuki Ike-uchi, 66–96. Tokyo: Kaitakusha.

Zushi, Mihoko. 2014b. Merge-based case valuation in Japanese. Paper presented at the 29th Annual Meeting of Sophia University Linguistics Society.

# Part II
# Development, processing and variations

# Part II
# Development, processing and variations

# 5 Structure dependence in child English
## New evidence*

*Koji Sugisaki*

## 1 Introduction

Since its conception, generative grammar stands on the fundamental assumption that the faculty of language is an organ of the body, on a par with "physical organs" such as the visual and immune systems and "mental organs" such as various kinds of memory. This "biolinguistic perspective", which regards human language as a biological object, focuses on the traditional problem of determining the specific nature of the faculty of language, and reinterprets it as the problem of discovering the genetic basis that underlies the acquisition and use of language.

Earlier attempts to identify the properties of this genetic endowment (called UG) attributed rich and complex properties to UG, to capture intricate and varied properties of each language while at the same time accounting for the rapidity and uniformity of language acquisition. A more recent framework called the Minimalist Program (Chomsky 1993 et seq.) emphasizes the efforts to reduce these richness and complexities of UG without sacrificing empirical adequacy, in order to gain insight into the evolution of language, which should be understood as the evolution of UG. Pushing this reduction of UG to the limit, Chomsky (2010) suggests that UG consists only of the simplest recursive operation *Merge*, which takes structures X and Y already formed and combines them into a new structure Z. Under this view, various linguistic phenomena should fall out of the interaction between this structure-building operation and the "third factor" (i.e. language-independent) principles of efficient computation, along with the requirements from the two interface systems (the conceptual-intentional systems and the sensorimotor systems) which access the structured expressions formed by Merge.

Chomsky (2012, 2013a, b) argues that, once we adopt such a framework and seek deeper explanations by eliminating unwanted stipulations, even a simple subject-predicate construction in (1a) and the corresponding interrogative in (1b) pose us a serious puzzle. In the simplest system, the sentence in (1a) has the structure [Subject-Predicate] ({XP, YP}), in which the nominal head of the subject and the T head of the predicate are equally prominent. Then, when the example in (1a) is subjected to a principle of efficient computation called the *Minimal Search Condition*, which locates the structurally closest element to the target position, why does that condition choose (1b) over (1c), fronting the auxiliary rather than the nominal head?

(1) a. Young children can write stories.
    b. Can young children write stories?
    c. *Children young can write stories?

Chomsky proposed that this puzzle would be resolved quite simply once we adopt the hypothesis that the subject must first be merged internally to the predicate (*the predicate-internal subject hypothesis*), a property which would probably derive from the requirements of the conceptual-intentional interface.

Building on Chomsky's (2012, 2013a, b) recent minimalist analysis of subject-auxiliary inversion in simple interrogatives, this chapter addresses the very simple question of whether young English-speaking children produce any ungrammatical *yes/no*-question that corresponds to (1c). If the ungrammatical status of (1c) follows from the structure-dependent condition of Minimal Search along with the predicate-internal subject hypothesis as envisioned by Chomsky, the results of this study would add a new piece of evidence for the influential claim made by Crain and Nakayama (1987) that children adhere to structure dependence from the earliest observable stages, as well as evidence for the claim made by Déprez and Pierce (1993) that the subject noun phrase originates and comes from inside the predicate even in the grammar of two-year-olds.

## 2 Subject-auxiliary inversion and structure dependence in English

In an early discussion of the phenomenon of subject-auxiliary inversion in English, Chomsky (1968: 61–2) analyzed the sentences with multiple instances of auxiliaries as in (2).

(2) a. The subjects [who will act as controls] will be paid.
    b. Will the subjects [who will act as controls] ___ be paid?
    c. *Will the subjects [who ___ act as controls] will be paid?

The formation of English interrogatives involves an operation that moves an auxiliary (more accurately, an inflectional element in the T position) to the sentence-initial position occupied by a complementizer C.[1] The contrast between (2b) and (2c) indicates that the distance to the matrix C position is measured structurally, not linearly. The well-formed status of (2b) suggests that the auxiliary that is raised to the C position is the one that is hierarchically closer to that position. The ungrammatical status of (2c), on the other hand, shows that the auxiliary that undergoes such raising is not the one that is linearly closer to that position, even though such an option might be plausible on grounds of easy parsing and simplicity of computation. Speculating about the reason for the reliance on structure-dependent operations, Chomsky (1968) argues that UG constrains syntactic operations to be structure-dependent, thereby excluding structure-independent ones from possible candidates.

More recently, Chomsky (2012) seeks a deeper explanation within the Minimalist framework and explores the issue of why human language relies on structural closeness rather than linear closeness. The best explanation for this choice, he argues, would be that linear order is a property of the system of production/perception (the sensorimotor system) and is imposed in the process of externalization.[2] The structures that enter into the system of thought (the conceptual-intentional system) do not have this property. As a consequence, linear proximity does not exist at the point when structures are formed by Merge, and hence is not accessible to the process of fronting an auxiliary.

Returning to the phenomenon of subject-auxiliary inversion, Chomsky (2012, 2013a, b) goes on to argue that even a simple subject-predicate construction in (1a) and the corresponding interrogative in (1b), repeated here as (3a) and (3b) respectively, also provide evidence for structure dependence, once subjected to a Minimalist analysis.

(3) a. Young children can write stories.
　　b. Can young children ____ write stories?
　　c. *Children young _____ can write stories?

The sentence in (3a) is traditionally described as a subject-predicate construction, with the subject *young children* and the predicate *can write stories*. In earlier analyses, it was stipulated that a sentence is a TP, with T the most prominent element, and that the subject noun phrase is in the *specifier* of TP, subordinate to T. However, if we are to abandon these stipulations and to regard the structure of (3a) simply as the set of the form {XP, YP} created by Merge, we are left without any argument for choosing (3b) over (3c), since the nominal head of the subject and the T head of the predicate are equally prominent. Quite importantly, the ill-formed status of (3c) demonstrates that the linear proximity to the C position is not at work, since the element that is linearly closer to the C head in the sentence-initial position is the nominal head *children*, not the T head of the predicate *can*.

As mentioned in the introduction, Chomsky (2012, 2013a, b) provides a very straightforward analysis which makes crucial use of the predicate-internal subject hypothesis (e.g. Fukui and Speas 1986, Kitagawa 1994, Koopman and Sportiche 1991, Kuroda 1988). Assuming that the full clause is headed by a complementizer C, which expresses its *force* (declarative, interrogative, etc.), the formation of interrogatives in English involves the search by C for the head to be raised. The "third factor" (i.e. language-independent) principles of efficient computation restrict the search to be minimal, which in effect leads to the search for the structurally closest head. This minimal search by C will necessarily select the auxiliary *can* as the target of raising, if the subject noun phrase has not yet been merged to the complement of C: the structure at this point is the set {C, predicate}, where the predicate is headed by T (to which the auxiliary *can* is attached). The relevant structure is shown in (4).

*Minimal Search*

(4) [C [$_{TP}$ can$_T$ [$_{vP}$ [young children] write stories]]]

Minimal search thus establishes a relation between C and T, which leads to the raising of T to C.[3]

Crucial to this analysis is the assumption that the subject noun phrase does not intervene between C and T when the relation is established: the subject is first merged internally to the predicate, and then moved to the surface position. In sum, under this Minimalist analysis, the contrast between (3b) and (3c) follows naturally from the interaction between (i) the structure-dependent condition of Minimal Search, which instantiates language-independent principles of efficient computation, and (ii) the predicate-internal subject hypothesis, which reflects interface requirements.

## 3 Subject-auxiliary inversion and structure dependence in child English: previous studies

If structure dependence in the formation of simple and complex *yes/no* questions in English reflects the properties of genetic endowment associated with our language faculty, it is expected that English-speaking children adhere to structure dependence from the earliest observable stages. In order to determine whether this expectation can indeed be borne out, Crain and Nakayama (1987) conducted an experiment on children's knowledge of the subject-auxiliary inversion involved in complex *yes/no* questions exemplified in (2). The subjects of their experiment were 30 English-speaking children, ranging in age from 3;02 (years;months) to 5;11 (mean age 4;07). The task for children was elicited production: in this task, children were shown a set of pictures and were invited to pose particular questions to a puppet, Jabba the Hutt. The requests to children from the experimenter contained a restrictive relative clause attached to the subject noun phrase, and hence involved two auxiliary verbs. For example, in one experimental trial, the picture depicted two dogs, one of which was asleep on a blue bench and another of which was standing up, and the experimenter provided children with the embedded question, "Ask Jabba if the dog that is sleeping is on the bench." The experiment was designed to determine whether children ask adult-like *yes/no* questions like (5a), or whether, instead, they would ask ungrammatical questions as in (5b), which should derive from an application of the structure-independent rule that moves the leftmost auxiliary of a declarative to the sentence-initial position.

(5) a. Is the dog that is sleeping _____ on the bench?
    b. *Is the dog that _____ sleeping is on the bench?

The results of Crain and Nakayama's experiment showed that English-speaking preschool children produced two types of ungrammatical utterances. One of them, which they classified as a 'prefix' error, was the kind of error in which an extra auxiliary verb is attached to the sentence-initial position, as illustrated in (6a). Another type of error, which they classified as a 'restarting' error, was the kind of error in which a well-formed fragment of a question is followed by a second question containing a pronoun, as illustrated in (6b).

(6) a. 'Prefix' error: *Is the boy who's watching Mickey Mouse is happy?
   b. 'Restarting' error: *Is the boy that is watching Mickey Mouse, is he happy?

Even though these two types of errors were observed to a certain extent, no child ever produced ungrammatical questions exemplified in (5b), which would have appeared if children had adopted the relevant structure-independent rule. A summary of the error data obtained in Crain and Nakayama's experiment is provided in Table 5.1.

These findings led Crain and Nakayama (1987) to conclude that even though children's *yes/no* questions were not fully adult-like, these questions did not clearly indicate that children applied structure-independent rules. Thus, according to Crain and Nakayama, children's ability to form complex *yes/no* questions is consistent with Chomsky's (1968) suggestion that syntactic operations are predetermined by UG to be structure-dependent.[4]

The evidence offered by Crain and Nakayama (1987) that structure dependence reflects properties of UG has been challenged by several different groups of researchers. For example, Pullum and Scholz (2002) argue that children are exposed to abundant examples that would allow them to reject the incorrect, structure-independent strategy for forming *yes/no* questions, as the quote in (7) shows.[5] One of their examples, which is taken from the *Wall Street Journal*, is given in (8).

(7) "Our preliminary investigations suggest that the percentage of relevant cases is not lower than 1 percent of the interrogatives in a typical corpus. By these numbers, even a welfare child [who has been exposed to only 10 million words of language use – KS] would be likely to hear about 7,500 questions

*Table 5.1* The number of children's errors in Crain and Nakayama's (1987) experiment

|  |  | Total | Prefix error | Restarting error | Structure-independent error |
|---|---|---|---|---|---|
| Group 1 | 15 children, from 3;02 to 4;07 (mean 4;03) | 50 (62%) | 30 (60%) | 10 (20%) | 0 |
| Group 2 | 15 children, from 4;07 to 5;11 (mean 5;03) | 17 (20%) | 9 (53%) | 5 (29%) | 0 |
| Total |  | 67 (40%) | 39 (58%) | 15 (22%) | 0 |

that crucially falsify the structure-independent auxiliary-fronting generalization, before reaching the age of 3."

(Pullum and Scholz 2002: 45)

(8) Is what I'm doing in the shareholders' best interest?
(Pullum and Scholz 2002: 43)

Legate and Yang (2002) point out an empirical problem in Pullum and Scholz's (2002) argument. Specifically, they suggest that, even though Pullum and Scholz might have shown the *existence* of disconfirming evidence which leads children to exclude the structure-independent rule of preposing the leftmost auxiliary, they failed to demonstrate its *sufficiency* to rule out this "first auxiliary" hypothesis. To evaluate its sufficiency, Legate and Yang (2002) constructed an independent yardstick from the subject-omission phenomenon in child English. Following the generalization that the availability of *there*-type expletives correlates with obligatory subjects, Legate and Yang adopt the hypothesis that *there*-type expletives constitute the evidence disconfirming an optional-subject grammar (e.g. Hyams 1986). Based on a random sample of 11,214 adult sentences in the CHILDES database (MacWhinney 2000), Legate and Yang estimated the frequency of *there*-type expletives to be around 1.2% (140/11214). This finding suggests that the frequency of critical sentences such as (9a) or (9b) that would rule out the first-auxiliary hypothesis also needs to be approximately 1.2%.

(9) a. Is [the boy who is in the corner] ___ smiling?
   b. How could [anyone that was awake] ___ not hear that?

Analyzing all the input data in the Nina corpus (Suppes 1973) available in CHILDES, Legate and Yang (2002) found that, among the 20,651 questions produced by adults, none were *yes/no* questions of the type in (9a), and that 14 were *wh*-questions of the type in (9b). This puts the frequency of the disconfirming evidence against the first-auxiliary hypothesis at approximately 0.068%, which is about 20 times lower than the amount of evidence needed to settle on the grammars with obligatory subjects. In light of this finding, Legate and Yang concluded that children adhere to structure dependence in the absence of sufficient evidence, which accords only with the view that structure dependence should follows from the innate predisposition for language acquisition.

Chomsky (2012: 20) also briefly addresses input-driven learning approaches that attempt to derive the structure-dependent nature of subject-auxiliary inversion solely from the input data, and suggests that "it would be of virtually no interest if one of them succeeded". The reason is that it would leave unanswered the following basic question: "Why does this principle [principle of structure dependence – KS] hold for every language, and every relevant construction?"

To summarize this section, evidence from the acquisition of complex *yes/no* questions in English is consistent with Chomsky's (1968) assumption that UG restricts syntactic operations to be structure-dependent: any attempt to derive the property of structure dependence solely from the input data would face serious conceptual and empirical problems.

## 4 Subject-auxiliary inversion in child English revisited: transcript analysis

The seminal acquisition study by Crain and Nakayama (1987) demonstrated that English-speaking preschool children never produced ungrammatical *yes/no* questions as in (10b), and argued that the absence of this type of error constitutes evidence that syntactic operations are predetermined by UG to be structure-dependent.

(10) a. Is the dog that is sleeping ____ on the bench?
     b. *Is the dog that ____ sleeping is on the bench?

Developing this line of research, I now address the question of whether young English-speaking children produce any incorrect *yes/no* questions as in (11b), in which the nominal element in the subject, rather than the auxiliary, is preposed to the sentence-initial position.

(11) a. Can young children ____ write stories?
     b. *Children young _____ can write stories?

As we have seen in the previous section, Chomsky (2012, 2013a, b) suggests that the sentence in (11b) is ruled out by the interaction between the minimal search by C for the structurally closest head and the base-generation of the subject within the predicate. Then, if this analysis is on the right track, the absence of errors like (15b) would be an indication that children's grammars conform to the structure-dependent condition of minimal search as well as to the predicate-internal subject hypothesis.

### 4.1 Subjects and method

An analysis was conducted on three longitudinal corpora for English available in the CHILDES database (MacWhinney 2000), which provided a total sample of more than 84,000 lines of child speech. The corpora examined in this study are summarized in Table 5.2.

Using the CLAN program KWAL, all the utterances containing a question mark ("?") were searched and checked by hand. In order to take account of the context of children's utterances and to exclude imitations and formulaic routines, each utterance was extracted from the transcript in its discourse window, which consisted of two conversational turns before and after the child's utterance.

76  Koji Sugisaki

Table 5.2 Corpora analyzed

| Child | Collected by | Age span | # of files | # of child utterances |
|---|---|---|---|---|
| Adam | Brown (1973) | 2;03:04–5;02:12 | 55 | 44,034 |
| Eve | Brown (1973) | 1;06–2;03 | 20 | 10,626 |
| Sarah | Brown (1973) | 2;03:05–5;01:06 | 139 | 29,467 |

(years;months:days)

For each child, I located the uses of *yes/no* questions with inversion and classified them into the following three types: (i) *yes/no* questions in which the subject noun phrase consists of a single word and an auxiliary is preposed, (ii) *yes/no* questions in which the subject noun phrase consists of multiple words and an auxiliary is preposed, and (iii) *yes/no* questions in which the subject noun phrase consists of multiple words and the nominal element in the subject is preposed.[6] Below, I will classify the first type of *yes/no* questions as *simple-subject yes/no questions with a fronted auxiliary*, the second type as *complex-subject yes/no questions with a fronted auxiliary*, and the third type as *complex-subject yes/no questions with a fronted nominal*. Examples for each of these types of interrogatives are given in (12).

(12) a. *Simple-subject yes/no questions with a fronted auxiliary:*
      Can you ___ write stories?
   b. *Complex-subject yes/no questions with a fronted auxiliary:*
      Can young children _____ write stories?
   c. *Complex-subject yes/no questions with a fronted nominal:*
      *Children young _____ can write stories?

### 4.2 Results

The results of my transcript analysis are summarized in Table 5.3.

Among the three children analyzed in this study, Adam and Sarah showed uses of *yes/no* questions with a subject noun phrase that consists of multiple words. However, none of these *yes/no* questions involved movement of the nominal element to the sentence-initial position. The representative utterances are presented in (13) and (14).

(13) Adam's complex-subject *yes/no* questions:
   a. *CHI: does this top go on here ?    (3;03:18, adam27.cha: line 4816)
   b. *CHI: can my train go in the water ? (3;05:29, adam32.cha: line 4474)

Table 5.3 The number of children's *yes/no* questions

| Child | Simple-subject yes/no questions with a fronted auxiliary | Complex-subject yes/no questions | |
|---|---|---|---|
| | | With a fronted auxiliary | With a fronted nominal |
| Adam | 1345 | 50 | 0 |
| Eve | 34 | 0 | 0 |
| Sarah | 570 | 18 | 0 |
| Total | 1949 | 68 | 0 |

(14) Sarah's complex-subject *yes/no* questions:
    a. *CHI: does that one get a button ? (3;07:23, sarah070.cha: line 1222)
    b. *CHI: is this one blue ? (3;10:16, sarah082.cha: line 275)

## 4.3 Analysis of the input data

Given the finding that children do not make any mistake of the type illustrated in (12c), let us now address the question of whether there is any possibility that children can deduce the ban on fronting the nominal as in (12c) solely from the surface analysis of the input data. One possible scenario for input-driven learning may proceed as follows. Encountering simple-subject *yes/no* questions with inversion as in (12a), children infer that some element must be fronted to form *yes/no* questions in the target language. However, these questions in the input are compatible with (at least) two strategies: (i) fronting the T head (to which an auxiliary is attached), or (ii) fronting the second element in the clause. The latter, structure-independent strategy eventually yields ungrammatical questions with a fronted nominal exemplified in (12c). However, children rule out this strategy by hearing abundant examples like (12b), in which an auxiliary, rather than the nominal in the second position, undergoes preposing.

This acquisitional scenario crucially depends on the assumption that the input data for children contain frequent use of complex-subject *yes/no* questions with a fronted auxiliary as in (12b). In order to determine whether this is actually the case, I analyzed the child-directed speech in the corpora for Adam and Sarah. Specifically, using the CLAN program KWAL, all the mother's utterances containing a question mark ("?") were searched, in which the number of *yes/no* questions with inversion were counted. The transcripts analyzed were limited to those before children's first clear uses of complex-subject *yes/no* questions.

The results of the analysis of the input data are summarized in Table 5.4.

Table 5.4 The number of *yes/no* questions in the child-directed speech

|  | Files analyzed | Total number of mother's utterances | Yes/No questions with inversion | |
|---|---|---|---|---|
|  |  |  | Simple-subject yes/no questions | Complex-subject yes/no questions |
| Adam's Mother | 01–27 | 10420 | 732 (97.60%) | 18 (2.40%) |
| Sarah's Mother | 001–070 | 15270 | 533 (99.07%) | 5 (0.93%) |

As we can see, the majority of adult *yes/no* questions with inversion contained a simple subject (as in (12a)), and the interrogatives whose subject noun phrase consists of multiple words (as in (12b)) were quite rare. Especially in the case of Sarah, the rate of the mother's complex-subject questions was limited to 0.93% of all the inverted *yes/no* questions. Let us now recall the findings by Legate and Yang (2002) we discussed in Section 3: they found that the frequency of *there*-type expletives, which should act as the disconfirming evidence against an optional-subject grammar, was around 1.2% in the child-directed speech. The frequency of complex-subject *yes/no* questions in the child-directed speech for Sarah is lower than this number. This finding casts doubt on the possibility that children are provided sufficient amount of critical evidence which help them decide on the option of fronting an auxiliary over the structure-independent option of fronting the second element in the clause.

## 4.4 Discussion

The results of the transcript analysis of three English-speaking children revealed that these children never produced any ungrammatical *yes/no* question in which the nominal element within the subject is preposed, rather than the auxiliary. The results of the analysis of child-directed speech suggested that these children should have been tempted to adopt a simpler, structure-independent rule of moving the second element in the sentence, since almost all of the *yes/no* questions in the mothers' speech contained a subject noun phrase which consists of a single word. Thus, the absence of such ill-formed *yes/no* questions in children's utterances add a new piece of evidence that children are genetically predisposed to rule out structure-independent options.

According to Chomsky's (2012, 2013a, b) minimalist analysis of simple interrogatives, the choice of the T head over the nominal as the raised element stems from the interaction between the third-factor principle of Minimal Search (which is structure-dependent) and the assumption that subjects are first merged internally to the predicate (the predicate-internal subject hypothesis). If this analysis of subject-auxiliary inversion in simple *yes/no* questions is on the right track, then the data from child English reported in this study provide support for the view that the subject noun phrase originates and comes from inside the verb phrase even in the earliest observable stages of syntactic development.

The claim that young children conform to the predicate-internal subject hypothesis has already been made by Déprez and Pierce (1993). Analyzing children's negation-initial (Neg-initial) sentences in English, German, and other languages illustrated in (15), Déprez and Pierce argued that the pre-sentential negative element is an instance of sentential negation (a variant of *not*), not an instance of anaphoric negation, in which the negative element negates a prior utterance. Then, since subject noun phrases in these utterances appear to occupy the position lower than the sentential negation, Déprez and Pierce concluded that children go through an early stage of acquisition during which subjects may optionally stay in their original position located internal to the predicate. Thus, according to Déprez and Pierce, Neg-initial utterances exemplified in (15) provide direct evidence that children's grammar conform to the predicate-internal subject hypothesis, thereby lending acquisitional support to this theoretical assumption.

(15) a. English: No Leila have a turn        (Nina, 2;01)
     b. German: Nein    Batsch   Hunger
                no      uncle    hunger
                'The uncle is not hungry.'       (Kathrin, 25–26 months)

However, various studies challenged the claim made by Déprez and Pierce (1993) that children do not distinguish between *no* and *not*. For example, Drozd (1995) examined the pre-sentential negations of 10 English-speaking children and found that most of these negative sentences can be paraphrased as exclamatory negation, like *No way Leila have a turn*. Building on this observation, Drozd argued that pre-sentential negations are not sentential negatives but an early form of *metalinguistic exclamative negation*, which is the use of idiomatic phrases like *no way* to express objection to a previous utterance. Thus, according to Drozd, pre-sentential negation has nothing to do with and hence cannot be derivationally related to sentential negation.

Similarly, Stromswold and Zimmermann (1999/2000) analyzed the negative utterances produced by five German-speaking children, and found that these children systematically distinguish between *nicht* 'not' and *nein* 'no'. The results of their transcript analysis indicated that children used *nicht* exclusively in sentence-medial position for sentential negation and *nein* exclusively in sentence-initial position for anaphoric negation. Thus, neither the data from child English nor the data from child German are consistent with the view that children go through an early stage during which subjects can remain in the predicate-internal position.

The results of the present study suggest that, even though the evidence from children's Neg-initial sentences presented by Déprez and Pierce (1993) may not be valid, their central claim still holds true for child English. If Chomsky (2012, 2013a, b) is correct in assuming that the fronting of the T head (rather than the nominal) in simple *yes/no* questions crucially relies on the base-generation of the subject within the predicate, the absence of nominal fronting in children's interrogatives indicates that the grammar of young English-speaking children conforms to the predicate-internal subject hypothesis.

## 5 Conclusion

Within the Minimalist framework, Chomsky (2012, 2013a, b) explored the relation between a simple subject-predicate construction and the corresponding *yes/no* question in English and contended that there is an issue in this phenomenon that has not been sufficient puzzled about: why is it the case that the auxiliary undergoes raising, not the nominal element in the subject noun phrase (e.g. *young children*)? Chomsky provided a very straightforward answer, in which the fronting of the auxiliary follows from the interaction between the structure-dependent condition of Minimal Search and the predicate-internal subject hypothesis. Building on this analysis, this study addressed the very simple question of whether English-speaking children produce any ungrammatical sentences which clearly contain fronting of a nominal in the subject rather than an auxiliary. The results of my transcript analysis revealed that no such error can be observed in children's speech. This finding suggests that children never employ the structure-independent option of preposing the second element in the sentence, even though the adult utterances directed to these children should tempt them to do so. These findings provide a new piece of evidence that young children conform to structure dependence, as well as to the predicate-internal subject hypothesis. Under the assumption that structure dependence follows from UG (more specifically, the structure-building operation of Merge) and the assumption that the predicate-internal subject hypothesis reflects requirements from the conceptual-intentional system, the findings of this study lend acquisitional support to the existence of a genetic endowment that underlies our faculty of language.

## Notes

* This paper is based on my presentation at EVOLANG IX: Workshop on Theoretical Linguistics/Biolinguistics (March 13, 2012). I would like to thank Cedric Boeckx, Koji Fujita, Masayuki Ike-uchi, Hisatsugu Kitahara, Miki Obata, Hajime Ono, Yukio Otsu, and Noriaki Yusa for valuable comments and suggestions. All the remaining errors are, of course, my own. This study was supported in part by the JSPS Grant-in-Aid for Scientific Research (A) (23242025, Principal Investigator: Koji Fujita, Kyoto University).
1 The displacement of an inflectional element under T can be observed directly when a vacuous dummy element is introduced to bear the inflection, as in (ib):

   (i) a. Young children wrote stories.
       b. Did young children write stories?

2 This is expected on the grounds that the human sensorimotor system is constructed in such a way that a human cannot speak in parallel (and hence has to speak linearly).

3 An important question remains as to whether any part of the continuation of the derivation permits the nominal element within the subject to be the structurally closest lexical item to C. See Kitahara (2011) for a detailed discussion of this issue.
4 See Gualmini and Crain (2005) for evidence from children's knowledge of the interaction between the semantic property of downward entailment and the structure-dependent notion of c-command.
5 Other major challenges include the study by Lewis and Elman (2002), who constructed a simple recurrent network to model question formation in English. For a critical discussion of Lewis and Elman's statistical-learning approach, see Gualmini (2004) and Gualmini and Crain (2005).
6 Stromswold (1990) analyzed the transcripts of 14 English-speaking children available in the CHILDES database and found that the overall inversion rate for *yes/no* questions was 93.7%. This high rate of inversion, according to Stromswold, suggests that no child had a grammar which prohibited inversion.

# References

Brown, Roger. 1973. *A first language: The early stages*. Cambridge, MA: Harvard University Press.

Chomsky, Noam. 1968. *Language and mind*. New York: Harcourt Brace Jovanovich.

Chomsky, Noam. 1993. A minimalist program for linguistic theory. In *The view from Building 20: Essays in linguistics in honor of Sylvain Bromberger*, ed. Kenneth Hale and Samuel Jay Keyser, 1–52. Cambridge, MA: MIT Press.

Chomsky, Noam. 2010. Some simple evo devo theses: How true might they be for language? In *The evolution of human language: Biolinguistic perspectives*, ed. Richard K. Larson, Vivian Déprez and Hiroko Yamakido, 45–62. Cambridge, UK: Cambridge University Press.

Chomsky, Noam. 2012. Introduction. In *Gengo kiso ronsyu* [Foundations of biolinguistics: selected writings], ed. Noam Chomsky, and trans. Naoki Fukui, 17–26. Tokyo: Iwanami Shoten.

Chomsky, Noam. 2013a. Poverty of the stimulus: Willingness to be puzzled. In *Rich languages from poor inputs*, ed. Massimo Piattelli-Palmarini and Robert C. Berwick, 61–67. Oxford: Oxford University Press.

Chomsky, Noam. 2013b. Problems of projection. *Lingua* 130:33–49.

Crain, Stephen, and Mineharu Nakayama. 1987. Structure dependence in grammar formation. *Language* 63:522–543.

Déprez, Viviane, and Amy Pierce. 1993. Negation and functional projections in early grammar. *Linguistic Inquiry* 24:25–67.

Drozd, Kenneth F. 1995. Child English pre-sentential negation as metalinguistic exclamatory sentence negation. *Journal of Child Language* 22:583–610.

Fukui, Naoki, and Margaret Speas. 1986. Specifiers and projection. In *Papers in theoretical linguistics: MIT working papers in linguistics 8*, ed. Naoki Fukui, Tova R. Rapoport and Elizabeth Sagey, 128–172. Cambridge, MA: MIT Working Papers in Linguistics.

Gualmini, Andrea. 2004. *The ups and downs of child Language: Experimental studies on children's knowledge of entailment relationships and polarity phenomena*. New York: Routledge.

Gualmini, Andrea, and Stephen Crain. 2005. The structure of children's linguistic knowledge. *Linguistic Inquiry* 36:463–474.

Hyams, Nina. 1986. *Language acquisition and the theory of parameters*. Dordrecht: D. Reidel.

Kitagawa, Yoshihisa. 1994. *Subjects in Japanese and English*. New York: Garland Publishing.

Kitahara, Hisatsugu. 2011. Can eagles that fly swim? Guaranteeing the simplest answer to a new puzzle. In *The proceedings of the twelfth Tokyo Conference on Psycholinguistics*, ed. Yukio Otsu, 1–15. Tokyo: Hituzi Syobo Publishing.

Koopman, Hilda, and Dominique Sportiche. 1991. The position of subjects. *Lingua* 85:211–258.

Kuroda, Shige-Yuki. 1988. Whether we agree or not: A comparative syntax of English and Japanese. *Linguisticae Investigationes* 12:1–47.

Legate, Julie Anne, and Charles D. Yang. 2002. Empirical re-assessment of stimulus poverty arguments. *The Linguistic Review* 19:151–162.

Lewis, John, and Jeffrey Elman. 2002. Learnability and the statistical structure of language: Poverty of stimulus arguments revisited. In *Proceedings of the 26th annual Boston University Conference on Language Development*, ed. Barbora Skarabela, Sarah Fish and Anna H.-J. Do, 359–370. Somerville, MA: Cascadilla Press.

MacWhinney, Brian. 2000. *The CHILDES project: Tools for analyzing talk*. Mahwah, NJ: Lawrence Erlbaum Associates.

Pullum, Geoffrey, and Barbara Scholz. 2002. Empirical assessment of stimulus poverty arguments. *The Linguistic Review* 19:8–50.

Stromswold, Karin. 1990. *Learnability and the acquisition of auxiliaries*. Doctoral dissertation, MIT.

Stromswold, Karin, and Kai Zimmermann. 1999/2000. Acquisition of *nein* and *nicht* and the VP-internal subject stage in German. *Language Acquisition* 8:101–127.

Suppes, Patrick. 1973. The semantics of children's language. *American Psychologist* 88:103–114.

# 6 Make a good prediction or get ready for a locality penalty
## Maybe it's coming late

*Hajime Ono, Kentaro Nakatani and Noriaki Yusa*

## 1 Introduction

We all make predictions. We are not fortune-tellers, but we can still hope to make good predictions. It could be that good predictors are also effective communicators. In a naïve conception of predictions, one may suggest that having more information in our hand leads to a good prediction. Now, let us consider those ideas about predictions in the realm of sentence processing. Suppose you have the following sentence fragments in Japanese.

(1) a. Taro-ga
       Taro-NOM
    b. Taro-ga      tsukue-o
       Taro-NOM    desk-ACC
    c. Taro-ga      tsukue-o    zookin-de
       Taro-NOM    desk-ACC    dustcloth-with

As soon as a reader encounters a nominative NP at the beginning of the sentence, such as (1a), we can predict that there must be a predicate (a verb or a predicative adjective) coming in, but we do not know yet what type of predicate (argument structure and/or lexical information). Then, after one or two more constituents are added to the context, as in (1b) and (1c), we can narrow down the number of possible continuations. When a nominative NP is followed by an accusative NP, as in (1b), it becomes clear that an intransitive verb will not appear in the structure, assuming that the reader predicts that those two constituents have a dependency on a single verb. Still, there are a number of verbs that could follow, and it may not be easy to predict the specific lexical content of the verb at this point. Note that, for (1c), after the reader sees another constituent, we can pick out a relatively specific predicate such as *huku* 'wipe' to be the best continuation with confidence (see an explanation of the 'surprisal' model in Hale (2001) and Levy (2008)). Then, exactly what kind of information will lead to a good prediction (in fact, it is called an 'expectation' in the literature) is at issue in the literature; obviously, a lot of different kinds of information could contribute to the prediction-making (Frazier and

Rayner, 1982, Crain and Steedman, 1985, Altmann and Steedman, 1988, Trueswell, Tanenhaus, and Garnsey, 1994, Tanenhaus, Spivey-Knowlton, Eberhard, and Sedivy, 1995, Bader, Meng, and Bayer, 2000, Miyamoto, 2002, among many others). In this chapter, however, we focus on a very restricted set of "useful" information, so the kinds of information that will be discussed here are by no means comprehensive. We review some relevant work in the following section.

Assuming that having more information is beneficial, we can still easily imagine the cases in which having many constituents before seeing a verb would not help us predict the identity of the verb. One such case is the one in which the intervening constituents are not arguments or adjuncts directly associated with the verb. In fact, having too many constituents seems to cause problems for processing a verb. It is known in the sentence-processing literature that there is an increase of the memory load, called locality effects, when the parser has to process a dependency that spans a long distance (Yngve, 1960, Chomsky and Miller, 1963, King and Just, 1991, Gibson, 1998). The parser may slow down due to having too many constituents in such a case.

In this chapter, we are interested in the relation between expectations and locality effects that seem to be counteracting each other. Having a lot of information helps the reader make a good prediction about the verb, but a large amount of information involved in an incomplete dependency may impose a lot of memory load. Previous work on expectations shows that not all types of information lead to the facilitation of the processing of the verb. We review Konieczny and Döring (2003) in the next section, who reported that a preverbal constituent contributed to the facilitation only when it was an argument of the verb. In section 3, we review relevant observations about locality effects. After discussing expectations and locality effects, we will discuss some recent empirical findings that locality effects are not independent of expectations, but are influenced by expectations. Manipulating the phrase structure in German embedded sentences, Levy and Keller (2013) found that locality effects sometimes emerged with the facilitation by expectations suppressed, and locality effects sometimes disappeared with the facilitation by expectations observed. Furthermore, Husain, Vasishth, and Srinivasan (2014) found that the locality effects were observed only when expectations were weak.

Then, one question arises: what kind of information leading to expectations influences the emergence of the locality effects? We will introduce results from our experiment in which the locality and expectations are manipulated in *wh*-interrogative sentences in Japanese. The results show that the locality effects disappeared when there was a specific expectation at work, which is quite similar to the results in Husain et al. (2014). Also, we will see that an expectation-related slowdown occurred later than the locality effects. This finding is particularly noteworthy because no previous work reported that locality effects and the expectation-related effects occurred at the same time in a single sentence. We argue that the way expectations relate to the locality effects is different depending on the type of prediction involved.

## 2 Expectations

Konieczny and Döring (2003) investigated whether having an additional constituent dependent on the verb leads to a faster reading time on the verb. In German, a tensed verb is placed clause-finally in the embedded clause, and thus all constituents, both arguments and adjuncts, appear before the verb. In example sentences in (2), a clause was embedded in the noun complement structure with an NP *die Einsicht* 'the insight'. In (2a), the embedded verb *verkaufte* 'sold' was preceded by three constituents (a subject, a dative object and a direct object with a modifying PP), while in (2b), there are two constituents (a subject with a genitive NP and a direct object with a modifying PP) before the verb. The only difference between the two examples was the case-marker of the determiner *dem/des* 'the-dative/the-genitive' of the second NP.

(2) a. NP-dative, NP-modifying PP
 Die Einsicht, dass der Freund dem Kunden das Auto aus Plastic
 the insight that the friend to the client the car out.of plastic
 verkaufte, . . .
 sold, . . .
 "the insight that the friend sold the client the car made from plastic . . ."
 b. NP-genitive, NP-modifying PP
 Die Einsicht, dass der Freund des Kunden das Auto aus Plastic
 the insight that the friend of the client the car out.of plastic
 verkaufte, . . .
 sold, . . .
 "the insight that the friend of the client sold the car made from plastic . . ."

In an eye-tracking study, Konieczny and Döring found that the mean reading time (regression path durations[1]) for the embedded verb in (2a) was about 200ms shorter than in (2b).

These observations indicate that having more information helps the reader make more precise predictions about the upcoming verb. In the NP-dative condition, the reader may predict the verb to be a ditransitive verb compatible with the dative NP. When the reader saw the right kind of verb that the reader predicted, the processing cost for the verb was low. In the NP-genitive condition, on the other hand, the reader was able to predict that, given the presence of the accusative NP, the verb must be one of the transitive verbs, but the number of possible verbs was still large, compared with the NP-dative condition. The verb in the example was compatible with the prediction by the reader, but there were other types of verbs that were also compatible with the prediction. It is plausible that the processing cost of the verb was higher when the reader did not have a specific verb in mind and was open to other possibilities.

In the same experiment, there were in fact two other conditions in which the V-modifying PP *aus Fraude* 'just for fun' was used instead of the NP-modifying PP. No statistically reliable effect of the PP-type was found. It is then clear that, although adding a V-modifying PP makes a more complex VP structure, this had no major impact on the reading time of the verb, suggesting that the lexical or grammatical information that lowers the processing cost of the verb is selective. Having more materials diminishes the reading time only if the presence of those materials can reduce the number of possible outcomes and contribute to a better prediction, possibly by providing more specific information about the argument structure of the verb (see also Konieczny, 2000). It is certainly at issue exactly what kind of lexical or grammatical (and presumably many other) information is connected to expectations discussed so far (Vasishth and Lewis, 2006, Levy, 2008, Nakatani and Gibson, 2010, Staub and Clifton, 2006).

## 3 Locality effects

Although it has been reported in the literature that more information leads to processing facilitation, there are opposite cases in which adding more materials in the structure slows down the parser. For instance, Grodner and Gibson (2005) observed reading-time slowdowns, called locality effects, when the length of the dependency between the two linguistic elements becomes larger. Dependency Locality Theory (DLT, Gibson, 2000) and Similarity-Based Interference (Gordon, Hendrick and Johnson, 2001, 2004) are representative accounts in the literature for these distance-based "locality" effects (Warren and Gibson, 2002).

Grodner and Gibson (2005) compared the following set of materials, in which the complexity of the embedded subject in the relative clause is manipulated. In (3a), the subject is a simple NP, *the nurse*; in (3b), a PP, *from the clinic*, is attached to the subject; in (3c), the subject is followed by a relative clause.

(3) a. The administrator who [the nurse] supervised scolded the medic while . . .
   b. The administrator who [the nurse *from the clinic*] supervised scolded the medic while . . .
   c. The administrator who [the nurse *who was from the clinic*] supervised scolded the medic while . . .

They measured the reading times of the embedded verb *supervised* and observed that the verb in (3c) was read the slowest, and there was a gradual increase in the reading times from (3a) to (3c). According to DLT, the reading time of the verb *supervised* is partially determined by the number of discourse referents and events intervening the verb and its subject. Compared with (3a), (3b) and (3c) had a discourse referent *clinic* between *the nurse* and *supervised*, the presence of which incurs a cost for integration at the verb. In (3c), the tensed verb *was*, representing an event, incurs another cost, thus the reading time for the verb in (3c) was the greatest.

The locality effects are not limited to the dependency between the subject and its predicate, but also observed in various long-distance dependency relations. For instance, Phillips, Kazanina, and Abada (2005) compared the following two sentences with different dependency lengths. The wh-phrase *which accomplice* originates as the object of the verb *recognize* (noted with the line) in both sentences. In (4a), the wh-phrase is at the edge of the embedded clause where it originates, and in (4b), the distance between the wh-phrase and the verb *recognize* is larger.

(4) a. The detective hoped that the lieutenant knew [which accomplice [the shrewd witness would recognize __ in the lineup]].
   b. The lieutenant knew [which accomplice [the detective hoped that [the shrewd witness would recognize __ in the lineup]]].

Complexity rating for those sentences indicate that native speakers of English judge (4b) as more difficult than (4a), showing that it is more costly for readers to process sentences with long dependencies. If we use a memory-based account, this locality effect observed at the verb shows that it is more costly to reactivate the element stored in working memory at the point of integration, if the activation level of the element in working memory has decayed (Lewis, 1996, Lewis, Vasishth, and Van Dyke, 2006, Van Dyke and McElree, 2006, Wagers, 2008).

## 4 Expectation and locality: their connections

In previous studies, facilitations of processing caused by expectations are mostly observed in head-final languages, such as German, Hindi, or Japanese. The increase of the processing cost caused by locality is often reported for head-initial languages like English. However, there are some recent suggestions that it is not quite accurate to consider that locality effects are totally absent from head-final languages, or expectations do not play a substantial role in head-initial languages. We review some recent studies of head-final languages that observed expectations and locality effects in the same types of constructions in the same language.

Levy and Keller (2013), critically reviewing Konieczny (2000) and Konieczny and Döring (2003), investigated the verb-final structures in German. They used materials in which the position of the critical verb, as well as the plausibility of the sentences, was controlled in their eye-tracking experiments. The templates of their materials are shown below. They created four versions of target sentences using the same phrases, by putting an adverbial phrase and a dative NP either in the temporal subordinate clause (with *nachdem* 'after' in the example below), or in the relative clause that modifies the subject in the matrix clause (*der Mitschüler* 'the classmate"). The adverbial phrase and the dative NP used in the target sentences were not a simple phrase but contained several words; thus, the placement of those phrases altered the length of the clause a lot. According to Levy and Keller (2013), verbs in the target sentences (both in the temporal clause and the relative clause) were all optionally ditransitives; the dative NP can appear with the verb as an argument, but it may be absent.

(5) The template for the target sentences

   a. adverbial and dative NP in subordinate clause
      After NP-NOM [*Adv*] [NP-DAT] NP-ACC verb, NP-NOM [who NP-ACC verb], . . .
   b. dative NP in subordinate clause; adverbial in relative clause
      After NP-NOM [NP-DAT] NP-ACC verb, NP-NOM [who [*Adv*] NP-ACC verb], . . .

c. adverbial in subordinate clause; dative NP in relative clause
After NP-NOM [*Adv*] NP-ACC verb, NP-NOM [who [NP-DAT] NP-ACC verb], . . .

d. adverbial and dative NP in relative clause
After NP-NOM NP-ACC verb, NP-NOM [who [*Adv*] [NP-DAT] NP-ACC verb], . . .

(6) A target sentence used in the condition (a)
Nachdem der Lehrer [zur zusätzlichen Ahndung des mehrfachen
After the teacher as additional payback for multiple
Fehlverhaltens] [dem ungezogenen Sohn des fleißigen
wrongdoings the.DAT naughty son the.GEN industrious
Hausmeisters] den Strafunterricht verhängte hat, der
janitor the.ACC detention.classes imposed has the.NOM
Mitshchüler, [der den fußball verstecht hat],
classmate who.NOM the.ACC football hidden has
die Sache bereinigt.
the affair corrected

"After the teacher imposed detention classes on the naughty son of the industrious janitor as additional payback for the multiple wrongdoings, the classmate who hid the football corrected the affair."

The critical region is the verb in the relative clause *verstecht* 'hidden', and Levy and Keller (2013) observed a statistically reliable interaction between the positioning of the adverbial and the dative NP. More specifically, there was a reading-time slowdown in (5a) compared with (5c), and also a slowdown in (5d) compared with (5c) in the second-pass time measure[2]. A similar pattern was also observed in the spillover region. They argued that the patterns of reading-time slowdowns they observed contain both expectation and locality effects. For example, the difference between (5a) and (5c) shows that the added constituent before the verb in (5c) facilitated the processing of the verb. In (5c), the critical verb was preceded by a dative NP, and the verb in (5c) was read faster than the verb in (5a) in which there was no dative NP before the verb, which is similar to the finding in Konieczny and Döring (2003). Note also that adding an adverb in the relative clause did not facilitate the processing of the verb in (5b). Although the reading time of the verb in (5b) was numerically shorter than that in (5a), the difference was not statistically significant. The lack of clear facilitation with an adverbial further suggests that the expectations are primarily driven by information about argument structure of the predicate that was made clear from the set of preverbal NPs in the structure. Even though it seems intuitive to witness that the combined semantic information of an adverbial and an accusative NP helps the reader predict the lexical content of the upcoming predicate, the results in Levy and Keller (2013) suggest that a facilitation based on the semantic information of this kind did not occur, or at least was not strong enough to make an impact on the reading time of the verb.

In this experiment, they observed the locality effects in (5d), compared with (5c). When there were both adverbial and dative NPs in the relative clause structure as in (5d), the reading times for the verb in the relative clause were longer. Levy and Keller argue that having a dative NP leads to the facilitation while also adding an adverbial cancels out such a facilitation effect. They also suggest that using a structure that requires a sufficient amount of working memory seems to be an important factor for obtaining both effects at the same time. In their paper, they reported another experiment in which they used simpler structures, and they did not observe a slowdown such as the one in (5d). Another consequence of their results is that a distance-based theory such as DLT is not sufficient to account for the whole pattern of results.

Related to Levy and Keller (2013), there is another study, Husain, Vasishth and Srinivasan (2014), showing the emergence of expectations and locality effects. In their experiment, they manipulated the strength of the expectation and investigated how it influences the (anti-)locality effects in Hindi. Just like in German, it has been shown previously that the additional constituent before a verb reduces the processing cost of the verb in Hindi (an anti-locality effect, e.g., Vasishth and Lewis, 2006). Husain et al. (2014) manipulated the strength of expectations not by the presence of constituents in the structure, but by the use of different object NPs, which varied in the degree of the expectations they would invoke for a particular verb. They prepared high and low expectation conditions, crossed with a distance factor (long vs. short). The short version of the high expectation condition is illustrated below (only the crucial part is shown). According to Husain et al. (2014), in the example of the high expectation condition, when placed in an appropriate context, the object *khayaal* 'care' provides a strong clue for the identity of the verb (in this case, *rakhe* 'keep'), due to the high collocations of the noun-verb sequence in Hindi. In the low expectation conditions, they used another noun, *gitaar* 'guitar', as the object that does not have any strong relation with the verb *rakhe* 'keep' although they are semantically fully compatible (the meaning of the sentence will be "she properly keeps the guitar"). Furthermore, they prepared the long version of the target sentences where an extra adverbial, *binaa kisi laaparvaahi ke* 'without any carelessness', was inserted after the object, so that the object was separated from the verb by several words. They were interested in the reading-time patterns of the verb, *rakhe* 'keep'.

(7) High expectation, short
    ... vah    apnaa    khayaal    achC se      rakhe, ...
        she    her      care       properly     keep
    "... she keeps good care of herself, ..."

In their self-paced reading experiment, they observed a main effect of expectations at the crucial verb region; the high expectation conditions were in general read faster than the low expectation conditions. Also, they observed a significant interaction between the expectation factor and the distance factor. Specifically, while the presence of an extra adverbial did not increase or decrease the reading time of the verb in the low expectation conditions, a speedup was observed in the long × high expectation condition, compared to the short × high expectation condition. In other words, the verb was read faster when the distance between the object and the verb was long. One thing that should be paid attention to in their results is that they increased the length between the object and the verb with an adverbial phrase. Recall from the German experiments of Konieczny and Döring (2003) and Levy and Keller (2013) that adding an adjunct phrase in the structure did not facilitate the processing of the verb. It seems unlikely that the adverbial greatly contributes to the prediction of the verb because, in the high expectation conditions in this experiment, the verb was already predicted. It could be that such a specific sentential context somehow boosts the impact of adding an adverbial, which is usually not strong enough to facilitate the processing of the verb. At this point, it is unclear what exactly causes the difference between German and Hindi cases, but further investigation is necessary about the role of an adverbial phrase with respect to the locality effects in those languages.

When we return to the main findings of their experiment, the results indicate that the impact of locality (i.e., the distance between elements to be integrated) is highly modulated by expectations. The Hindi case shows that the locality effects (slowdowns) appear only when there is no strong expectation at work. Husain et al. (2014) argued that such complementarity arises because the parser can build a detailed structural representation when the verb is highly expected given the preverbal constituents. In the high expectation conditions above, the parser not only builds a detailed phrase structure for VP, but also integrates the expected verb before the parser actually encounters the verb. Note that the locality effects are assumed to be, at least in a substantial part, a cost of integrating the verb and the dependent elements in the structure (Gibson 1998 among others); therefore, no locality effects should be found at the verb region when the verb is fully predicated in advance, evoking no need for integration at the point of the actual encounter. Husain et al. found a distance-based expectation facilitation (i.e., the more intervening elements, the easier) only in the high expectation conditions because the counteracting locality effects (the more intervening elements, the harder) were absent.

## 5 An interim summary and forward

The findings in Levy and Keller (2013) and Husain et al. (2014) show that the facilitation effects based on expectations and the processing costs based on locality are two forces that counteract each other. In one situation where the parser can make a strong prediction, the facilitation effects based on expectation are more pronounced, and in another situation where properties of the verb are not highly predictable, locality effects do appear more strongly. As we discussed above, Husain et al. (2014) further argue that such complementarity between expectations and locality arises because, when an item is highly predictable, the parser can integrate the item into the structure before the parser actually encounters that item. In the Hindi case they discussed, the crucial cue for the parser to make a prediction is information about the collocation of the object and the verb. On the other hand, the crucial cue for the prediction in the German cases of Levy and Keller (2013) and Konieczny and Döring (2003) is the presence of NPs in the preverbal position that helps the parser predict the argument structure of the verb to follow.

Now we would like to investigate further other types of prediction by the parser that may lead to the modulation of the locality effects. If Husain et al.'s (2014) argument is applicable to more general cases, then it seems that a wide range of information about the verb, including highly grammatical cues, may impact the locality effects. On the other hand, it could be that their argument may be applicable only to certain types of expectation, such as information about the argument structure of the verb. In the following experiment, we used *wh*-interrogatives in Japanese as target sentences. The use of *wh*-interrogatives provides a particularly interesting testing ground for further investigating the interaction between the locality effects and expectation-based facilitations, because they involve two types of grammatical dependencies at the same time: thematic dependencies and *wh*-Q dependencies. The presence of a *wh*-phrase strongly invokes an expectation for a Q-particle to come, which may or may not be attached to its theta-assigning verb. If the prediction for the Q-particle suppresses the locality effects, then the suggestion by Husain et al. (2014) is shown to be applicable quite generally, in that information other than the argument structure of the verb has an influence on the locality effects.

## 6 Experiment

We conducted a self-paced reading experiment, examining whether information other than the argument structure of the verb would influence the emergence of locality effects. As briefly discussed above, we set up a paradigm of expectation triggered by *wh*-phrase, which obligatorily forms a dependency relationship with clause-final Q-particle *ka* in Japanese. The Q-particle is a scope-marking verbal suffix, and it may appear with the verb assigning a thematic role to the *wh*-phrase. This is the earliest possible position where the required Q-particle may be placed. Alternatively, the Q-particle, being a scope marker, may be attached to a verb in a clause higher than the one containing the *wh*-phrase.

(8) a. Early, clause-mate Q-particle:
    *dare-ga*      kita-*ka*    sitteiru.
    who-NOM        came-Q       know
    "(I) know who came."
  b. Late, higher Q-particle
    *dare*-ga      kita-to      omoimasu-*ka*?
    who-NOM        came-C       think-Q
    "Who do (you) think came?"

In the experiment, we varied the position of Q-particle *ka*, crossed with the manipulation of the distance between the *wh*-phrase and the verb. Note that the presence of *wh*-phrase made the test materials just like the Hindi high-expectation conditions. In Hindi, the specific lexical content of the verb is predicted by the object, while in our Japanese sentences, what is predicted by the *wh*-phrase is the encounter with the Q-particle at the earliest possible position. Thus, it is worth testing whether the locality effects would be incurred by this type of grammatical expectation. Additionally, in the conditions in which the Q-particle is *not* attached to the theta-assigning verb, a surprisal effect might be incurred, because the parser would prefer to resolve incomplete dependencies as soon as possible (Miyamoto and Takahashi, 2002; more on this below). In a sense, these conditions can be considered counterparts of Hindi low-expectation conditions. We would like to see if stronger locality effects are observed in these "less expected" conditions than in the "expected" conditions in which the Q-particle is attached to the clause-mate, theta-assigning verb.

Another important point we should pay attention to with respect to the findings in Levy and Keller (2013) and Husain et al. (2014) is that they observed locality effects when the structure being processed requires a lot of working memory resources. In fact, in an experiment by Grodner and Gibson (2005), the locality effects were not so pronounced when the structure was simple; their locality effects clearly appeared when the relevant part of the sentence was embedded inside of relative clauses. In Hindi experiments by Husain et al. (2014), the sentences were rather long, which obviously consumes much working memory. It has been shown in various measurements that sentences with *wh*-dependency require more working memory resources to process in general (King and Kutas, 1995, Kluender and Kutas, 1993, Kaan et al. 2000, Phillips, Kazanina and Abada, 2005). We examine the locality effect using *wh*-interrogative sentences in Japanese.

## 6.1 *Method*

Thirty undergraduate students participated in a self-paced reading experiment run by Linger software (written by Douglas Rohde). Twenty-four sets of target sentences were prepared and distributed into four lists according to the 2 × 2 factorial design (Q-position × Locality). Fifty-four filler sentences were also prepared in order to mask the purpose of the experiment from the participants. Sentences were presented pseudo-randomly. Reading times for each *bunsetsu* phrase were measured, and each sentence was followed by a comprehension question about the content of the sentence to make sure that the students were paying enough attention to the content of the sentences while reading.

The template of the target sentences is shown in (9), and a sample set of the target materials in (10). We manipulated the distance between the *wh*-phrase *dono soodanin-ga* 'which counselor-nom' and the verb, e.g., *sinziteiru* 'believe', by switching (scrambling) the positions of the *wh*-phrase and the embedded clause. In the distant conditions, the embedded clause was placed between the *wh*-phrase and the verb, while in the local conditions, it was placed before the *wh*-phrase. In the local conditions, the *wh*-phrase and the verb that had the thematic relation with the *wh*-phrase were placed next to each other. We also manipulated the position of the Q-particle *ka* which determined the scope of the *wh*-phrase. In the Mid-Q conditions, the Q-particle was attached to the verb that assigns a thematic role to the *wh*-phrase (it was called "Mid-Q" because among the three clauses involved, it appeared in the middle one). This position of the Q-particle was the grammatically closest position for the *wh*-phrase, in the sense that the *wh*-phrase and the Q-particle were in the same clause and the *wh*-phrase was c-commanded and licensed by the Q-particle. In the High-Q conditions, on the other hand, the Q-particle appeared with the matrix verb, and the whole sentence became a *wh*-question. In the High-Q conditions, the *wh*-phrase and the Q-particle were not in the same clause, and the Q-particle was attached to the verb in a higher (superjacent) clause. See Ono and Nakatani (2014) for the use of similar paradigm.

(9) Mid-Q conditions

    a. [which NP-NOM [...comp clause...] verb-Q ] Loc NP-TOP V DISTANT
    b. [[...comp clause...] which NP-NOM verb-Q ] Loc NP-TOP V LOCAL
    High-Q conditions
    c. [which NP-NOM [...comp clause...] verb-C ] Loc NP-TOP V-Q DISTANT
    d. [[...comp clause...] which NP-NOM verb-C ] Loc NP-TOP V-Q LOCAL

(10) A sample set of target sentences

    a. dono soodanin-ga   [sensee-ga   tyapatu-no   namaikina
       which counselor-NOM teacher-NOM dyed.hair-GEN impertinent
       gakusee-o   hidoku sikatta-to] sinziteiru-ka
       student-ACC harshly scolded-C] believe-Q
       kyoositu-de  hogosya-wa kikimasita.
       classroom-at parents-TOP asked

b. [sensee-ga    tyapatu-no    namaikina    gakusee-o    hidoku sikatta-to]
   teacher-NOM dyed.hair-GEN impertinent student-ACC harshly scolded-C]
   dono soodanin-ga    sinziteiru-ka kyoositu-de hogosya-wa kikimasita.
   which counselor-NOM believe-Q   classroom-at parents-TOP asked
   "The parents asked which counselor believed that the teacher had harshly
   scolded the impertinent student with dyed hair."

c. dono soodanin-ga    [sensee-ga    tyapatu-no    namaikina
   which counselor-NOM teacher-NOM dyed.hair-GEN impertinent
   gakusee-o    hidoku sikatta-to]  sinziteiru-to
   student-ACC harshly scolded-C] believe-C
   kyoositu-de hogosya-wa iimasita-ka.
   classroom-at parents-TOP said-Q

d. [sensee-ga    tyapatu-no    namaikina    gakusee-o    hidoku sikatta-to]
   teacher-NOM dyed.hair-GEN impertinent student-ACC harshly scolded-C]
   dono soodanin-ga    sinziteiru-to kyoositu-de hogosya-wa iimasita-ka.
   which counselor-NOM believe-C   classroom-at parents-TOP said-Q
   "Which counselor did the parents say believed that the teacher had harshly
   scolded the impertinent student with dyed hair?"

As is clear from the sample set of the target materials and discussed above, we manipulated not only the linear distance between the *wh*-phrase and its predicate, but also the position of the licensing Q-particle. This manipulation was independently motivated by the findings by Miyamoto and Takahashi (2002). They investigated the *wh*-dependency formation in Japanese and compared the following paradigm (see also Aoshima, Phillips and Weinberg, 2004; Ono and Nakatani 2010, 2014). In a series of self-paced reading experiments, they found that, given the presence of the *wh*-phrase in the embedded clause, the embedded verb with the non-interrogative complementizer *to* 'that' in (11b) was read slower than the same verb with the Q-particle. This observation was widely recognized as Typing Mismatch Effects (TMEs). They argued that the parser prefers the structure in (11a) because the length of the dependency formation in (11a) is shorter than that in (11b). In other words, the parser makes prediction that the embedded verb appears with the Q-particle, based on the presence of the *wh*-phrase.

(11) a. kakarityoo-wa [senmu-ga        dono pasokon-o
        supervisor-TOP the director-NOM which computer-ACC
        tukatteiru-ka] kikimasita
        is.using-Q    asked
        "The supervisor asked which computer the director is using."

b. kakarityoo-wa [senmu-ga      dono pasokon-o
   supervisor-TOP the director-NOM which computer-ACC
   tukatteiru-to] iimasita-ka
   is.using-C     said-Q
   "Which computer did the supervisor say that the director is using?"

In terms of expectations, TMEs are an indication that the presence of the *wh*-phrase leads the parser to make a fine-grained prediction. More specifically, when the parser encounters a non-*wh* NP with an accusative Case-marker, the parser can predict there must be a transitive verb, but it cannot make any further specification of the verb: the verb may appear with the non-*wh* declarative complementizer or the Q-particle *ka*. By contrast, if the parser encounters an accusative *wh*-NP, it makes a fine-grained, non-thematic prediction about the verb region; it should preferably come with *ka*, rather than *to*.

## 6.2 Results

Data from three participants were eliminated due to their low accuracy rates for the comprehension questions (less than 66.6 per cent). We used reading time data only from the trials in which the participants answered the comprehension question correctly. The reading-time data whose z-scores were 3 or greater in each condition × region cell were eliminated. Because we were interested in locality effects and TMEs at the mid-verb and its spillover regions, reading time data from the regions 8–10 were submitted to fit linear mixed effects models, in which the Q-position factor and the locality factor were taken as fixed effects, and participants and items factors as random effects. The two fixed effects were centered such that for the locality factor, the distance condition was encoded as 1, and the local condition as –1; similarly, for the TME-associated factor, the Mid-Q was encoded as 1, and the High-Q as –1. The models were fit using the `lmer` command in the lme4 package in R; the p values were computed using the `mixed` function in the afex package. Table 6.1 lists the results from the LMER analysis for region 8–10. Where appropriate, we compared the conditions pairwise, the results of which are reported in the text.

In region 8 (the "mid" verb), there was a main effect of locality, showing that the verbs in the distant conditions were read slower than those in the local conditions. There was no main effect of the Q-position. Although there was no interaction between the two factors, a planned pairwise comparison showed that there was a locality effect in the High-Q conditions, indicating that the distant × High-Q condition was read slower than the local × High-Q condition ($\beta$ = 55.05, SE = 25.91, t value = 2.12).

In region 9 (locative PP, spillover region 1), there was also a main effect of locality as well as a main effect of the Q-position. The main effect of locality is in fact in the opposite direction to the one in the region 8. In region 9, the verbs in the local conditions seemed to be read slower, but it looked like this

*Table 6.1* Results of the statistical analyses based on the linear mixed effects models fit for the reading time in critical region 8 and spillover regions 9 and 10

|  | Estimate | Std. Error | t value | p value |
|---|---|---|---|---|
| **Region 8** | | | | |
| (Intercept) | 1168.47 | 78.69 | 14.85 | |
| locality | 58.72 | 18.68 | 3.14 | 0.00 |
| Q-position | −23.03 | 18.66 | −1.23 | 0.22 |
| locality × Q-position | −15.45 | 18.68 | −0.83 | 0.41 |
| **Region 9** | | | | |
| (Intercept) | 1141.58 | 69.34 | 16.46 | |
| locality | −40.74 | 18.98 | −2.15 | 0.03 |
| Q-position | −46.06 | 18.95 | −2.43 | 0.02 |
| locality × Q-position | 21.40 | 18.96 | 1.13 | 0.26 |
| **Region 10** | | | | |
| (Intercept) | 1090.77 | 78.06 | 13.97 | |
| locality | −25.07 | 17.52 | −1.43 | 0.15 |
| Q-position | −48.54 | 17.47 | −2.78 | 0.01 |
| locality × Q-position | −13.02 | 17.49 | −0.74 | 0.46 |

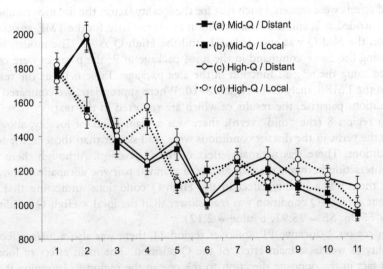

*Figure 6.1* Mean reading times (millisecond) of the target sentences for the four conditions. Error bars show standard errors.

effect emerged due to the slow reading time in the local × High-Q condition. In order to examine the slow reading time in the local × High-Q conditions, further planned comparisons were conducted. We found that there was a simple main effect of the Q-position in the local conditions ($\beta$ −75.08, SE = 27.73, t value = −2.71), showing that the local × High-Q condition was read slower than the local × Mid-Q condition. On the other hand, such an effect of the Q-position was absent in the distant conditions.

Finally, in region 10 (a nominative NP, a spillover region 2), there was a main effect of the Q-position, showing that the High-Q conditions were read slower than the Mid-Q conditions.

## *6.3 Discussion*

One of the main interests in this experiment was to examine whether and how the locality effects and expectations influence each other. Given the findings in Levy and Keller (2013) and Husain et al. (2014), we were specifically interested in testing whether one would suppress the other, and whether the locality effects would be observed only when there was no strong expectation. Below, we will briefly summarize the findings and discuss implications we can draw for the connection between the locality effects and expectations.

In region 8 (critical region), there was a main effect of locality, indicating that the critical verb in the distant conditions was read slower than that in the local conditions. Although it is unfortunate that the interaction of the two fixed effects was not significant, a further planned pairwise comparison suggested that the slow reading time of the distant × High-Q condition was driving the main effects of locality. The reading time data showed that there was no locality effect in the Mid-Q conditions where the Q-particle appeared with the critical verb. Furthermore, the comparison between the two High-Q conditions where the Q-particle was missing from the verb that was in the same clause as the *wh*-phrase showed that there were actually locality effects observed between the two.

As discussed above, if we take the Mid-Q conditions as equivalent to Husain, et al's Hindi high expectation conditions, and the High-Q conditions as the Hindi low expectation conditions, the current pattern of results in this region turned out quite similar to what has been observed in Husain et al. (2014): locality effects showed up only when expectations were not strong. In other words, when expectations were in fact strong (our Mid-Q conditions), locality effects did not show up. Such a correspondence between Hindi data and Japanese data leads us to claim that the types of expectations that may suppress locality effects are not restricted to information about argument structure of the verb, but are more generally found. In the current experiment, what was predicted by the *wh*-phrase was the presence of the Q-particle appearing at the critical verb as a suffix, and obviously this *wh*-licensing morpheme had no direct relation to the argument structure of the verb. Recall that Husain et al. (2014) argue that locality effects are not observed in the high expectation conditions because the reader can expect a specific verb to appear in the high expectation conditions, and the reader can integrate the verb into the structure before actually seeing it. Then, when the reader sees the actual verb, the fully elaborated phrase structure has already been built, and there is not much to do in terms of integration. In our current results, the Q-particle was fully predicted by the presence of the *wh*-phrase. The reader can build the structure with the Q-particle that is yet to be seen. When the reader actually sees a verb with the Q-particle, the representation of the Q-particle has already been integrated into the structure, presumably without the lexical content of the verb by itself.

Turning to the High-Q conditions, where the verb was followed by a declarative complementizer *to*, we observed locality effects: the distant × High-Q condition was read slower than the local × High-Q condition. The fact that locality effects showed up in this comparison also fits well with the findings from Husain et al. (2014). In the target sentences, the verb without the Q-particle was not expected, given the existence of the *wh*-phrase in the clause. We could interpret the finding along the line of the analysis elaborated in Husain et al. (2014): because the verb with the declarative complementizer *to* was not expected at all, the representation for the declarative complementizer could not be supposed or built in before the reader actually saw the declarative complementizer. Because no benefit was available from the expectation-based preprocessing, the verb and the complementizer in the High-Q conditions must be processed upon the reader seeing the verbal complex. It has been assumed that the integration of the verb requires the reactivation of the arguments, and it is quite costly to reactivate linearly distant elements because the activation of such an element has already decayed (Lewis 1996, Gibson 1998). Therefore, the distant × High-Q condition was read slowly, compared with the local × High-Q condition.

Note that in addition to the locality effect in region 8, we also observed slowdowns of High-Q conditions in region 9 and 10, though the detailed patterns are slightly different. Let us first look at the effects we observed in region 9. In this region, the local × High-Q condition was read slower, compared with the local × Mid-Q condition, and it is plausible that this is a TME (Miyamoto and Takahashi 2002; we will come back to the lack of TMEs in the distant conditions shortly). This finding is quite crucial because as far as we are aware of, no timing difference has ever been observed with respect to the emergence of locality effects and expectation-related effects, such as TMEs in our case (see Ono and Nakatani 2010, 2014 for some related discussions). Also, the current results showed that locality effects were observed earlier in the sentence than the expectation-related effects. We believe that there is a principled reason why locality effects must be observed earlier than TMEs. Recall that locality effects are claimed to reflect the processing costs associated with the integration of the verb into the structure (Gibson 1998). TMEs, on the other hand, could be a slowdown effect involving the checking process with respect to the feature compatibility between the *wh*-phrase and the verb plus the particle. If such a checking process can take place only after the verb and the relevant morpheme are actually integrated into the structure, then TMEs must be observed after the locality effect, which is a cost associated with the integration of the verb. The above interpretation can also account for why no other previous work has observed a timing difference with respect to the locality effects and expectations. As we suggest, checking the compatibility of the relevant feature must take place after the integration of the verb has been completed. Note that what previous studies have dealt with concerning expectations was the lexical content of the verb itself; for instance, Levy and Keller (2013) used information on the possible argument structures of the verb, and Husain et al. (2014) used the collocation of the verb and certain object NPs. Therefore, upon integrating the verb into the structure, all the relevant checks about the lexical content of the verb were done at that step. In our current case, the difference in timing was observed because we dealt with complementizers that appeared as a suffix, which may be an independent syntactic projection in the structure.

Finally, we observed another timing difference between the High-Q conditions. It seems that the emergence of TMEs was slightly faster in the local × High-Q condition than the distant × High-Q condition. Between the two local conditions, TMEs were observed in region 9, and probably the effects remain through region 10. On the other hand, TMEs in the distant × High-Q condition were clearly observed only in region 10. Although we have to wait for some further investigation into the timing issue with respect to TMEs, we could hypothesize that the difference was due to the dependency length between the verb with the complementizer and the *wh*-phrase in the structure. Again, assuming that part of TME involves the checking process for the compatibility of the relevant features, it is reasonable to consider that it is more costly to check the feature compatibility that spans a long distance. For instance, Sprouse, Fukuda, Ono, and Kluender (2011) showed that *wh*-interrogatives in Japanese were judged worse when the distance between the *wh*-phrase and the licensing Q-particle was long. Then, we could say that TMEs for the distant conditions were found later than that for the local conditions because the checking process for the feature compatibility was distance-sensitive.

## 7 Conclusion

We started our discussion in this chapter by raising a question about how to make a good prediction. Focusing on the sentence-processing mechanism, we observed that there is a certain benefit for having more constituents. A reader can expect a very specific verb to appear and can process the verb faster when case-marked NPs narrow down the choices of verbs. At the same time, slowdowns at the verb are observed due to the working memory cost incurred by processing long-distance grammatical dependencies. Based on recent works by Levy and Keller (2013) and Husain et al. (2014), it is now clear that expectations and locality effects influence each other; when there are specific expectations, the locality effects are suppressed, but locality effects show up when expectations are not strong. In this chapter, we provided data showing that such a mutually exclusive situation arises not only where predictions are based on thematic properties of a verb (such as the argument structure and the object-verb collocation), but also where the reader can predict the *wh*-licensing

Q-particle, which appears as a suffix for the verb. Furthermore, we showed that an expectation-related slowdown occurred later in the sentence than locality effects. Such a timing difference between the two factors has never been observed in the literature, and we suggest that the different types of expectations are responsible for the timing difference. Finally, in our experiment, the TMEs in the distant conditions were delayed compared with the TMEs in the local conditions. One possible account for the delay has to do with the backward search mechanism suggested in Sprouse et al. (2011). Taken together, the present study has shown some new findings related to the locality and expectations, which calls for further investigation on the relationship between these two memory-based components of sentence processing and grammatical properties such as scope-marking, head-finality, and suffixation.

## Notes

1 There are several measurements used in most of the eye-tracking experiments in the literature. Regression path durations, also known as regression path time, are the sum of fixation durations in the region before the reader makes a saccade to the region on the right.
2 Second-pass time is the sum of all fixation durations of the region except for first-pass time, which is the sum of all fixation durations before there is a saccade either to the right or the left. Total time is the sum of first-pass time and second-pass time.

## References

Altmann, Gerry T. M., and Mark J. Steedman. 1988. Interaction with context during human sentence processing. *Cognition* 30:191–238.

Aoshima, Sachiko, Colin Phillips, and Amy Weinberg. 2004. Processing filler-gap dependencies in a head-final language. *Journal of Memory and Language* 51:23–54.

Bader, Markus, Michael Meng, and Josef Bayer. 2000. Case and reanalysis. *Journal of Psycholinguistic Research* 29:37–52.

Chomsky, Noam, and George A. Miller. 1963. Introduction to the formal analysis of natural languages. In *Handbook of mathematical psychology (Vol. 2.)*, ed. Robert Duncan Luce, Robert R. Bush and Eugene Galanter, 269–321. New York: Wiley.

Crain, Stephen, and Mark J. Steedman. 1985. On not being led up the garden path: The use of context by the psychological parser. In *Natural language parsing: Psychological, computational, and theoretical perspectives*, ed. David R. Dowty, Lauri Karttunen, and Arnold Zwicky, 320–358. Cambridge, UK: Cambridge University Press.

Frazier, Lyn, and Keith Rayner. 1982. Making and correcting errors during sentence comprehension: Eye movements in the analysis of structurally ambiguous sentences. *Cognitive Psychology* 14:178–210.

Gibson, Edward. 1998. Linguistic complexity: Locality of syntactic dependencies. *Cognition* 68:1–76.

Gibson, Edward. 2000. Dependency locality theory: A distance-based theory of sentence processing difficulty. In *Image, language, brain*, ed. Alec Marantz, Yasushi Miyashita, and Wayne A. O'Neil, 95–126. Cambridge, MA: MIT Press.

Gordon, Peter C., Randall Hendrick, and Marcus Johnson. 2001. Memory interference during language processing. *Journal of Experimental Psychology: Learning, Memory, and Cognition* 27:1411–1423.

Gordon, Peter C., Randall Hendrick, and Marcus Johnson. 2004. Effects of noun phrase type on sentence complexity. *Journal of Memory and Language* 51:97–114.

Grodner, Daniel, and Edward Gibson. 2005. Consequences of the serial nature of linguistic input for sentential complexity. *Cognitive Science* 29:261–290.

Hale, John. 2001. A probabilistic Earley parser as a psycholinguistic model. In *Proceedings of the 2nd Conference of the North American Chapter of the Association for Computational Linguistics (Vol. 2)*, 159–166. Pittsburgh, PA: Association for Computational Linguistics.

Husain, Samar, Shravan Vasishth, and Narayanan Srinivasan. 2014. Strong expectations cancel locality effects: Evidence from Hindi. *PLoS ONE* 9(7):e100986. doi:10.1371/journal.pone.0100986

Kaan, Edith, Anthony Harris, Edward Gibson, and Phillip Holcomb. 2000. The P600 as an index of integration difficulty. *Language and Cognitive Processes* 15:159–201.

King, Jonathan W., and Marcel Adam Just. 1991. Individual differences in syntactic processing: the role of working memory. *Journal of Memory and Language* 30:580–602.

King, Jonathan W., and Marta Kutas. 1995. Who did what and when? Using word- and clause-level ERPs to monitor working memory usage in reading. *Journal of Cognitive Neuroscience* 7:376–395.

Kluender, Robert, and Marta Kutas. 1993. Bridging the gap: Evidence from ERPs on the processing of unbounded dependencies. *Journal of Cognitive Neuroscience* 5:196–214.

Konieczny, Lars. 2000. Locality and parsing complexity. *Journal of Psycholinguistic Research* 29:627–645.

Konieczny, Lars, and Philipp Döring. 2003. Anticipation of clause-final heads: Evidence from eyetracking and SRNs. In *Proceedings of the 4th International Conference on Cognitive Science and the 7th Conference of the Australasian Society for Cognitive Science*, 330–335. University of New South Wales, Sydney.

Levy, Roger P. 2008. Expectation-based syntactic comprehension. *Cognition* 106:1126–1177.

Levy, Roger P., and Frank Keller. 2013. Expectation and locality effects in German verb-final structures. *Journal of Memory and Language* 68:199–222.

Lewis, Richard L. 1996. Interference in short-term memory: the magical number two (or three) in sentence processing. *Journal of Psycholinguistic Research* 25:93–115.

Lewis, Richard L., Shravan Vasishth, and Julie A. Van Dyke. 2006. Computational principles of working memory in sentence comprehension. *Trends in Cognitive Sciences* 10:447–454.

Miyamoto, Edson T. 2002. Case markers as clause boundary inducers in Japanese. *Journal of Psycholinguistic Research* 31:307–347.

Miyamoto, Edson T., and Shoichi Takahashi. 2002. The processing of wh-phrases and interrogative complementizers in Japanese. In *Japanese/Korean Linguistics 10*, ed. Noriko Akatsuka and Susan Strauss, 62–75. Stanford, CA: CSLI Publications.

Nakatani, Kentaro, and Edward Gibson 2010. An on-line study of Japanese nesting complexity. *Cognitive Science* 34:94–112.

Ono, Hajime, and Kentaro Nakatani. 2010. Integration of wh-phrases and predicates in Japanese sentence processing. In *Technical report of IECIE (TL2010)* 110:99–104. [downloadable at https://sites.google.com/site/hajimeonoling/]

Ono, Hajime, and Kentaro Nakatani. 2014. Integration costs in the processing of Japanese wh-interrogative sentences. *Studies in Language Sciences* 13:13–31.

Phillips, Colin, Nina Kazanina, and Shani Abada. 2005. ERP effects of the processing of syntactic long-distance dependencies. *Cognitive Brain Research* 22:407–28.

Sprouse, Jon, Shin Fukuda, Hajime Ono, and Robert Kluender. 2011. Reverse island effects and the backward search for a licensor in multiple wh-questions. *Syntax* 14:179–203.

Staub, Adrian, and Charles Clifton, Jr. 2006. Syntactic prediction in language comprehension: Evidence from either . . . or. *Journal of Experimental Psychology: Learning, Memory, and Cognition* 32:425–436.

Tanenhaus, Michael K., Michael J. Spivey-Knowlton, Kathleen M. Eberhard, and Julie C. Sedivy. 1995. Integration of visual and linguistic information in spoken language comprehension. *Science* 268:1632–1634.

Trueswell, John C., Michael K. Tanenhaus, and Susan M. Garnsey. 1994. Semantic influences on parsing: Use of thematic role information in syntactic disambiguation. *Journal of Memory and Language* 33:285–318.

Van Dyke, Julie A., and Brian McElree. 2006. Retrieval interference in sentence processing. *Journal of Memory and Language* 55:157–166.

Vasishth, Shravan, and Richard L. Lewis. 2006. Argument-head distance and processing complexity: Explaining both locality and antilocality effects. *Language* 82:767–794.

Wagers, Matthew. 2008. *The structure of memory meets memory for structure in linguistic comprehension*. Doctoral dissertation, University of Maryland, College Park.

Warren, Tessa, and Edward Gibson. 2002. The influence of referential processing on sentence complexity. *Cognition* 85:79–112.

Yngve, Victor H. 1960. A model and a hypothesis for language structure. *Proceedings of the American Philosophical Society* 104:444–466.

# 7 Some things to learn from the intersection between language and working memory

*Gonzalo Castillo*

Working memory is a set of cognitive models that attempt to capture information processing and the short-term storage function that supports it. This rapidly expanding field of research started by proposing an encapsulated system with no commitments regarding the nature of language, but it has now reached a stage in which fruitful theoretical exchanges with linguistics are possible when an effort is made by both disciplines to approach a biologically grounded discourse. Here, I intend to conflate a specific view of WM and syntax that I believe already matches that requirement: state-based models of WM and Boeckx's (2013) Unbounded Merge. This exercise will hopefully reveal a common theme about how to look for the neurocognitive underpinnings of language, which I take to be one of the most challenging and important problems of modern linguistic science. More specifically, this chapter is organized in two parts: first, the main concepts of WM will be introduced; second, it will be discussed to what extent Unbounded Merge can accommodate these concepts by providing a computational description for them. I will finally argue that the results of this exchange open new avenues of research for biologically minded linguists.

## 1.1 The original multicomponent model

Working memory (WM) models were born to explain the necessary connection that exists between paying attention, understanding and keeping things in short-term memory. The most influential WM account is the original multicomponent model proposed by Baddeley and Hitch (1974), which I will introduce here by taking into account its most recent incarnations (Baddeley and Logie 1999, Baddeley 2000, Baddeley 2012, cf. Figure 7.1) to define a set of concepts that will be useful to keep in mind throughout this paper.

An important fact about the multicomponent model is that the function of storing content for short-term recall is performed in separate buffers. To handle the visual and auditory modalities, a *visuospatial sketchpad* and a *phonological loop* are respectively proposed. The phonological loop can maintain auditory content for approximately two seconds unless *verbal rehearsal* (repeating what one has just heard) is used to prevent decay (Baddeley and Ecob 1970).

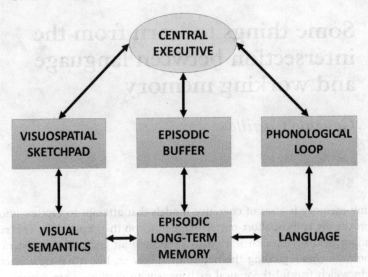

*Figure 7.1* The multicomponent model of WM (adapted from Baddeley 2012). "Visual semantics", "Episodic long-term memory" and "Language" correspond to 'crystallized' information stored in long-term memory, while WM itself comprises the rest of the systems.

The visuospatial sketchpad binds visual features to objects files, indexes that mark potential objects to which visual properties are attached (Kahneman et al. 1992). Their separation is motivated by the fact that it is possible to find modality-dependent performance when simultaneously carrying out a visual and a language comprehension task (Baddeley and Hitch 1974).

The complexity and noise of the environment are automatically filtered through these perceptual buffers, providing the material on which processing mechanisms can act. Within the model, these processing mechanisms are summarized under the notion of the *central executive*, a domain-general capacity to implement goals. In order to achieve those goals, the central executive can switch between tasks and integrate perceptual information into coherent multimodal episodes, effectively supporting what we call cognition. This integration and coordination is assumed to be based on the allocation of *top-down attention* (Treisman and Gelade 1980), which can be considered as the fuel of the central executive. Top-down attention results from the existence of volition, and is opposed to bottom-up attention, which refers to the automatic shifts of attention that are triggered by the sudden presence of unexpected stimuli in our field of perception. From a WM perspective, triggers of bottom-up attention need to be avoided to effectively carry out the resolution of a specific cognitive task, so the central executive also has the function of shielding itself from interferences (*inhibition*) (Miyake et al. 2000).

Even in the absence of interferences, the contents stored in short-term memory decay with time. Verbal rehearsal has already been mentioned as way of preventing the loss of phonological content, but the central executive also has the domain-general capacity of going back to the contents stored in any buffer and re-allocate attention to them in order to maintain them for longer. This process is known as *refreshing* (Johnson and Hirst 1993).

The last component that completes the standard WM model is the *episodic buffer* (Baddeley 2000), a domain-general storage system where the central executive integrates the information from the perceptual buffers into coherent episodes known as *chunks* (Miller 1956). Chunks are considered as complex units of thought updated by content stored in long-term memory, so their formation marks the dividing line between perception and cognition. One of the reasons that motivated the introduction of the episodic buffer was the domain-specificity and limited capacity of the perceptual buffers, which could not explain how it was possible that humans could produce and understand linguistic utterances. Even if the phonological buffer's capacity is exceeded during the presentation of the stimulus, and verbal rehearsal is blocked by a concurrent task, understanding a list of words as a meaningful sentence increases short-term storage dramatically, from 5 unrelated words up to more than 12 (Potter and Lombardi 1990). This is known as the *sentence superiority effect*, and the way it was captured by the model was by suggesting that long-term memory has a boosting effect on multimodal integrations, including the associations between sound and meaning that characterize language (Baddeley and Wilson 2002). The episodic buffer can be considered to have an approximate limit of 4 simultaneous chunks, a limit that has been termed "the magical number 4" (Miller 1956; Cowan 2001, 2010).

Although at first it was assumed that the central executive uses attention to produce chunks, more recently, studies like Baddeley et al. (2009) and Allen et al. (2006) have shown that chunk formation is not significantly affected by the realization of concurrent tasks that are designed to tax attentional resources. This has forced a redefinition of chunk formation as a mostly automatic process, while central executive functions are left for flexible chunk manipulation (Baddeley et al. 2009), a central characteristic of language that in linguistics and philosophy of language is known as compositionality.

One of the main criticisms of the multicomponent model is that its central executive is little else than a homunculus, as it has the same capacities as the system that it is trying to explain (e.g. Verbruggen et al. 2014). But as Baddeley (2012) correctly argues, the introduction of homunculi can be useful if they are not used as explanations, but as a way of constraining the set of things that are in need of explanation.

From this introduction to the multicomponent model a basic vocabulary of concepts has emerged: task switching, inhibition, attention, refreshing, chunk formation, chunk manipulation, the processing/storage distinction, and the short-term/long-term memory distinction. We can ask ourselves how these concepts connect with our language capacities, but in order to find satisfactory answers and establish linking hypotheses between linguistics and working memory, we need first to find a neurocognitive model that can extract them from the specificity of the multicomponent model and translate them into a more neutral code. The next three subsections will try to further develop these concepts by adapting them to what we know about how processing works at the level of the brain.

## 1.2 Working memory as an emergent state

The multicomponent model of WM is a top-down approach, meaning that it is a model that was designed to account for behavioral performance in a series of experiments. Based on what we know about how the brain functions, it is also possible to propose bottom-up approaches to WM. The bottom-up cognitive model that will be introduced here, Baars's (1988) Global Workspace theory, was not originally designed to explain WM functions per se, but to provide a functionalist, brain-based interpretation of the phenomenon of conscious awareness. In spite of that, since both models deal with how processing functions, recent cognitive and AI models based on the global workspace already incorporate a full account of WM functions (e.g. Franklin et al. 2012), so it is worth dedicating a few paragraphs to understand how the concepts that we have extracted from the multicomponent model can start to approach the neural substrate, and how the exercise of reconstructing them from the bottom up can significantly change them.

According to Global Workspace theory, cognition and behavior result from the ever-changing, complex interaction between localized processors that are in charge of representing the world. These processors range from low-level visual feature detectors to more complex systems that integrate them, producing conceptual structures dedicated to specific domains of cognition, such as the detection of objects, the animate vs. inanimate distinction or basic counting mechanisms (Spelke and Kinzler 2007). Faced with the myriad of stimuli that constantly bombard us, these processors will activate in parallel, trying to match an input with the specialized function that they can perform best, producing cascades of activation that will necessarily yield multiple, conflicting interpretations of the world. An important question in cognitive science, known as the frame problem, is how these localized processors get prioritized and integrated into a coherent whole to create a mind that can navigate the environment in an efficient, flexible, and appropriate manner. In short, how does a modular mind become central cognition (Fodor 1987, Dennett 2006)?

Global workspace theory (Baars 1988, 2007; cf. also Dehaene and Changeux 2011) describes the way in which this happens as an emergent property of the architecture of the brain, which is designed so that localized processors can communicate through a virtual space with limited bandwidth, known as the global workspace. The global workspace is used by localized processors to form various long-distance coalitions whenever their interpretations can complement each other, getting stronger by isolating competing coalitions until these are shut out from accessing the global workspace and only one winner emerges, imposing its contents on the global workspace (Shanahan 2012). This moment corresponds to the end of the interpretative part of the process and is known as a global broadcast: a unified, disambiguated conception of the world that corresponds to an instant of conscious awareness. A global broadcast entails a coherent episode of widespread activation that provides a new input for localized processors to start the whole interpretative process anew, making the regime of the winning coalition collapse whenever another coalition gets strong enough to occupy the global workspace (Figure 7.2).

Central cognition, therefore, works in a cyclic fashion, an observation that is captured by the concept of the cognitive cycle, a unit of consciousness that has been purported to last between 260–390ms (Madl et al. 2011). Within the model, it is assumed that higher-order cognitive tasks are carried out by the coherent integration of multiple cognitive cycles. The cognitive control functions of WM, therefore, can be reduced to the enabling of this integration, transforming parallel, localized activation into a serial, interconnected mode of thought that can be sustained long enough to solve cognitive tasks, i.e. that is adaptive. WM, in short, is a *selective serializer*.

How well do the concepts that we extracted from the multicomponent model resist this bottom-up approach? One of the first things that should be obvious is that within the global workspace account there is not a clear inclusion of attention, much less of top-down or volitional attention. Solving the problem of free will is beyond the scope of this short paper, but it can be assumed that the setting of goals also works in the same bottom-up fashion as the rest of cognition, forming coalitions that compete and finally broadcast their contents

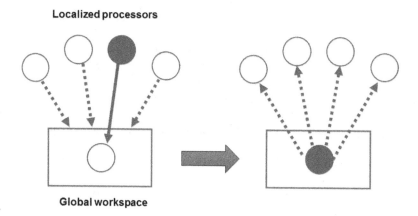

*Figure 7.2* The global workspace model (adapted from Franklin et al. 2012). A localized processor forms a coalition strong enough to occupy the global workspace and broadcast its contents to all other processors.

to influence the development of subsequent cognitive cycles. If that is the case, attention and central executive functions such as sustainment of goals, task switching and inhibition can be understood as the boosting and suppression (Gazzaley et al. 2005) of localized processors that is triggered by a global broadcast, meaning that there is no little man shouting orders from the rooftop, but only emergent behavior arising from the complex interaction of multiple systems that have been fine-tuned by evolution.

## 1.3 A neurocognitive approach to processing and storage

The reduction of attention to global broadcast-dependent boosting and suppression does not mean, nonetheless, that some regions of the brain cannot be thought of as responsible for cognitive control: they would have to be those that are well situated to inhibit or boost across long distances, and to receive bottom-up feedback that will regulate their functioning. The most studied 'control region' regarding WM is the prefrontal cortex, which fulfills those requirements by being highly interconnected with subcortical regions, and also part of the network that engages during the performance of WM-related tasks, the Multiple Demand (MD) network (Duncan 2010). The MD network is the physical realization of the process described in the global workspace model, as it connects with parietal regions of the cortex dedicated to sensory, cognitive and memory processing. Such long-distance interconnectivity should be the product of synchronization of neural oscillations (Buszáki 2006; Kitzbichler et al. 2011). A key finding by Duncan (2010) is the presence of independently evolving patterns of activation in the MD network (orthogonality), which would be useful in sustaining identity and content of stimuli across cognitive cycles. From the perspective of the MD network, prefrontal cortex can be understood as an enhancer of capacities that are performed somewhere else in the brain, and only in this interconnected scenario can be said to support higher-order cognition, including learning and memory processes, but also the implementation of goals and abstract representations (cf. Fedorenko and Thompson-Schill 2014). This enhancement can be understood as an increase in the time window in which a localized processor remains active (a quantitative measure that can correspond to the operation of refreshing, cf. Rowe et al. 2000), and also as a qualitative increase in the granularity of represented content (Mercado 2008). As an instance of the latter, Miller et al. (2011) discovered that disrupting prefrontal cortex activity blurs the distinctiveness with which occipital visual areas activate to face versus scene stimuli.

Since Goldman-Rakic (1987), however, a theoretical assumption about the standard WM model was that the two perceptual buffers could be localized in the prefrontal cortex, correlating with sustained activity of some of its neurons after the presentation of a stimulus. Indeed, the literature on areas of the prefrontal cortex supporting short-term memory functions is large. Postle (2006) lists some of the buffers that have been proposed: egocentric and allocentric spatial working memory, including hand-centered vs. eye-centered vs. foot-centered spatial working memory; working memory for manipulable vs. nonmanipulable objects, working memory for faces vs. houses; working memory for olfaction . . . In the domain of language, it is possible to dissociate between working memory for phonological vs. semantic vs. syntactic content (Shivde and Thompson-Schill 2004), and, within syntax, some of the subregions of Broca's area have been proposed to be specialized for processing basic syntax, while others would handle the short-term memory requirements of complex syntactic structures such as *wh*-movement (Fiebach et al. 2005, Fedorenko et al. 2012). How can we reconcile these findings? Where does short-term storage happen within state-based accounts of WM?

The domain-generality of prefrontal cortex functions does not necessarily preclude the emergence and measurement of domain-specific subregions that guarantee an efficient cognitive control of specific tasks. Language in particular could benefit from the existence of dedicated channels of control, since it requires the precise combination across multiple cognitive cycles of very distinct material (widely distributed instances of sound and meaning). Fedorenko et al.'s (2012) language-specific regions of Broca's area show individual-specific variability, preferentially activate to linguistic input, and recruit more domain-general areas in the prefrontal cortex whenever processing demands increase. Instead of constituting an actual buffer where linguistic information is put on hold, they can be considered as a case of *neuronal recycling* by extensive exposure to language since infancy (Matchin 2014). Neuronal recycling was first proposed by Dehaene and Cohen (2007) as a way to account for the existence of visual regions of the brain that seem to be specialized for the recognition of letters when reading one's native script. Given the recent invention of writing systems, the researchers concluded that their existence has to be a result of the environment shaping the brain. According to Matchin (2014), the term neuronal *recycling* should perhaps be replaced by neuronal *retuning*, stressing the idea that the areas that specialize do so only if they are functionally related to the specialized task to perform. For language-specific regions, this would indicate that the study of cognitive control functions, which take place adjacently to these language-specific areas, can still inform us when trying to assign their function.

All in all, the question of where short-term storage takes place is ultimately a question about the format of those short-term storage representations. The current view is that modality-specific storage capacities are an automatic result of processing modality-specific information. Applied to the multicomponent model, this view entails that the visuospatial sketchpad and the phonological loop can no longer be considered dedicated buffers localized in a distinct short-term memory area, but the visual and auditory systems themselves, which activate either in the presence of stimuli or as a result of top-down signals. What we perceive from the outside as the distinct categories of perception, imagination and memory actually take place in a much related manner, using the same physical locations (posterior brain regions) to create phenomenological experiences through the MD network.

Jonides et al. (2008) provide a detailed model of short-term memory encoding and maintenance in which three different stimuli get serially encoded and sustained until the first one reappears and is matched to its memory (cf. Figure 7.3). Distributed patterns of sustained activity represent each stimulus during the encoding phase, which simultaneously and within the same location trigger different plasticity mechanisms (Zucker and Regehr 2002) that are responsible for short-term retention. After the stimulus is gone, and before presenting a new one, sustained activity would start to decrease while plasticity mechanisms decay much more slowly. This decrease of sustained activity should take a few milliseconds after the stimulus is gone, so it could come back more quickly if the same stimulus was presented again instead of a new one. This would explain why there is a difference in reaction time between retrieving the most recently presented item as compared to those that are supposed to be stored in a magical number four format (e.g. McElree and Dosher's 1989). When the second and third stimuli are presented, synaptic weight changes are affected by the same nature of some of their visual features, which may interfere with the retrieval of the first item (the objective of the task). The last presented item, which should have the same perceptual identity as the first and be recognized as such (a process known as updating), would trigger the same pattern of sustained activity by means of the pattern-completion property of attractor networks (Hopfield 1982).

The model could be almost directly translated into a dynamic neuronal dimension by proposals such as Dipoppa and Gutkin (2013) or Roux and Uhlhaas (2014), in which different kinds of oscillatory activity would synchronize to perform the encoding, sustainment and updating operations. The difference is that here the sustainment operation is not performed by plasticity, but by synchronization that is strong enough to deflect potentially irrelevant encodings (interferences). This idea is in line with McElree (2006), who reviews a series of experiments in which the distinction between short-term and long-term memory is blurred by the fact that no qualitatively different reaction times are observed during the retrieval process from each purported component. If short-term memory and long-term memory form a continuum, capacity limitations such as Cowan's (2001) magical number four may be better described as processing constraints that can reduce to rapid periods of refreshing. Instead of interference phenomena overriding encoded material in short-term memory (e.g. Oberauer 2013), what we would find is that performance is reduced by interference as a result of new encodings that are not related to the task that needs to be performed. Similarly blurring the dividing line between processing and storage, Barrouillet et al. (2011) propose the notion of cognitive load, a function of the amount of time it takes to process a specific stimulus, which has been shown to be negatively correlated with the amount of information that can be sustained in WM. The debate is still open regarding the distinctiveness of short-term storage and processing functions, and the modeling of interference phenomena.

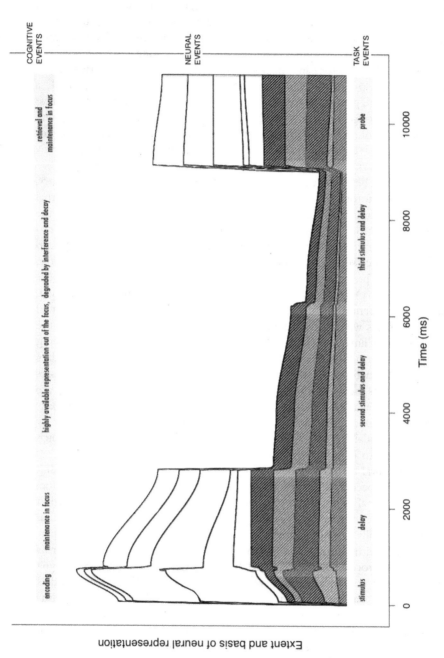

*Figure 7.3* The encoding, maintenance and retrieval of a stimulus in short-term memory while two other stimuli are presented in a sequence (adapted from Jonides et al. 2008). The white areas represent widespread sustained activity, while the grey areas represent plasticity mechanisms that keep the encoding past the act of perceiving.

## 1.4 Working memory and long-term memory

From the perspective of an organism, the physical world is ever-changing but causal, as patterns and relationships can be extracted when gathering perceptual data, and these patterns can be used in the future to obtain reliable outcomes. WM and learning are two interconnected processes, as long-term memory can be considered a repository in which the processing of present experiences can be stored to help the processing of future ones. In the context of WM, a description of the processes that operate as part of it cannot be complete unless the act of chunking (producing complex, multimodal structures for processing the present in connection with the past) is accounted for, including how processing generates chunks, how it is possible to alter them to arrive at new meaningful patterns, and how their formation maximizes WM capacities.

As mentioned before, the use of long-term memory activates the same sensory, cognitive and motor areas that are present during the processing of external stimuli, and a functional description of memory cannot be divorced from the description of processing that we have established so far. Taking that into account, the literature on long-term memory divides the object of its study into two distinct processes that are specifically subserved by two different networks: declarative and procedural memory. Declarative memories comprise episodes (episodic memory) and facts (semantic memory) that are consciously recalled, and their recollection mainly involves ventral pathways that connect prefrontal cortex structures with medial temporal lobe structures, crucially including the hippocampus (Ullman 2004), which seems to be especially relevant in keeping multimodal information together and bound to a context during the first days of memory formation, but that gradually loses importance as time progresses and connections between the cortical regions that sustain that memory are strengthened (Takashima et al. 2009). Rubin et al. (2014) go further, suggesting that the hippocampus plays a crucial role in sustaining flexible behavior by making multimodal representations available to control networks like MD, which would connect to it through ventro-medial prefrontal cortex.

The procedural memory system, on the other hand, consists of the learning of habits and procedures, defined as "sequential, repetitive, and motor behaviors elicited by external or internal triggers that, once released, can go to completion without constant conscious oversight" (Graybiel 2008). Examples of procedural behaviors are skills like riding a bicycle or playing the piano, in which automaticity develops with practice, but also the implicit acquisition of habits. In terms of the brain, procedures are acquired by a prefrontal-thalamic-basal ganglia network, which connects dorsolateral prefrontal cortex with striatum, and is modulated by dopaminergic inputs that can produce two main states of activation in the network depending on where dopamine is expressed (cf. D'Esposito and Postle 2015 for a review). In terms of WM, the striatum would favor task switching, while the prefrontal cortex would be responsible for sustainment or the deployment of the MD network, in a model analogous to the one by Jonides et al. (2008) that was previously presented to explain short-term encoding.

The analogy is not just a coincidence: as it was said before, the activity of prefrontal cortex as a control mechanism together with the MD network cannot be understood without appealing to the feedback it receives from subcortical structures. The fact that the prefrontal cortex and striatum are connected in a loop would indicate a mutual modulation that has the capacity to flexibly and emergently alter the flow of processing in the brain without the need for a prime mover or homunculus.

The process of generating a procedure would be based on repeated exposure to the same activation patterns throughout the course of various trials, which would generate a chunk. In experiments by Fujii and Graybiel (2003, 2005) it is shown that after training monkeys to make complex sequences of saccadic eye movements, some neurons in dorsolateral prefrontal cortex and striatum start to show increased activation at the boundaries of those sequences, signaling that a chunk has been completed. The process of forming a chunk is correlated with decreased activation in the prefrontal cortex, whose sustainment capabilities are not needed anymore to connect the dots that would constitute an action sequence. The process of acquiring a chunk can therefore be understood as a transfer between the declarative and the procedural memory components, or as a gradual decrease in cognitive control and MD network activation. Buschman and Miller (2014) propose a model in which the development of procedures would reduce cognitive load by decreasing the amount of time in which the MD network needs to be sustained to correctly solve cognitive tasks. Within this model, fast associations of a stimulus-response nature would be handled by the faster, more primitive basal ganglia, while the development of abstract categories and generalizations would require the slower processing of the prefrontal cortex, with repeated exposures that guarantee a link between the different associations.

Chunking is a widespread phenomenon in animals, which show differing degrees of flexibility in the process of retrieving and modifying chunks. Usually, once a behavior has been proceduralized, the learning process is considered as finished and, in many animals, behavior crystallizes. Crystallization has been thoroughly studied in the domain of vocal learning birds. Some species of birds have males that learn their courtship and territorial songs by imitating a tutor's song during a period of babbling known as the subsong stage (Jarvis 2007). During this period, the song patterns that are produced are very variable, with a degree of accuracy that seems to go back and forth until it starts to stabilize and the bird acquires its song. In some species like the zebra finch, this song gets fixed for life and is considered to be crystallized. However, in other species, such as the canary, crystallized songs are renewed for each breeding season. Finally, other species of birds such as the lyrebird are open-ended learners, as they can use their vocal components to acquire new vocalizations throughout life (Petkov and Jarvis 2012). These distinctions seem to depend on species-specific striatal capacities to regulate plasticity. In humans, unchangeable or crystallized action patterns are rare, save perhaps for cases of mental illness like OC-spectrum disorders (Graybiel 2008), and the acquisition of skilled behavior like playing a musical instrument shows that flexibility can persist past the formation of chunks, which would correspond to cognitive control capabilities exerting their power on retrieved chunks.

As we can see, both memory systems are highly integrated with the MD network through ventral or dorsal pathways, and they should support both long-term memories and processing functions, blurring the dividing line between the system to draw a picture that comes back to the global workspace model, in which different brain areas contribute to processing only as part of a large-scale network that handles the basic operations of integration of information along a time dimension. In this sense, the declarative/procedural distinction is useful in signaling that learning does not reduce to a clear-cut opposition between conscious and unconscious ways of learning, or between content and action, but that these distinctions can ultimately be traced back to a *gradient of cognitive control* (Botvinick and Cohen 2014) that will determine the complexity and degree of flexibility during the process of extracting meaning from the world. The implementation of syntax will necessarily have to navigate through the altering flows of information that I have so far described. The remaining of this paper will try to put forth some insights that could help to achieve this quest someday.

## 2.1 Working memory and syntax

Poeppel and Embick (2005) provide an overview of the state of the art of the interface between linguistic theory and cognitive neuroscience, arriving at the conclusion that reasonable linking hypotheses between the two disciplines are non-existent. They identify the cause for this lack of dialogue in the different granularity and incommensurability of the concepts that each discipline deals with. They also propose a solution: each field of study should look at the computational side, producing computational models of their respective phenomena that can be associated with one another. In the case of linguistic theory, this entails a decomposition of its concepts into sets of distinct, smaller processes that could be performed by neuronal populations. This strategy should not only ground linguistic concepts in biology, but also inform us about how the brain functions to produce the hallmarks of human cognition, entailing a real interdisciplinary exchange.

Throughout this paper, I have listed a series of classic concepts in WM and provided a description of how they are redefined in computational and neurocognitive accounts, usually by being split into different stages or suboperations that together produce what from the outside we perceive as a distinct category. Such was the case of attention, which was reduced to localized, parallel encoding during processing, distinguishing between processes that boost specific neural assemblies while suppressing others, and a process of bottom-up encoding of goals that may influence what gets boosted or suppressed across multiple cognitive cycles. If WM and linguistic theory can initiate an interdisciplinary dialogue, they will have to do it by agreeing on a common ground in which the concepts of each discipline are grounded and reduced. There are many obstacles for this to happen, perhaps the largest being the conception of linguistic knowledge as an innately specified universal grammar that is taken by many to be the explanation for how something as complex as language can be *grown* by the individual, instead of just learned. I will leave aside most of these concerns and difficulties for the purposes of this paper and instead focus on the essentials of linguistic theory: hierarchical combination. It is my belief that the action of splitting linguistic concepts in order to get to a lower floor in the causal chain should start by distinguishing grammar from the process of hierarchical combination that supports it. As I will show along these lines, connecting hierarchical combination with WM is a relatively straightforward task that seems to be a more fruitful approach than taking language as a complex object and contrasting it with the equally complex multicomponent model just to see that nothing matches. So let's start from there and expand.

According to Chomsky (2008), "The most elementary property of language – and an unusual one in the biological world – is that it is a system of discrete infinity consisting of hierarchically organized objects." In most recent theoretical accounts, the weight of building syntactic structures is carried by the operation Merge, which takes two lexical items and combines them under a label that is taken from one of them, thus producing a structure that can be captured by the notion of hierarchy. Syntax is thus, following the inverted-Y model (Chomsky 1995), a connector of three dimensions: a pool of lexical items, and conceptual and sensorimotor structures that constitute the interfaces, and its mission is to link the three of them to support linguistic behavior. Regarding discrete infinity, discreteness refers to the combination of atoms or lexical items, which are understood as roots of words instead of full-fledged words, morphology also resulting from combination. Infinity is the property that emerges from recursively applying Merge over the structures that it builds, a process that would, if left unconstrained, reach unbounded levels of embeddings that would nonetheless remain part of the linguistic system.

Of course, there are boundaries that make this just an abstraction. Merge is implemented in brains, a type of hardware that necessarily imposes multiple constraints, which have been summarized by the notion of the three factors in language design (Chomsky 2005): structures that are impossible because of universal grammar (the "genetic endowment" of the language faculty), which may arbitrarily restrict combination to favor or deny specific outputs; structures that are impossible due to constraints on how the interfaces are built ("language- or even organism-independent" factors), which would crucially include working memory limitations that make center-embedding a difficult task, and that make the combinatory process collapse and reset whenever utterances surpass storage limitations; and, finally, factors related to experience, including, for example, the acquisition of Spanish over Chinese, which would also determine what gets merged and when.

This distinction between Merge and constraints on Merge can also be useful in applying Merge to WM accounts, because it constitutes a dividing line between grammar and syntax. If syntax is not defined as a set of rules, but as the process that handles integration, the implementation of Merge acquires a clear, converging direction towards WM accounts that are based on capturing how information integrates at the neuronal level. An example of a cognitive model that gets closer to Merge is Oberauer's (2009), summarized in Figure 7.4, which I introduce here so that we have a summary of everything we have learned about WM in this chapter. The model distinguishes between the focus of attention and the fringe of attention, following the idea that the cognitive cycle is characterized by a moment of unification (the global broadcast) in which only one item (or one item and its context) influences processing, identified in the figure as B, which slowly descends with dotted lines towards the focus. Simultaneously, activated representations in long-term memory descend to a region of direct access known as the fringe of attention, indicating that they are more easily retrievable but do not occupy the focus of attention within that cognitive cycle. Note that long-term memory is characterized as a network of increasing degrees of activation, indicated by the opposition between white and grey dots, and by the inclusion of letters in the fringe of attention to indicate an even greater degree of activation. If we wanted to translate Merge into this account, Merge would have to occupy the focus of attention, linking two items (or an item and its context) at each cognitive cycle. Merge, in short, would have to be identified with the processing operation, not with the whole of WM, nor with grammar.

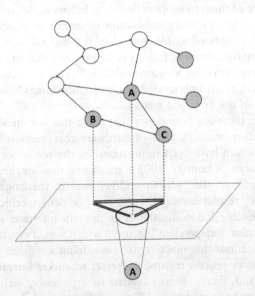

*Figure 7.4.* Oberauer's model of Declarative Working Memory (adapted from Oberauer 2009). A, B, C correspond to the "fringe of attention", grey circles are activated long-term memory items, the "focus of attention" has selected A as the target for processing.

This conflation of Merge with the focus of attention would forcefully imply that Merge is not a linguistic operation per se, but a fundamental process that is shared across species, so further clauses would have to be added to the model to explain what in Merge, or in grammar (for example, the use of labels), or WM, may constitute the human-specificity that supports the acquisition of language. For now, it should be noted that the building of trees, that is, the production of structures that can be defined as hierarchical, may not be one of those human-specific features. If integration of information is subject to the goal of processing, it would be necessary to select an order in which the items in the fringe of attention enter the focus of attention, completing subtasks in order to fully encode the item and the context in which a response has to be selected. Thus, an argument that follows this reasoning states that domains like motor planning, which is phylogenetically prior to language, have, despite significant differences (e.g. Moro 2014, cf. also Boeckx and Fujita 2014), much in common with the hierarchical structures that Merge builds in language. As an example, Pulvermüller (2014) shows how the action of walking to the bathroom to brush one's teeth can be described by a center-embedding operation that not only is hierarchical in nature, but also exceeds the usual limit of three embeddings characteristic of language use.

It should be noted that this domain-general capacity is not Merge per se, as Merge as it has so far been described produces only unordered sets that would then need to be connected to the interfaces to linearize and become speech. What we have here is the notion of sequential hierarchy (Fitch and Martins 2014), in which the building of hierarchy, like integration in WM, is subject to a sequence of time in which, crucially for language and many types of actions, the order matters. Even if all processing is sequential, we should distinguish between hierarchical information that is presented sequentially, such as the act of recognizing a face, in which the order of the components does not alter the product, and hierarchies that can become meaningful only if they follow a specific order of operation, that is, if they are serialized in a fixed order. If Merge is to be equated with processing, it cannot remain being a producer of unordered sets unless we propose a complementary mechanism that serializes in a similar way as the global broadcast serializes localized processors in the global workspace framework. The problem with this option is that it would amount to saying that Merge is only a parallel mechanism that is not capable of producing the type of hierarchies that characterize language, in which order matters.

A more parsimonious solution than adding a complementary serializer is to implement the property of forming sequential hierarchies in Merge itself, but crucially leave outside of syntax any judgments about the rightness of the sequences that Merge produces, keeping in place the separation between syntax and grammar. If Merge is to be identified with the focus of attention, it also has to be deployed along a full cognitive cycle, which is a process that is parallel and serial at different stages, so it would make sense that Merge is also defined as a process that is parallel and serial at different stages. The difference would be that, while the cognitive cycle captures all localized processors trying to reach the global workspace, Merge would be a very specific subset of that, describing the process by which the winner of the competition gets to form, occupy the global workspace and globally broadcast. Merge, in sum, enables the sequence, but constitutes only one frame of the movie. A formalization of all of this is offered in Boeckx (2013), which divides Merge into four subtasks that may ultimately have distinct neural realizations (here identified with different kinds of oscillatory activity at the level of the brain). Assume that elements in capital letters mean 'already labeled' and 'serialized', while small caps mean hierarchical structures that are still in the process of forming.

a) Find mergeable elements:
   N, v
b) Combine:
   v(v, N)
c) Label:
   V(V, N)
d) Forget:
   V

Recursion (a new cognitive cycle):

e) Find mergeable elements:
   V, t
f) Combine:
   t(t, V)
g) Label:
   T(T, V)
h) Forget:
   T

The steps of finding mergeable elements and combining them would amount to the forming of the winning long-distance coalition through the global workspace. We assume in the diagram that we have captured the merging operation in the middle of its functioning, which should have started prelinguistically, whenever cognition developed within the individual's brain, so the $N$ component is already interpreted, and the objective here is to merge it with something new, $v$. This assumption is not trivial, as it connects linguistic Merge with the extralinguistic instances of Merge that led to the acquisition of language, including the acquisition of words and categories, which should participate from the same syntactic processes as the building of a sentence (cf. Carey 2009).

Importantly, the labeling operation comes from the interfaces, as it corresponds to the global broadcast in which information is integrated and serialization takes place. This entails that structure building is a separate process from structure interpretation, a principle known as exoskeletality which opposes views in linguistics that state that structure takes precedence over content and is cognitively impenetrable (e.g. Fodor and Pylyshyn 1988). Here, the labeling of the constituents is handled by information that can come from the morphophonological or conceptual components, with Merge subserving only the function of hierarchizing that content. This also means that structure building can be affected by interference phenomena, which I will return to below.

There are at least two different ways of interpreting the labeling process. Merge can here take just roots and hierarchize them at the syntactic level, generating an unordered set that is then serialized by a unification operation, which would add morphosyntactic, morphophonological and phonological information in three parallel steps that would constitute the operation of linearization, giving rise to labeling those categories as a serial structure (Boeckx et al. 2014). A different interpretation would consider words as chunks in long-term memory that were originally acquired as roots that were fed into the merging operation, but that in normal language use are retrieved automatically without the need to reconstruct them through unification unless they become the object of attention, in which case they could be flexibly manipulated as any other chunk. The difference between these two views also entails a bigger difference that has to do with the question of whether there really is a system of linguistic knowledge that is separate from a system of linguistic processing, that is, whether there is one single tree or two independently generated trees during the moment of processing (Phillips 2013). The value of an account based on two trees is that it allows syntax to be separate from the morphosyntactic differences of each independent language, constituting a universal conceptual structure. An account based on a single tree, on the other hand, would have to see syntax, not as the linking component between the three interfaces, but as the builder of conceptual structure, which can operate with or without language-specific features depending on the nature of the information that gets integrated at each cognitive cycle.

Regarding the forgetting step, it signals the closing of a cognitive cycle and of the structure-building operation: the information that is forgotten is actually transferred to long-term memory, and cannot be modified anymore as part of that cognitive cycle. Only through an operation of refreshing in subsequent cognitive cycles can it be retrieved and manipulated. Thus, Merge amounts to a cyclic rhythm in which the structure-building process is delineated by phases that always include a head and a complement.

## 2.2 Working memory and grammar

The time dimension in language is, of course, not foreign to classic generative linguistics, which recognized that natural languages displayed non-local dependencies that can be captured only by mildly context-sensitive grammars (Chomsky 1963, Joshi 1985). A simple sentence like "Kate builds it up" (Pulvermüller 2010) shows us how frequent non-local dependencies are in language, with the subject connecting with the verb's inflection, and the verb connecting with *it* and *up*. This takes us to a problem that any account of language based on WM has to be able to answer: how does Merge know where to go? Or in other words, how is a selective gating mechanism such as Merge or the focus of attention able to select the most appropriate content to let in and sustain? The start of an answer was already given from a domain-general point of view by appealing to a self-sustaining mechanism of cognitive control, but the domain of language allows us to distinguish different dimensions (pragmatic, semantic . . .) that should influence processing, grammar being perhaps one of the most important.

The non-locality of grammar, and grammar as a whole, can be interpreted as a set of chunks or learned content that fulfill the specific role of guiding and constraining the merging operation during processing so that it becomes an efficient mechanism for extracting meaning out of the world. Regarding the non-locality of many syntactic objects, it can be understood as a trigger of specific refreshing operations across time that operate automatically (without the need for cognitive control). While the processing of an utterance is taking place, the labeling of some structures will generate expectations about how to proceed (e.g. a subject will want to find its verb) which will not depend on explicit rules or be encoded in any specific format, but emerge from the act of labeling itself, from processing an item. Crucially, these expectations would not need to carry on across cognitive cycles, so further processing resources will be made available as normal, and the operation of Forget will still operate. Following the model put forth by Jonides et al. (2008), what would happen is that the recently processed content will automatically refresh when, within a new cognitive cycle, a new item is processed that partially matches it. A way to represent this can be found in Figure 7.5.

The idea that non-locality is not necessarily a controlled process is in line with experiments by McElree and colleagues (McElree et al. 2003, Öztekin and McElree 2007, Van Dyke and McElree 2011) on memory retrieval during language comprehension. First, these experiments examined the search mechanism employed in language comprehension to detect and understand non-local relationships, reaching the conclusion that, since no time differences are observed in retrieval despite the amount of material that is placed in between the two elements, it is probably a direct-access mechanism in which information becomes readily available for interpretation without the need to go (Merge) step by step

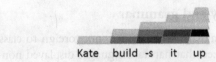

*Figure 7.5* A representation of the online establishment of dependencies as the parallel reactivation of already processed material. The processing of each word leaves a trace in short-term memory that is refreshed (represented here by the color getting darker) by the ensuing processing of specific items that have a grammatical relationship with it. The most active trace at each step is the one the syntactic relationship corresponds to.

through the whole sentence in order to find the target. Second, they examined the role of interferences within these retrieval operations, appealing to the hypothesis of cue overload (Watkins and Watkins 1975) to explain why, in some cases (for example, in center-embedding), retrieval produces slowdowns or even fails: if the retrieval cues become not distinctive enough, i.e. if it turns out that it is possible to associate them with many recently processed items, the probability of correct retrieval is diminished. In the language we are using here, if the activation of an item does not match the target item or matches more than one target item, the probability of Merge generating the target item again in an operation of refreshing will decrease.

If, as stated before, the prefrontal cortex within the MD network seems to be responsible for increasing the representational resolution of specific items (Mercado 2008), the acquisition and deployment of grammar would crucially involve MD network activity that would be in charge of reducing interferences and disambiguating. However, it should be noted that the acquisition of language is a gradual process that coincides with the protracted development of the prefrontal cortex in humans. Chrysikou and Thompson-Schill (2011) hypothesize that this protracted development may facilitate the acquisition of language, as acquisition would start by incorporating more basic structures, converting them into chunks by repeated exposure, creating in this way a bootstrapping process that, as long as the MD bandwidth allows it, will finally attain more complicated patterns. Interestingly, the authors link the stages previous to the acquisition of a full-fledged grammar to a higher frequency of out-of-the-box thinking, giving the example of German and Defeyter (2000), an experiment in which it was shown that children younger than five years old were more able to come up with the idea of using a box as a platform to reach an object than were older children, who tended to see the box as just a container. Within the picture being developed, out-of-the-box, creative thinking would be a result of stochastic activity entering the global workspace as a result of low prefrontal activation, while a combination of high and low prefrontal activation would render a focused but flexible way of thinking. The acquisition or proceduralization of grammar, therefore, should behave exactly as its deployment, being a combination of the basal-ganglia-related and prefrontal-cortex-related states within the prefrontal-thalamic-basal ganglia network, but taking place gradually as cognitive control abilities develop and bootstrap the emergence of the more complex structures.

Regarding the domain-specificity of the language subregions in Broca's area, special attention should be paid to the phenomenon of disambiguation, which would be produced when an item triggers multiple potential candidates. Taking into account that Broca's area can be divided into multiple subregions, probably with a different function for each (cf. Embick and Poeppel 2006), a general cognitive control role for this area can be to prevent Merge from falling into interferences that may disrupt the achievement of complex goals, a capacity that will vary according to the species, and that may find its maximum exponent in humans. The functioning of Broca's area, nonetheless, would still reduce to mechanisms that are shared across species and are also involved in non-linguistic tasks.

It should be noted, finally, that Merge cannot be said to be speeded up by grammar, as the multiple types of oscillations required to embed content that Merge should consist of are distinct because of their different temporal frequencies. What grammar could be doing instead is to make Merge waste less time merging irrelevant information for the completion of a cognitive task. If there is a system of chunks or frequent patterns of activation in place that is acquired through experience, this should make cognition efficient without speeding it up. This phenomenon should be of the same nature as symbol training, which seems to enhance the representational capacities of animals (e.g. Thompson et al. 2001, Srihasam et al. 2012): it would make them focus on specific aspects of processing that are relevant, while suppressing others. Also related to this, the external guiding of attention by placing visual clues in front of prelinguistic children seems to facilitate their chunking abilities, giving them higher-order chunking capacities that were once believed to be an exclusive domain of a fully developed grammar (Rosenberg and Feigenson 2013). Grammar, like the WM capacities that support it, is a selective serializer.

## Conclusion

Throughout this paper, the multicomponent model of Baddeley and Hitch (1974), which was designed as an independent system to capture the phenomenon of domain-specific short-term memory, has been reduced to an emergent property that models the general way in which neuronal networks interact, giving rise to a fruitful exchange of knowledge among multiple disciplines that were initially mute regarding this topic, such as developmental psychology, animal cognition, psycholinguistics and neurocognitive science. This exercise is still pending in many bastions of linguistic theory, where language is taken as a distinctive component or organ whose inevitable particularities seem to grant its separation from other areas of cognition, creating a bad tendency to label the discovery of similarities as little else than metaphors and to banish interdisciplinary dialogue.

The implementation of language should leave behind the traditional Broca-Wernicke model to include the interaction of multiple brain regions and networks. The study of language processing can be informed by the study of domain-general cognition, and the structure of language knowledge and the stages of language acquisition should be inextricably linked to the natural constraints imposed on processing mechanisms. Likewise, the phenomena of processing, short-term memory and long-term memory cannot be separated if our objective is to design a realistic cognitive model of how the brain works.

## Acknowledgements

Research funded by the FI-DGR 2015 Program from the Generalitat de Catalunya.

# References

Allen, R. J., Baddeley, A. D., & Hitch, G. J. (2006). Is the binding of visual features in working memory resource-demanding? *Journal of Experimental Psychology: General*, 135(2), 298.
Baars, B. J. (1988). *A cognitive theory of consciousness*. London: Cambridge University Press.
Baars, B. J. (2007). The global workspace theory of consciousness. In M. Velmans and S. Schneider (eds.) *The Blackwell Companion to Consciousness*. Malden, MA: Blackwell, 236–246.
Baddeley, A. (2000). The episodic buffer: A new component of working memory? *Trends in Cognitive Sciences*, 4(11), 417–423.
Baddeley, A. (2012). Working memory: Theories, models, and controversies. *Annual Review of Psychology*, 63, 1–29.
Baddeley, A. D., & Ecob, J. R. (1970). Simultaneous acoustic and semantic coding in short-term memory. *Nature*, 227, 288–289.
Baddeley, A. D., & Hitch, G. (1974). Working memory. *Psychology of Learning and Motivation*, 8, 47–89.
Baddeley, A. D., Hitch, G. J., & Allen, R. J. (2009). Working memory and binding in sentence recall. *Journal of Memory and Language*, 61(3), 438–456.
Baddeley, A. D., & Logie, R. H. (1999). Working memory: The multiple-component model. In A. Miyake & P. Shah (Eds.), *Models of working memory: Mechanisms of active maintenance and executive control*, 28–61. Cambridge, UK: Cambridge University Press.
Baddeley, A., & Wilson, B. A. (2002). Prose recall and amnesia: Implications for the structure of working memory. *Neuropsychologia*, 40(10), 1737–1743.
Barrouillet, P., Portrat, S., & Camos, V. (2011). On the law relating processing to storage in working memory. *Psychological Review*, 118(2), 175.
Boeckx, C. (2013). Merge: Biolinguistic Considerations. *English Linguistics*, 30(2), 463–484.
Boeckx, C. A., & Fujita, K. (2014). Syntax, action, comparative cognitive science, and Darwinian thinking. *Frontiers in Psychology*, 5, 627.
Boeckx, C., Martinez-Alvarez, A., & Leivada, E. (2014). The functional neuroanatomy of serial order in language. *Journal of Neurolinguistics*, 32, 1–15.
Botvinick, M. M., & Cohen, J. D. (2014). The computational and neural basis of cognitive control: Charted territory and new frontiers. *Cognitive Science*, 38(6), 1249–1285.
Buschman, T. J., & Miller, E. K. (2014). Goal-direction and top-down control. *Philosophical Transactions of the Royal Society B: Biological Sciences*, 369(1655), 20130471.
Carey, S. (2009). *The origin of concepts*. New York: Oxford University Press.
Chomsky, N. (1963). Formal properties of grammars. In R. D. Luce, R. R. Bush, & E. Galanter (Eds.). *Handbook of mathematical psychology*, vol. 2, 323–418. New York and London: Wiley.
Chomsky, N. (1995). *The minimalist program*. Vol. 1765. Cambridge, MA: MIT Press.
Chomsky, N. (2005). Three factors in language design. *Linguistic Inquiry*, 36(1), 1–22.
Chomsky, N. (2008). On phases. *Current Studies in Linguistics Series*, 45, 133.

Chrysikou, E. G., & Thompson-Schill, S. L. (2011). Dissociable brain states linked to common and creative object use. *Human Brain Mapping*, *32*(4), 665–675.

Cowan, N. (2001). The magical number 4 in short-term memory: A reconsideration of mental storage capacity. *Behavioral and Brain Sciences*, *24*(1), 87–114.

Cowan, N. (2010). The magical mystery four. How is working memory capacity limited, and why? *Current Directions in Psychological Science*, *19*(1), 51–57.

D'Esposito, M., & Postle, B. R. (2015). The cognitive neuroscience of working memory. *Psychology*, *66*(1), 115.

Dehaene, S., & Changeux, J. P. (2011). Experimental and theoretical approaches to conscious processing. *Neuron*, *70*(2), 200–227.

Dehaene, S., & Cohen, L. (2007). Cultural recycling of cortical maps. *Neuron*, *56*(2), 384–398.

Dennett, D. C. (2006). The frame problem of AI. In *Philosophy of Psychology: Contemporary Readings*, New York: Routledge. 433–454.

Dipoppa, M., & Gutkin, B. S. (2013). Flexible frequency control of cortical oscillations enables computations required for working memory. *Proceedings of the National Academy of Sciences*, *110*(31), 12828–12833.

Duncan, J. (2010). The multiple-demand (MD) system of the primate brain: Mental programs for intelligent behaviour. *Trends in Cognitive Sciences*, *14*(4), 172–179.

Embick, D., & Poeppel, D. (2006). Mapping syntax using imaging: Problems and prospects for the study of neurolinguistic computation. *Encyclopedia of Language and Linguistics*, *2*, 484–486.

Fedorenko, E., Duncan, J., & Kanwisher, N. (2012). Language-selective and domain-general regions lie side by side within Broca's area. *Current Biology*, *22*(21), 2059–2062.

Fedorenko, E., & Thompson-Schill, S. L. (2014). Reworking the language network. *Trends in cognitive sciences*, *18*(3), 120–126.

Fiebach, C. J., Schlesewsky, M., Lohmann, G., Von Cramon, D. Y., & Friederici, A.D. (2005). Revisiting the role of Broca's area in sentence processing: Syntactic integration versus syntactic working memory. *Human brain mapping*, *24*(2), 79–91.

Fitch, W., & Martins, M. D. (2014). Hierarchical processing in music, language, and action: Lashley revisited. *Annals of the New York Academy of Sciences*, *1316*(1), 87–104.

Fodor, J. A. (1987). Modules, frames, fridgeons, sleeping dogs and the music of the spheres. In Z. Pylyshyn (Ed.). *The robot's dilemma: The frame problem in artificial intelligence*. Norwood, NJ: Ablex, 139–149.

Fodor, J. A., & Pylyshyn, Z. W. (1988). Connectionism and cognitive architecture: A critical analysis. *Cognition*, *28*(1), 3–71.

Franklin, S., Strain, S. Snaider, J., McCall, R., & Faghihi, U. (2012). Global workspace theory, its LIDA model and the underlying neuroscience. *Biologically Inspired Cognitive Architectures*, *1*, 32–43.

Fujii, N., & Graybiel, A. M. (2003). Representation of action sequence boundaries by macaque prefrontal cortical neurons. *Science*, *301*(5637), 1246–1249.

Fujii, N., & Graybiel, A. M. (2005). Time-varying covariance of neural activities recorded in striatum and frontal cortex as monkeys perform sequential-saccade tasks. *Proceedings of the National Academy of Sciences of the United States of America*, *102*(25), 9032–9037.

Gazzaley, A., Cooney, J. W., Rissman, J., & D'Esposito, M. (2005). Top-down suppression deficit underlies working memory impairment in normal aging. *Nature neuroscience*, *8*(10), 1298–1300.

German, T. P., & Defeyter, M. A. (2000). Immunity to functional fixedness in young children. *Psychonomic Bulletin & Review*, *7*(4), 707–712.

Goldman-Rakic, P. S. (1987). Circuitry of primate prefrontal cortex and regulation of behavior by representational memory. In F. Plum (ed.). *Comprehensive Physiology*. Bethesda. MD: American Physiological Society, 373–417.

Graybiel, A. M. (2008). Habits, rituals, and the evaluative brain. *Annual Review of Neuroscience*, *31*, 359–387.

Hopfield J. J. (1982) Neural networks and physical systems with emergent collective computational abilities. *Proceedings of the National Academy of Science USA 79*(8), 2554–2558.

Jarvis, E. D. (2007). Neural systems for vocal learning in birds and humans: A synopsis. *Journal of Ornithology*, *148*(1), 35–44.

Johnson, M. K., & Hirst, W. (1993). MEM: Memory subsystems as processes. *Theories of Memory*, *1*, 241–286.

Jonides, J., Lewis, R. L., Nee, D. E., Lustig, C. A., Berman, M. G., & Moore, K. S. (2008). The mind and brain of short-term memory. *Annual Review of Psychology*, *59*, 193.

Joshi, A. K. (1985). Tree adjoining grammars: How much context-sensitivity is required to provide reasonable structural descriptions? In David R. Dowty, Lauri Karttunen, and Arnold M. Zwicky (Eds.) *Natural language parsing*, 206–250. New York: Cambridge University Press.

Kahneman, D., Treisman, A., & Gibbs, B. J. (1992). The reviewing of object files: Object-specific integration of information. *Cognitive Psychology*, *24*(2), 175–219.

Kitzbichler, M. G., Henson, R. N., Smith, M. L., Nathan, P. J., & Bullmore, E. T. (2011). Cognitive effort drives workspace configuration of human brain functional networks. *The Journal of Neuroscience*, *31*(22), 8259–8270.

Madl, T., Baars, B. J., & Franklin, S. (2011). The timing of the cognitive cycle. *PloS one*, *6*(4), e14803.

Matchin, W. G. (2014). *Investigations of the syntax-brain relationship*. PhD Dissertation. University of California, Irvine.

McElree, B. (2006). Accessing recent events. *Psychology of Learning and Motivation*, *46*, 155–200.

McElree, B., & Dosher, B. A. (1989). Serial position and set size in short-term memory: The time course of recognition. *Journal of Experimental Psychology: General*, *118*(4), 346.

McElree, B., Foraker, S. & Dyer, L. (2003). Memory structures that subserve sentence comprehension. *Journal of Memory and Language*, *48*, 67–91.

Mercado III, E. (2008). Neural and cognitive plasticity: From maps to minds. *Psychological Bulletin*, *134*(1), 109.

Miller, B. T., Vytlacil, J., Fegen, D., Pradhan, S., & D'Esposito, M. (2011). The prefrontal cortex modulates category selectivity in human extrastriate cortex. *Journal of Cognitive Neuroscience*, *23*(1), 1–10.

Miller, G. A. (1956). The magical number seven, plus or minus two: Some limits on our capacity for processing information. *Psychological Review*, *63*(2), 81.

Miyake, A., Friedman, N. P., Emerson, M. J., Witzki, A. H., Howerter, A., & Wager, T. D. (2000). The unity and diversity of executive functions and their contributions to complex "frontal lobe" tasks: A latent variable analysis. *Cognitive Psychology, 41*(1), 49–100.

Moro, A. (2014). On the similarity between syntax and actions. *Trends in Cognitive Science, 18,* 109–110

Oberauer, K. (2009). Design for a working memory. *Psychology of Learning and mMotivation, 51,* 45–100.

Oberauer, K. (2013). The focus of attention in working memory – From metaphors to mechanisms. *Frontiers in Human Neuroscience, 7,* 673.

Öztekin, I., & McElree, B. (2007). Retrieval dynamics of proactive interference: PI slows retrieval by eliminating fast assessments of familiarity. *Journal of Memory and Language, 57,* 126–149.

Petkov, C. I., & Jarvis, E. D. (2012). Birds, primates, and spoken language origins: Behavioral phenotypes and neurobiological substrates. *Frontiers in Evolutionary Neuroscience, 4,* 12.

Phillips, C. (2013). Parser-grammar relations: We don't understand everything twice. In San M., Laka I., Tanenhaus M. K. (eds.). *Language Down the Garden Path: The Cognitive and Biological Basis for Linguistic Structure,* 294–315. New York: Oxford University Press.

Poeppel, D., & Embick, D. (2005). Defining the relation between linguistics and neuroscience. In Cutler A. (ed.). *Twenty-First Century Psycholinguistics: Four Cornerstones,* 103–118. Mahwah: NJ: Four Cornerstones.

Postle, B. R. (2006). Working memory as an emergent property of the mind and brain. *Neuroscience, 139*(1), 23–38.

Potter, M. C., & Lombardi, L. (1990). Regeneration in the short-term recall of sentences. *Journal of Memory and Language, 29*(6), 633–654.

Pulvermüller, F. (2010). Brain embodiment of syntax and grammar: Discrete combinatorial mechanisms spelt out in neuronal circuits. *Brain Language, 112,* 167–179.

Pulvermüller, F. (2014). The syntax of action. *Trends in Cognitive Science, 18,* 219–220.

Rosenberg, R. D., & Feigenson, L. (2013). Infants hierarchically organize memory representations. *Developmental Science, 16*(4), 610–621.

Roux, F., & Uhlhaas, P. J. (2014). Working memory and neural oscillations: Alpha – gamma versus theta – gamma codes for distinct WM information? *Trends in Cognitive Sciences, 18*(1), 16–25.

Rowe, J. B., Toni, I., Josephs, O., Frackowiak, R. S., & Passingham, R. E. (2000). The prefrontal cortex: Response selection or maintenance within working memory? *Science, 288*(5471), 1656–1660.

Rubin, R. D., Watson, P. D., Duff, M. C., & Cohen, N. J. (2014). The role of the hippocampus in flexible cognition and social behavior. *Frontiers in Human Neuroscience, 8,* 742.

Shanahan, M. (2012). The brain's connective core and its role in animal cognition. *Philosophical Transactions of the Royal Society B: Biological Sciences, 367*(1603), 2704–2714.

Shivde, G. S., & Thompson-Schill, S. L. (2004). Dissociating semantic and phonological maintenance using fMRI. *Cognitive, Affective, & Behavioral Neuroscience, 4,* 10–19.

Spelke, E. S., & Kinzler, K. D. (2007). Core knowledge. *Developmental Science, 10*(1), 89–96.

Srihasam, K., Mandeville, J. B., Morocz, I. A., Sullivan, K. J., & Livingstone, M. S. (2012). Behavioral and anatomical consequences of early versus late symbol training in macaques. *Neuron, 73*(3), 608–619.

Takashima, A., Nieuwenhuis, I. L., Jensen, O., Talamini, L. M., Rijpkema, M., & Fernández, G. (2009). Shift from hippocampal to neocortical centered retrieval network with consolidation. *The Journal of Neuroscience, 29*(32), 10087–10093.

Thompson, K. R., Rattermann, M. J., & L Oden, D. (2001). Perception and judgement of abstract same-different relations by monkeys, apes and children: Do symbols make explicit only that which is implicit? *Hrvatska revija za rehabilitacijska istraživanja, 37*(1), 9–22.

Treisman, A. M., & Gelade, G. (1980). A feature-integration theory of attention. *Cognitive Psychology, 12*(1), 97–136.

Ullman, M. T. (2004). Contributions of memory circuits to language: The declarative/procedural model. *Cognition, 92*(1), 231–270.

Van Dyke, J. A., & McElree, B. (2011). Cue-dependent interference in comprehension. *Journal of Memory and Language, 65*(3), 247–263.

Verbruggen, F., McLaren, I. P., & Chambers, C. D. (2014). Banishing the control homunculi in studies of action control and behavior change. *Perspectives on Psychological Science, 9*(5), 497–524.

Watkins, O. C., & Watkins, M. J. (1975). Buildup of proactive inhibition as a cue-overload effect. *Journal of Experimental Psychology: Human Learning and Memory, 1*(4), 442.

Zucker, R. S., & Regehr, W. G. (2002). Short-term synaptic plasticity. *Annual Review of Physiology, 64*(1), 355–405.

# 8 Eliminating parameters from the narrow syntax
## Rule ordering variation by third-factor underspecification*

*Miki Obata and Samuel Epstein*

## 1 Parameters and language variation

The concept of parameters has been central to generative inquiry since the inception of Chomsky's (1981) principles and parameters approach. UG comprises a finite set of principles, while parameters express variant forms of (some of) those principles. Individual I-languages are acquired (grown) by setting parameters/fixing their values, as is triggered by primary linguistic data, i.e. externalized acoustic disturbances or retinal images of hand shapes being input into a parameterized UG, thereby determining experience. This approach enables us to capture a number of construction-specific and language(s) – specific rules or rule systems – non-explanatorily postulated for each individual language in pre-P&P theories. In this sense, with the P&P approach, the theory of UG becomes more restricted, yet concomitantly permitting (highly restricted) variation between I-languages, which can be understood as an earlier step toward the seamless quest of "minimalism", meaning nothing other than a commitment to generalization and explanation-seeking normal science.

While the P&P approach provided great advances in understanding the architecture of the language faculty, Chomsky's (1995) minimalist approach further propels the project forward. He attributes language variation to the lexicon, which is learned through exposure to externalized Primary Linguistic Data (PLD), so that the computational system/syntax is no longer equipped with parameters but works universally, based on instructions provided by linguistic features (selected from the lexicon) which can display (infinitely restricted) variation between I-languages. This "minimalist" advance sought to eliminate variation from the syntax, i.e. to capture what is human syntax, and accorded with the standard seemingly unavoidable and enduring view that the lexicon and morphology do indeed vary among I-languages.

> With regard to the computational system, then, we assume that $S_0$ is constituted of invariant principles with options restricted to functional elements and general properties of the lexicon.
>
> (Chomsky 1995)

Borer (1984) also argued that variation is limited to functional lexical elements, the so-called Borer-Chomsky conjecture. Although parameters are embedded in principles in the P&P approach, they are separated here. Parameters are recaptured through variation in the features of functional lexical categories, while principles/operations are applied with considerations of interface conditions and of computational efficiency. Under this approach, the narrow syntax operations are further simplified and universalized.

The shift to a more restricted minimalist design of the narrow syntax is motivated not only by (the scientifically standard) adherence to Occam's razor, but also from an evolutionary perspective. Chomsky (2005) hypothesizes that not enough time has elapsed to allow for the gradual evolution of a highly complicated UG. If recursive Merge is the defining mutation that endowed modern humans with the infinite capacity (the simplest and seemingly unavoidable assumption), then, as Chomsky notes, it (recursion) could not have developed gradually (a system either is or is not recursive), and the historical record, impoverished as it is, seems to indicate that the defining moment occurred as part of the (so-called) Great Leap forward approx. 75Kya, a split-second before the present time, in evolutionary terms. Thus, there is a convergence between principles of normal explanation-seeking science adhering to Occam's razor and the empirical demands of explaining the (known) facts regarding human linguistic evolution; both demand simplicity (for different reasons). While UG becomes as minimal as possible, we confront the 'opposing' issue of how linguistic variation – a form of complexity – which has been captured under syntactic parameters, can be recaptured by appeal to the lexicon. The next section discusses this issue by focusing on three factors argued in Chomsky (2005) and clarifies how these three factors contribute to the expression of linguistic variation.

## 2 Three factors and language variation: variation in externalization

Chomsky (2005) outlines how the following three factors determine the nature of human I-languages.

(1) a. Genetic endowment
    b. Experience
    c. Principles not specific to the faculty of language

The first factor, genetic endowment, is the species-specific UG. The second factor, experience, is exposure (of UG) to the PLD, which triggers variation in internalized I-languages. (Importantly, as we will discuss below, the second factor is not exactly an independent factor – that is "Experience" is a relational notion, and the experiences one can have are delimited by one's genetic endowment. The experiences one does have are determined not only by genetics, delimiting the class of "possible human experiences", but by countless other factors influencing human history, physics, social norms, traffic accidents, and myriad factors impacting the particulars of individual existence.) The third factor, (more general) principles not specific to the faculty of language, consists of general properties of biological, physical, and/or computational principles of two kinds: (i) data processing and (ii) architectural/computational and developmental constraints. By assuming (as must everyone) the existence of the third factor in addition to the other two factors, (see also Chomsky 1965) the language faculty can be understood as an organ constituting part of a biological computational system, conforming also to physical law, and as such must be constrained by the general constraints on all such systems. The trick (as in all empirical inquiry) is to determine the truth of the matter, i.e. for each aspect of knowledge, how did the three factors contribute to its internalization?

With respect to linguistic variation, Berwick and Chomsky (2011) hypothesize that variation is, perhaps entirely, attributed to externalization:

> Parameterization and diversity, then, would be mostly – possibly entirely – restricted to externalization. That is pretty much what we seem to find: a computational system efficiently generating expressions interpretable at the semantic – pragmatic interface, with diversity resulting from complex and highly varied modes of externalization, which, furthermore, are readily susceptible to historical change.
>
> (Berwick and Chomsky 2011)

To illustrate such variation in externalization, consider the availability of overt (main verb) V-movement famously discussed in Pollock (1989) and Chomsky (1995), variation which was ascribed to the feature strength of Agr: If Agr is strong (e.g. in French), verbs undergo overt movement to T. If Agr is weak (e.g. in English), verbs stay in situ. As mentioned in the Borer-Chomsky conjecture, features on functional heads (e.g. strong vs.weak Agr) differentiate individual I-languages. The deep (unifying) idea is to reduce apparent complex diversity, to 'small' analytical differences. Similarly in Chomsky (2014), T is distinguished in terms of its ability to label: Strong T can be a label (e.g. in Italian), while weak T does not have the capacity to provide a label by itself (e.g. in English). From this one variant property of one lexical category, crosslinguistically variant Extended Projection Principle (EPP) and Empty Category Principle (ECP) effects are deduced. "Italian and English appear radically different" but under analysis they are, by hypothesis, almost identical. These two analyses posit crosslinguistically variant syntactic derivations executed by an invariant universal syntax with the variant derivations triggered by language-specific strength of features on functional heads in the lexicon. The nature of the language-specific features is learned through experience, i.e. via exposure (of UG) to externalized input. That is, the locus of variation is not the narrow syntax but (a highly restricted part of) the lexicon. Consider another case, word order variation. One example is externalization of the symmetric relation generated by (set-forming) simplest Merge (cf. Richards 2008). For example, the orders of VO and OV, which were described by the Head-Parameter in the P&P framework, are not currently specified in the narrow syntax because the simplest Merge generates only the linearly unordered set {V, O}. After Transfer, the unordered set must be linearly externalized for the SM systems to 'execute' the representation (as temporally ordered successive articulatory gestures or hand shapes) as either VO (e.g. in English) or OV (e.g. in Japanese). That is, Merge generates order-indeterminacies or order-underspecified representations in the narrow syntax. (The revolutionary empirical claim is then that word-order is not syntactic, but rather phonological, see e.g. Kayne 1994, Chomsky 1995) Linearization of {V, O} can apply in either of the two logically possible ways: V-O or O-V. Word order variation is 'fixed' by the learner one way or the other, in the process of exposure to externalized PLD. (See Richards 2008 for a more detailed account based on a third-factor interpretation of the Precedence Resolution Principle proposed in Epstein et al. 1998.) Under the Borer-Chomsky conjecture, syntax is no longer equipped with parameters but rather variation in I-languages is attributed to variant features in the lexicon, internalized on the basis of linguistic experience (externalized acoustic or hand shape) input into UG.

The lexicon and the morpho-phonological component can be the loci of variation, so that externalized input plays the crucial role in triggering variation among I-languages (the strong interpretation of the Berwick and Chomsky 2011 hypothesis excerpted above). It is important to note that, if variation is restricted to externalization, it sounds like only the second factor (experience) is involved. But this is not correct, in that it is experience *input into UG* that determines a particular I-language/knowledge state. With respect to this point, Epstein (2014) proposes:

> if the organism is held constant (two human infants) and the exposure is varied (Tagalog vs. Japanese acoustic disturbances in the environment), then any differences in the development of the two infants, must be due to a species-level property by which these variant exposures are mapped to those particular developmental trajectories. In this sense, "language variation" (in humans) is, contra. much standard locution, innate (biologically constrained). That is, it is a defining property of the species that the possible class of variant developmental trajectories is determinable by variant experience. Contrary to the implication of the standard nature-nurture dichotomy, "nurture" is then itself, definable only in terms of nature, and "human language variation" is a species property or capacity (and not "that which is not innate").
> 
> (Epstein in press)

In the next section, we present some cases showing that the third factor can also be implicated in crosslinguistic variation (as well as externalization) and demonstrate how exactly the third factor can contribute to variation.

## 3 Ordering: third-factor underspecification

Third factors are "principles not specific to the faculty of language" according to Chomsky's (2005) definition. In other words, principles belonging to the third factor are obeyed not only within the language system but also in other biological, physical or computational systems. This means that explanation by appeal to the third factor can be regarded as having greater generality than one based on the first factor, in the sense that any appeal to UG, whether parameterized or not, is circumvented under third-factor explanation thereby contributing to the goals of (i) getting the facts right, e.g. a frog's DNA does not in fact determine the shape of a frog's cells, rather this frog-property is determined by physical law) (ii) explaining language evolution in a manner consistent with both (a) Occam's razor, and (b) the known facts indicating the existence of a sudden, recent, and simple mutation. Obata, Epstein and Baptista (2015) propose that variant "rule ordering" can underlie (cross)linguistic variation. How can linguistic variation be explained in terms of such (by hypothesis) third factor licensed ordering variation? Obata, Epstein and Baptista (2015) demonstrate that variant orders of applying (universal) syntactic operations can generate multiple grammatical outputs, each of which equally satisfies the interface conditions and each of which is derived in a computationally efficient manner. That is, given the *independently motivated* definition of "computationally efficient satisfaction of the interfaces" – not proposed in order to capture variation – there are cases in which more than one derivational rule ordering is optimal. In cases in which the third-factor content of the theory independently entails that more than one rule ordering is optimal, the possibility of (cross)linguistic variation is predicted. The question then is whether these predictions, once identified, are correct and, if any are, how much variation might be explained in these terms? Obata, Epstein and Baptista (2015) explain variation of C-agreement in Cape Verdean Creole and Haitian Creole and also agreement variation in English and Kilega by appeal to different timing/ordering of agreement and Merge. (See Obata, Epstein and Baptista (2015) for details of the analysis and also Obata and Epstein 2011, 2012 for relevant analysis.) We will now present additional cases which can be similarly captured under ordering variation permitted by third-factor underspecification of what constitutes computational efficiency.

Chomsky (2013) discusses Aux-inversion with matrix *wh*-movement in English and argues that C searches T by (the third-factor principle of) minimal search, *before* the subject moves to T-Spec as in (2).[1]

(2) C's minimal search before subject-movement finds T

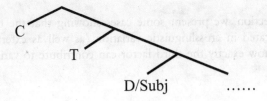

If, by contrast, C's search is executed *after* subject raising, as in (3), the subject/D is the goal of minimal search by C. Assuming that only a minimally searched goal can undergo attraction by C, T/Aux can undergo Internal Merge to C in the derivation of (2), but not (3).

(3) C's minimal search after subject-movement finds D

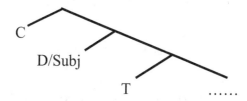

Although Chomsky does not address the derivation of (3) (except to note its hypothesized exclusion in English), note that this is an equally computationally efficient derivation. Convergence by hypothesis requires ONLY that (i) the subject raises (EPP) and (ii) that C attracts a head under minimal search. *Crucially, then, the order of application of these two operations is underspecified by the third-factor law of computational efficiency.* Two rules must apply, any more would be computationally inefficient/unnecessary for convergence, but their relative ordering is not specified. Computationally efficient satisfaction of the interfaces, therefore, in fact predicts the possibility of derivations like (2) and (3). In fact, there is a language which has D-to-C movement. Carstens and Shoaff (2014) propose that D-to-C movement takes place in relative clauses of Kisongo Maasai. This language has two word orders: SVO and VSO. If the direct object contains a relative clause, only the VSO order is allowed. If the subject contains the relative clause, on the other hand, either VSO or SVO is allowed depending on the presence of a left-edge morpheme *ore*. If the morpheme *ore* is included, both *ore* and the subject precede the verb (i.e. SVO) as in (5). If it is absent, the verb is initial (i.e. VSO) as in (4).

(4) a. ɛ-tadwaa      [ɛnkıtɛŋ n-atooko   ɛŋkarɛ] ɔlɛɛ
       3sSA-see.PST cow       AAE.fem-drink.PST water man
    b. *[ɛnkıtɛŋ n-atooko   ɛŋkarɛ] ɛ-tadwaa ɔlɛɛ
       cow   AAE.fem-drink.PST water 3sSA-see.PST man
       'The cow that drank the water saw the man'
(5) a. Ore [ɛnkıtɛŋ n-atooko    ɛŋkarɛ] ɛ-tadwaa ɔlɛɛ
       ORE cow AAE.fem-drink.PST water 3sSA-see.PST man
    b. *Ore ɛ-tadwa [ɛnkıtɛŋ n-atooko   ɛŋkarɛ] ɔlɛɛ
       ORE 3sSA-see.PST cow  AAE.fem-drink.PST water man
       'The cow that drank the water saw the man'

Carstens and Shoaff (2014) suggest that the morpheme *ore* is a relative determiner, which corresponds to demonstrative *that* in English, and moves from inside of the subject DP *ɛnkıtɛŋ* (cow) to C in the relative clause. Based on this assumption, they explain the complementary distribution of the morpheme *ore* and V-movement to C. If their analysis is applied to Chomsky's (2013) analysis, the following derivations are carried out (although Carstens and Shoaff present other derivations based on a different movement system.)[2]

(6) VSO
 $[_{CP}\ C_{rel}\ [_{TP}\ V+v+T\ [_{vP}\ [_{DP}\ NP]\ \ldots$
 $\rightarrow [_{CP}\ V+v+T+C_{rel}\ [_{TP}\ <V+v+T>\ [_{vP}\ [_{DP}\ NP]\ \ldots$

(7) Ore SVO
 $[_{CP}\ C_{rel}\ [_{TP}\ [_{DP}\ ore\ NP]\ V+v+T\ [_{vP}\ \ldots$
 $\rightarrow [_{CP}\ ore+C_{rel}\ [_{TP}\ [_{DP}\ <ore>\ NP]\ V+v+T\ [_{vP}\ \ldots$

In (6), C minimally searches the V+v+T amalgam before the subject DP moves to T-Spec. (as schematized in (2)). As a result, the V amalgam moves to C. In (7), on the other hand, the subject DP is moved to T-Spec before C minimally searches. Under minimal search, C searches and finds D within the subject DP and D-to-C movement takes place. (as schematized in (3)). That is, both of these orders generate convergent outputs and the two derivations are equally optimal/efficient, computationally, applying all and only the (universal syntactic) operations necessitated by convergence.

## 4 A note on INTRA and INTER I-language variation

Notice that variation between I-languages and variation among "constructions in a given I-language" must both satisfy the same constraints; UG and third factor under experience. The theory therefore predicts, in principle, a correspondence between (A) the variation displayed by different I-languages and (B) the variation displayed within a particular I-language. For example Obata and Epstein (2011, 2012) argue that the enduring oddity of English *tough*-constructions is 'nothing more' than a type of lexical item inducing the same kind of (computationally efficient) object agreement as is seen characteristically in Kilega *wh*-movement (Carstens 2005, 2008). That is, English *tough*-constructions are "just Kilega". The prediction that intra-language constructional (lexical) variation comports with inter-I-language variation is predicted by the theory – all such systems are constrained by UG and third factor, with third factor leaving certain aspects of derivations underspecified, thereby allowing variants triggered by externalized outputs (generated by variant systems, allowed by third-factor underspecification, generating variable externalized outputs. . . . etc.!) By hypothesis, the orders in (2) and (3) both apply "within English", thereby generating variant outputs in English subject wh-movement cases:

(8) a. Which man visited New York?
 b. Which man did visit New York?
 (with emphatic intonation of *did*; Koopman 1983)

With the order of (2), (8b) can be derived. When C searches, the subject *which man* is still within vP, so that T undergoes Internal Merge to C. After that, the subject *which man* undergoes Internal Merge to Spec-T (counter-cyclically, see Epstein, Kitahara and Seely (2012)) and to Spec-C. With the order of (3), on the other hand, (8a) can be derived. The subject *which man* moves to Spec-T before C's minimal search. In the case of Kisongo Maasai, D *ore* can be morphophonologically separated from its complement NP *cow*. However, English D can never be isolated because of its morphological properties (a form of morphophonological variation) which are analyzed in the process of externalization, so that D pied-pipes the whole DP for SM-convergence. As a result, DP *which man* moves to Spec-C. Thus the exceptional vexing (8a) is simply the Kisongo-Maasai type of derivation, just like the mysterious *tough*-construction (i.e. mysterious only "in English"!) is seen as nothing more than the Kilega type of derivation (Obata and Epstein 2011, 2012).

The important points worth emphasizing are that the orders in (2) and (3) are each computationally efficient, in that C finds a goal (as required by hypothesis for convergence) and in each derivation the exact same operations are applied (subject-raising and C minimally searching) and something moves (overtly) to C by EF. Crucially, then, the theory predicts that there is greater than one computationally efficient type of derivation: convergence requires that C search and attract some category, but as far as optimal *computation* goes, EF on C can be satisfied by any category moving to C-Spec. EF on T can be similarly satisfied. But optimal computation leaves underspecified the ordering of these two operations, each of which is necessary for convergence. Both orders are computationally efficient. Equally optimal distinct derivations generate predicted variation. Prediction (the theory of computationally efficient satisfaction of the interfaces predicts the possibility of two distinct orderings) via underspecification in third-factor machinery is arguably the furthest we can get from ad hoc stipulation, including the stipulating syntactic parameters as part of UG. That is, if this approach is viable, intra and inter "language" non-morphological variation are unified and in part captured by what is NOT stated (in general, third-factor laws NOT specific to the human language faculty.)

The empirical domain and viability of the application of such an approach remain to be seen. But the idea that language-specific 'syntactic phenomena' (like *tough*-constructions "in English" or optional Aux-inversion "in English" matrix subject *wh*-movement) might be partially accounted for, by what is NOT stated in laws NOT specific to the human language faculty, was until very recently inconceivable. If scientific explanation is one's goal, this form of third-factor analysis seems very worthy (in our view) of further exploration, no matter where it might lead.

## 5 Discussion: what's an (I-) "language"?

We reviewed some approaches to crosslinguistic variation and further explored third-factor underspecification as a locus of I-language variation (Richards 2008, Boeckx 2011, Obata and Epstein 2011, 2012, Obata, Epstein and Baptista

2015) triggered by (variant) externalized outputs (Berwick and Chomsky 2011) which constitute the input to the learner, based on the Borer-Chomsky conjecture. Under the third-factor underspecification approach, the options of underspecified rule order are entailed by an independently motivated formulation of the SMT "computationally efficient satisfaction of the interfaces" without any ad hoc descriptive stipulations. Based on Obata, Epstein and Baptista (2015), we further extended the argument in favor of ordering as a third factor. As a case study, we discussed Chomsky's (2013) analysis of Aux-inversion in terms of Carstens and Shoaff's (2014) analysis of Kisongo Maasai and English matrix subject *wh*-movement, and suggested that both T-to-C and D-to-C are equally possible computationally efficient derivations satisfying the interface conditions, each of which is derivable by allowable ordering variation entailed by what is NOT formally stated, within the third-factor underspecification approach. The theory thus predicts that computationally efficient satisfaction of the interfaces can be attained by variant rule orderings, giving rise to variant externalized outputs which serve as inputs to the learner. As is perhaps obvious, 'constructions' (or, more primitively, lexical items or their features) in a given I-language are subject to the exact same constraints which apply 'across' I-languages. The resulting prediction is that any 'construction' in a particular language (e.g. English *tough*-constructions) are in principle possible constructions ("IN" another "language"). The particular languages that have existed and do exist are an accident of many historical factors, but the class of possible (human) I-languages (our object of inquiry) and hence the class of possible (human) derivations is a single set of what we might call "human". Hence it is not particularly surprising (nor is it necessary) to find that "English *tough*-constructions are 'simply' Kilega", and "English matrix subject *wh*-movement lacking auxiliary inversion is 'simply' Kisongo Maasai". This, of course, further blurs (if not obliterates) the common sense (non-scientific) notions of languages, such as English, Kisongo Maasai and Kilega, i.e. if we are on track here, the English *tough*-construction 'is' Kilega and the English subject *wh*-movement in matrix clauses lacking T to C, 'is' Kisongo Maasai. Scientifically speaking, the I-language English is neither a T-to-C "language"; nor is it a D-to-C "language". Rather, "it" is both, i.e. my I-language performs both operations, each licensed by third factor and triggered by input. Similarly, under Obata and Epstein (2011), we cannot say that English involves subject agreement. Rather in *tough*-constructions, T agrees with a moved/moving/shifted object first merged in object position. (i.e. "John IS easy to please" originates as "Be easy to please John". *John* then shifts to embedded C-Spec where matrix T can probe and agree with (what was a direct object) *John*. Then *John* raises ultimately to matrix subject position. See Obata and Epstein 2011, 2012) Thus "English" is neither a subject agreement nor object agreement "language". My I-language generates both subject agreement derivations and derivations in which a shifted object appearing in embedded C-Spec is the goal of a higher T probe. Binary parameters of the form "Subject agreement or Object agreement" seem not only descriptive, but (descriptively) inadequate. The class of possible human derivations is a single set. The set my I-language generates is determined by UG (including the Borer-Chomsky hypothesis), third factor, and the particular inputs I received, which were themselves determined by a host of factors including historical accident, invasions, etc.

Within the Minimalist framework (as in biology more generally) *humanly possible I-language* is the object of *explanatory* inquiry (by contrast, *predicting* what forms happen to exist on earth today or in the past is a quite different enterprise). If on track, the I-language internalized by the second author of this chapter has Kilega-type derivations (Object agreement in *tough*-constructions, not just Subject agreement), as well as having Kisongo Maasai-type derivations (D to C, not just T to C) and other derivational types, too, all falling within the constraints imposed on what can be a humanly possible I-language, as determined by UG, the third factor and the infinite class of possible externalized inputs (including none) into such infinitely and multiply constrained, linguistic, biological, physical and computational mental organs.

## Notes

\* We are very grateful to Marlyse Baptista, Vicki Carstens, Noam Chomsky, Hisatsugu Kitahara, Acrisio Pires and Daniel Seely for extremely valuable and helpful discussion, suggestions and comments.
1 If Internal Merge involves probing, matching and Agree (thereby determining the syntactic objects to be attracted) and therefore conforms to minimal search, while External Merge does not, then we lose the unification of Internal Merge and External Merge, as proposed by Chomsky (2013).
2 The analysis presented here does not explain why D-to-C movement is allowed only in the case of relative clauses modifying the subject, nor why the morpheme *ore* always appears in the sentence initial position. See Carstens and Shoaff (2014) for more detailed analysis.

## References

Berwick, Robert, and Noam Chomsky. 2011. The biolinguistic program: The current state of its development. In A.-M. Di Sciullo and C. Boeckx (eds.). *The Biolinguistic enterprise: New perspectives on the evolution and nature of the human language faculty*. 19–41. Oxford: Oxford University Press.
Boeckx, Cedric. 2011. Approaching parameters from below. In C. Boeckx and A-M Di Sciullo (eds.), *The Biolinguistic enterprise: New perspectives on the evolution and nature of the human language faculty*. 205–221. Oxford: Oxford University Press.
Borer, Hagit. 1984. *Parametric syntax: Case studies in semitic and romance languages*. Dordrecht: Foris Publications.
Carstens, Vicki. 2005. Agree and EPP in Bantu. *Natural Language & Linguistic Theory* 23:219–279
Carstens, Vicki. 2008. Raising in Bantu. Unpublished manuscript, University of Missouri, Columbia
Carstens, Vicki, and Cassady Shoaff. 2014. D-to-C and VSO/SVO alternation in Kisongo Maasai: Evidence from relative clauses. Unpublished manuscript, University of Missouri, Columbia.

Chomsky, Noam. 1965. *Aspects of the theory of syntax*. Cambridge, MA: MIT Press.
Chomsky, Noam. 1981. *Lectures on government and binding*. Dordrecht: Foris.
Chomsky, Noam. 1995. *The minimalist program*. Cambridge, MA: MIT Press.
Chomsky, Noam. 2005. Three factors in language design. *Linguistic Inquiry* 36:1–22.
Chomsky, Noam. 2013. Problems of projection. *Lingua* 130:33–49.
Chomsky, Noam. 2014. *Problems of projection: Extensions*. Unpublished manuscript, MIT.
Epstein, Samuel D. in press. *Why nurture is natural (too)*. Unpublished manuscript, University of Michigan, Ann Arbor.
Epstein, Samuel, Erich Groat, Ruriko Kawashima, and Hisatsugu Kitahara. 1998. *A derivational approach to syntactic relations*. Oxford: Oxford University Press.
Epstein, Samuel D., Hisatsugu Kitahara and Daniel Seely. 2012. Structure building that can't be. In M. Uribe-Etxebarria and V. Valmala (eds.) *Ways of structure building*. 253–270. Oxford: Oxford University Press.
Kayne, Richard. 1994. *The antisymmetry of syntax*. Cambridge, MA: MIT Press.
Koopman, Hilda. 1983. ECP effects in main clauses. *Linguistic Inquiry* 12:93–133.
Obata, Miki, and Samuel D. Epstein. 2011. Feature-splitting internal merge: Improper movement, intervention and the A/A'-distinction. *Syntax: A Journal of Theoretical, Experimental and Interdisciplinary Research* 14:122–147.
Obata, Miki, and Samuel D. Epstein. 2012. Feature-splitting internal merge: The case of *tough*-constructions. In M. Uribe-Etxebarria and V. Valmala (eds.) *Ways of structure building*. 366–384. Oxford: Oxford University Press.
Obata, Miki, Samuel D. Epstein, and Marlyse Baptista. 2015. Can crosslinguistically variant grammars be formally identical?: Third factor underspecification and the possible elimination of parameters of UG. *Lingua* 156:1–16.
Pollock, Jean-Yves. 1989. Verb movement, universal grammar and the structure of IP. *Linguistic Inquiry* 20:365–424.
Richards, Marc. 2008. Two kinds of variation in a minimalist system. In varieties of competition. In F. Heck, G. Muller and J. Trommer (eds.). *Varieties of Competition. Linguistische Arbeits Berichte* 87: 133–162. Universität Leipzig.

# Part III
# Conceptual and methodological foundations

# 9 On certain fallacies in evolutionary linguistics and how one can eliminate them

*Koji Fujita*

## 1 Introduction

Studies of the biological origins and evolution of the human language faculty (*evolutionary (bio)linguistics*) are a highly inter-/cross-disciplinary field of research in which a huge variety of researchers, including not only linguists and biologists but psychologists, philosophers, anthropologists, archaeologists, primatologists, computer scientists and many more need to gather together and collaborate in order to explore how language first came into being (more precisely, how the new genus *Homo* with the faculty of language, *H. sapiens*, came into being).

As is often the case with other inter-/cross-disciplinary areas, evolutionary linguistics has suffered from a lack of enough agreement among researchers on the target of inquiry and some fundamental ideas. Language evolution is itself a very tricky notion of which different people have different understandings. To begin with, there is almost no consensus on what *language* and *evolution* exactly refer to when people talk about language evolution. Typically, many researchers are, in fact, studying the evolution of linguistic communication when they purport to study language evolution, the two of which must be sharply distinguished.

This chapter takes a close look at four major fallacies which can be easily detected in current evolutionary linguistics. A possible scenario of language evolution which helps eliminate these fallacies will be suggested. The goal is to eliminate these fallacies in order to establish the underpinnings for a truly inter-/cross-disciplinary development of evolutionary linguistics.

## 2 Fallacy of communication (or fallacy of thought)

There is one recurring statement one often hears in the context of language evolution: "Language evolved for the purpose of communication." This apparently innocent claim is the best illustration of what I call the fallacy of communication and is plainly wrong in two important respects (see Balari and Lorenzo (2010) for relevant discussions). Firstly, biological evolution is a blind process and has no purpose in sight; nothing evolves for any specific function. *X evolved for Y* is only a metaphorical description of *X evolved and then began to be used for Y*.

In evolutionary biology, it is useful and customary to draw a line between *original function* and *current utility* (see Bateson and Laland 2013), the importance of which was already recognized by Darwin himself in his discussion of *preadaptation* (*exaptation* will be more appropriate). An oft-quoted example is bird feathers, which originally had an adaptive value in relation with thermoregulation, sexual dimorphism, locomotion and the like before they became useful for flight (see Clarke 2013 for new findings on this issue).

In the case of human language, there is no denying that its current utility includes communicative function among many others, though whether communication is a major function of language or whether language is a major tool of communication even today remains to be seen (think of the vast area of our daily communication where nonverbal means are far more efficient), in part due to the vague nature of communication itself. The question to be asked in the context of language evolution is, then, whether communication was the original function or rather an extended, later co-opted function.

There is one well-known piece of argument against the view that communication was the original function, which goes as follows. The first individual who happened to acquire language by mutation would have had no one else around to use this new faculty to communicate with. This "lone mutant" puzzle cannot be solved by kin selection (Fitch 2005) either, because it would have taken at least two generations (parent-offspring) before kin selection could work to make communication by language adaptive. The first mutant had to remain all alone in any event, no matter how short the period was. The alternative view, that the original function of language was thought, internal to the individual, therefore deserves credit.

Within the minimalist framework of generative grammar, this latter view has been corroborated by the (theory-internal) observation that the computational system of human language (syntax) is optimally designed for *internalization*, not for *externalization* (Chomsky 2013, Berwick et al. 2013; see also Bolhuis et al. 2014, 2015). That is to say, mapping from syntax to the Conceptual-Intentional (CI) interface for thought takes place in a uniform and straightforward manner across languages, while mapping to the Sensory-Motor (SM) interface for communication requires many language-particular adjustments, including specification of diverse morphophonological realization, linearization of hierarchical structure and deletion of copies both created by the recursive combinatorial operation *Merge*, itself generating only unordered sets, all of which would be unnecessary if human language were exclusively an instrument of thought. Language as an instrument of communication has to be more complicated and therefore is very likely to be a later innovation, a matter of *cultural evolution*, to speak metaphorically (see Figure 9.1).

The problem with this minimalist consideration to support the view that communication initially had nothing to do with language evolution is, of course, that it is totally theory-dependent; a different way of looking at language will necessarily lead to a different picture of language evolution. This is where

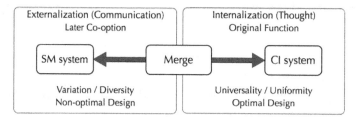

*Figure 9.1* The minimalist design of human language, with an interface asymmetry.

Jackendoff's (2010) caveat "Your theory of language evolution depends on your theory of language" must be seriously reckoned with.

Highly relevant here is the question of whether *protolanguage* existed before the advent of human language. In fact, there is suggestive evidence for its existence from a variety of animal cognition experiments and also from studies of human pathological conditions. Primates or non-primates, several non-human species have been shown to have the capacity for non-hierarchical, linear grammar. Dolphins, for example, can discriminate different meanings of a pair of sentences depending on the order of the words in them ("words" here are represented by gestures and electronic sounds); they understand how *take the ball to the hoop* differs from *take the hoop to the ball* (Herman 2010). This is, of course, not to say that dolphins have Merge, since Merge-based syntax will yield a grammar that is dependent on hierarchy and not on linearity.

In humans, Broca's aphasics from time to time exhibit a similar behavior. They correctly understand *the dog chased the cat* but *the cat was chased by the dog* only at chance level. In the past, this situation was linked to the idea that the patients had some trouble processing a structure generated by movement (Y. Grodzinsky's *trace-deletion hypothesis* being a concrete proposal), an idea which is no longer an option. For one thing, given the VP-internal subject hypothesis, deriving an active sentence involves movement, too. For another, movement is now understood to be a case of Merge (internal Merge) and has no special status in the minimalist theory of syntax.

The correct interpretation of this degraded performance by Broca's aphasics, then, seems to be this. Merge-based hierarchical structure processing ceases working, and instead a simple processing strategy based on word order comes to the fore, according to which the subject (agent) must precede the object (theme or patient). This explains not only the active-passive contrast above, but also why a relative clause like *the cat that chased the dog (was small)* can be correctly understood but *the dog that the cat chased (was big)* cannot, or even why Chinese aphasics perform in the opposite way (because in Chinese the relative clause precedes the head nominal) (see Grodzinsky 2004).

Taken together, these observations point to the possibility that non-hierarchical linear grammar is an evolutionary precursor of hierarchical grammar, and that

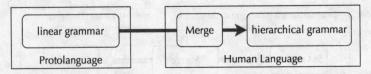

*Figure 9.2* The shift from protolanguage to full human language triggered by Merge.

the former was a component of protolanguage and still exists in the brain of modern humans as a vestige of language evolution. The emergence of Merge was the crucial event in the (nearly abrupt) shift from protolanguage to full human language (Figure 9.2).

So suppose there was protolanguage before human language, very probably at some point in the lineage of *H. erectus*. Chomsky is well-known for his persistent opposition to this idea, but this opposition is based on his belief that Merge was an instantaneous innovation (as a result of brain rewiring caused by mutation) and did not evolve gradually, and that no form of language was possible until Merge emerged (Chomsky 2005 and elsewhere; see also Bickerton 2009). But we have seen the possibility that Merge-less non-hierarchical grammar operates in the modern human and non-human brains as a reflex of protolanguage. Given protolanguage, and the discrepancy between protolanguage and human language, the "lone mutant" argument against communication as an original function of language now needs to be rethought and weakened.

It is only natural that protolanguage, to the extent that it was shared among a group of individuals, was already adaptive as an instrument of communication, thought, and more. Then arrived a new individual equipped with Merge and hierarchical grammar in addition to the old non-hierarchical grammar. Was this individual totally unable to use this new capacity to communicate with other members of the group? The situation might be compared to an intact person surrounded by Broca's aphasics. Not a very good comparison, to be sure, but the point should be clear. Just as this intact person can map part of his/her configurational grammar to non-configurational grammar for ease of communication ([ S [ V O ]] can be safely modified to [ S V O ]), the first individual with human language was likely to be able to utilize it to communicate with other individuals, though of course in a restricted way.

This observation is not to deny the primary importance of language as an instrument of thought brought about by Merge and the CI interface, but it does prompt us to take a generic, pluralistic view towards the original function(s) of language, by departing from the simplistic *thought vs. communication* debate and taking all other possibilities into consideration. Who could deny the possibility that its original function was neither thought nor communication (both being later co-options) but was something much more fundamental, such as hierarchical perception/reorganization of the image of the world, so that it would become more easily manipulable for both thought and communication?

## 3 Fallacy of a single origin

Human language is a complex system which combines several independent capacities (modules), most notably the Merge-based computational system plus the CI and SM systems with which it interfaces. This modular architecture is what guarantees the *evolvability* of language; in general, non-modular monolithic systems hardly evolve. Accordingly, the problem of language evolution needs to be decomposed into two distinct problems: (A) How did each of these modules evolve? What are their precursors? (B) How did these modules get integrated into the complex system we call language? Problem (A) can be loosely called *the origin problem*; problem (B) *the interface problem* (Figure 9.3).

In this perspective, it no longer makes sense to simply ask what human language evolved from, but we need to ask which components of language evolved from what kind of preexisting capacities. Any statement of the form *Language evolved from X* (X being a single trait, be it gestures, speech, songs, music or whatever else) is doomed to be false, and this wrong idea that language must have evolved from a single preexisting capacity well represents the fallacy of a single origin.

M. Corballis once stated that "language evolved, not from the vocal calls of our primate ancestors, but rather from their manual and facial gestures" (Corballis 2002: ix). Why not *both vocal calls and gestures,* each responsible for different aspects of language? What is called for is again a pluralistic view, in which every candidate must be considered as a possible precursor of a particular module, not of the whole language suite, without necessarily precluding the possibility that a single module had a multiple precursor.

Very famously, Hauser et al. (2002) proposed the dichotomy between a faculty of language in the narrow sense (FLN) and a faculty of language in the broad sense (FLB), FLN being the component of language which is both species-specific and domain-specific, while FLB includes FLN and every other component which is not specific in one way or another. They further suggested that only recursion, which can be safely equated with Merge in the present discussion (Hauser et al. did not refer to Merge per se at all), belongs to FLN, whereas the CI and SM systems do not. In other words, these latter two systems may easily find their evolutionary precursors in the realm of non-human cognition, including primate conceptual systems and phonological syntax of songbirds.

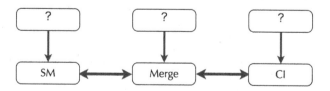

*Figure 9.3* Two major problems of language evolution. The vertical arrows represent the origin problem; the horizontal arrows the interface problem.

To the extent that FLN is species-specific and evolutionarily discontinuous, its existence presents a barrier to studies of language evolution. Below I will argue that the FLN/FLB dichotomy is an illusion and that Merge also had an evolutionary precursor, as already implicated in Figure 9.3.

## 4 Fallacy of continuity (anthropomorphism)

So suppose every component of human language is evolutionarily continuous with some cognitive faculties which can be found in the animal kingdom. Human language is unique not in any of its constituents but in their combination. This conforms to the general picture of biological evolution in which a novel trait appears as a result of recombination of old traits and is good news for every evolutionary linguist endeavoring to understand the evolution of human language in a naturalistic way on a par with other instances of evolution.

It is for this reason that studies of animal cognition, communicative or not, play an important role in today's evolutionary linguistics. Now we read reports after reports of a wide range of animals, not limited to primates, mammals and birds, having mental/cognitive faculties to a greater or lesser degree which were once believed to be monopolized by humans (see, for example, Stoop et al. 2013 for the surprising observation that *Drosophila*'s courtship behavior is grammatically more complex than human language).

It is important to note that these reports, to the extent that they are correct, do not conflict with our understanding that language is a uniquely human trait. As just mentioned, the uniqueness of human language lies in its combination of distinct modules and not in any one of these modules taken in isolation (including Merge, I believe). Instead, these studies offer an invaluable asset to the development of evolutionary linguistics in that they help us investigate how this species-specific faculty of language may have evolved from what kind of non-species-specific faculties through the process of recombination and reorganization.

This much said, we need to realize that some human faculty X and its animal counterpart Y, no matter how closely they are evolutionarily linked, are not the same after all and that any gap, however small, which may exist between X and Y is the key to understanding species-specificity. We know that higher primates like chimpanzees have rudimentary forms of conceptual capacity and a theory of mind, and we study these in order to see how these mental faculties may have evolved in the hominin lineage. So far, so good. But troubles begin when one goes further and believes that chimpanzees' theory of mind (X) is the same as the humans' (Y), or that studies of the evolution of Y can safely be replaced by studies of the evolution of X.

The fallacy of continuity, or anthropomorphism, has been very dangerous in comparative cognitive/psychological studies, so that as early as 1903 C. L. Morgan published a caveat against such an anthropomorphic bias (see Buckner 2013): "In no case may we interpret an action as the outcome of the exercise of higher psychological processes, if it can be fairly interpreted in terms of processes which stand lower in the scale of psychological evolution and development." Later to be known as Morgan's Canon, this criterion has served as a good antidote for excessive assimilation or identification of human and non-human cognition.

A fairly recent example of violating this canon may be found in the so-called Integration Hypothesis of human language evolution (Miyagawa et al. 2014). This hypothesis maintains that the non-finite state nature of human language results from the combination of two separate systems, E(xpressive) and L(exical), each being finite-state in isolation and found in other species (E in songbirds and L in monkeys, etc.). The conceptual advantage of this hypothesis is that it allows us to understand language evolution in perfect conformity to the general pattern of evolution as creation of something new by combining old things.

The empirical disadvantage is that birds' E-system and monkeys' L-system, to the extent that they exist, differ greatly from the human counterparts (roughly, the systems of functional categories and lexical categories). In particular, the human system of lexical categories is extremely rich and generative and always mind-dependent when making reference to the outside world. As Chomsky has stressed on many occasions, human language has no direct reference relation, in sharp contrast to animal communication systems, which are related to mind-independent entities (Chomsky 2007). This gap we really need to mind.

## 5 Fallacy of the FLN/FLB dichotomy

The above discussion is directly connected to the reevaluation and rejection of Hauser et al.'s FLN/FLB dichotomy. This dichotomy is based on the supposition that many but not all aspects of language are neither species-specific nor domain-specific. But exactly in what sense are they not species-specific or domain-specific?

Take the SM system as an example of the non-FLN part of FLB and compare the vocal learning abilities of humans and songbirds. They look very similar not only behaviorally but in terms of the neural and genetic mechanisms involved, and they represent a good example of homoplasy. So this aspect of the SM system is evolutionarily continuous and the human version is not species-specific. It is not domain-specific, either, because the same vocal learning ability can be put to use outside the linguistic domain. In this weak sense, it is certainly not part of FLN. Still, human vocal learning is not exactly the same as bird vocal learning, and it does not work quite the same for linguistic and non-linguistic purposes. In this strong sense, the human linguistic vocal learning can be argued to belong to FLN, too, because it has properties that cannot be found anywhere else.

Now consider Merge (recursion) as the sole candidate of FLN. The species-specificity and domain-specificity of Merge is secured by the fact that it is the elementary computational operation dedicated to human language only. Still, outside the linguistic domain, Merge-like combinatorial operations can be found in human and non-human physical/metal capacities, including locomotion, tool using and tool making, counting, the theory of mind, mental time travel, etc. In addition, as will be argued below, Merge is likely to have its evolutionary precursor in the action domain of both humans and non-humans. If so, Merge does have certain evolutionary continuity with a non-human, non-linguistic function. In this weak sense, then, even Merge does not belong to FLN: FLN is virtually an empty set.

In short, the FLN/FLB dichotomy is an illusion, arising from applying different criteria to Merge and other components. In fact, Merge is uniquely human only to the same extent that the CI and SM systems can be judged so, and all of them are evolutionarily continuous with similar capacities of other species. Merge is unique to human language, but then the CI and SM systems are equally unique when these systems work together with Merge for specifically linguistic functions.

It seems reasonable to conclude that every component of language is equally (non-)unique, and that the FLN/FLB dichotomy is a hindrance to our natural understanding of language and its evolution (see Boeckx 2013 for related discussion).

## 6 Motor control origin of Merge and Merge-only evolution

There have already been some discussions of the evolutionary continuity of Merge with other biological traits, on how this computational capacity may have evolved from preexisting non-linguistic functions shared by humans and non-humans. Here I review the main proposals of Fujita (2014) with some amendment (see also Boeckx and Fujita 2014 and the references therein).

Both humans and non-humans have the capacity to combine physical objects hierarchically and sequentially, as typically observed in tool using and tool making. Chimpanzees' nut cracking behavior is one classical example, but other mammals, birds, and even insects are now known to be good tool users. What deserves special mention is the metatool use by New Caledonian crows, where the corvids ingeniously invent new ways of combining three distinct steps of tool use to get food (Taylor et al. 2010; more recently it has been reported that they can combine as many as eight steps), though of course Morgan's Canon dictates us to ask whether these feats evidence the existence of the birds' human-like insight (see Taylor et al. 2012).

The hypothesis of the motor control origin of Merge maintains that Merge is an exaptation of this object-combining function (call this *Action Merge*, adapting Greenfield's (1991, 1998) *Action Grammar*), which was later expanded to manipulate abstract symbols in humans. The possible evolutionary links between stone tool technology, language and cognition have often been discussed in the field of cognitive archaeology, based on the observation that language and tools share certain cognitive and neurological underpinnings (see Stout 2010 for example).

As noted, however, language is a complex trait, and each of its components is very likely to have evolved from distinct precursors. The proposal here is that the evolution of stone tool technology (as a form of Action Merge) should be

## Fallacies in evolutionary linguistics  149

specifically linked to the evolution of Merge, not language at large. More concretely, the domain-specific Action Merge first evolved into a domain-general combinatorial capacity (General Merge), from which many domain-specific capacities, including Merge, derived in the manner of descent with modification (Figure 9.4).

Merge then applied to the already existing rudimentary forms of the lexicon (protolexicon), the CI system (proto-CI) and the SM system (proto-SM) at the stage of protolanguage, reorganizing them into the kind of hierarchical systems which characterize modern human language. This, then, is the Merge-only model of human language evolution, a radical instantiation of the minimalist approach to language evolution, which says that Merge is the only innovation you need for creating human language, with everything else following from Merge applying to these preexisting faculties (Figure 9.5).

The two problems of language evolution mentioned in section 3 (see Figure 9.3) can now be addressed in a systematic way. Every component of language, including Merge, had a precursor originally unrelated to language, and the emergence of uniquely human Merge allowed other components to evolve into equally unique systems. The formation of the interfaces took place almost automatically, as a reflex of the Merge-triggered evolution of the relevant systems; since Merge transformed these systems into what they are, it inevitably interfaces with them.

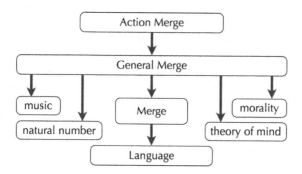

*Figure 9.4* Motor control origin of Merge.

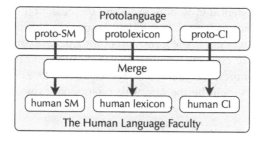

*Figure 9.5* Merge-only evolution of human language.

Perhaps surprisingly, what is implied here is a syntax-first model of language evolution. Contrary to what is often taken for granted, in this model, syntax was not the finishing touch of language evolution but rather its point of entry. So syntax must have evolved where there was no corresponding (compositional) semantics, for example, and there is some indication that something similar indeed takes place in the evolution and development of birdsong (phonological) syntax (Berwick et al. 2011). The birth of the human lexicon may also be dependent on Merge; as proponents of Distributed Morphology argue (Marantz 1997 et seq.), words are syntactically derived objects and not something prestored in the lexicon before syntactic computation takes place.

Let me stress that the virtue of these hypotheses reviewed here lies not in their novelty or radicalism but in that they show the possibility and necessity of considering a wider range of conceivable scenarios, without any bias towards hitherto proposed popular hypotheses. Pluralism is the essence of evolutionary studies.

## 7 Conclusion

This chapter has pointed out four major fallacies commonly found in evolutionary linguistics (sections 2–5) and briefly presented a conceivable scenario concerning the evolution of Merge and human language (section 6), which I believe can carefully avoid falling victim to these fallacies.

It is important to note that the Merge-only hypothesis is not an example of the fallacy of a single origin: It says that the evolution of human language required only Merge as a new ingredient together with all other preexisting capacities, which Merge allowed to evolve into distinct components of language. This is totally different from saying that language evolved from Merge alone. Merge itself evolved from its own precursor (Action Merge) through the process of descent with modification and is therefore no exception to the orthodoxy of Darwinian evolution. To complete this picture, it is also necessary to clarify the adaptive value(s) of Merge per se, because Merge must have evolved first before it became useful as a component of human language, which I hope to address in future work.

In addition to those discussed in this chapter, one can easily detect a variety of fallacies which today's evolutionary linguistics suffers from, including strong adaptationism (hyper-selectionism). Adaptation and selection continue to be a major factor in biological evolution, including the evolution of Merge and language, but they can no longer be the only factor. Modern biological thinking reflects this understanding in the formation of *the Extended Synthesis*, which adds many new ideas like evo-devo, epigenetic modification and niche construction to the Modern Synthesis (Pigliucci and Müller 2010). Pluralism has started to prevail. The same should be the case with evolutionary (bio)linguistics, too.

## References

Balari, Sergio, and Guillermo Lorenzo. 2010. Communication: Where evolutionary linguistics went wrong. *Biological Theory* 5:228–239.

Bateson, Patrick, and Kevin N. Laland. 2013. Tinbergen's four questions: An appreciation and an update. *Trends in Ecology & Evolution* 28(12):712–718.
Berwick, Robert C., Angela D. Friederici, Noam Chomsky, and Johan J. Bolhuis. 2013. Evolution, brain, and the nature of language. *Trends in Cognitive Sciences* 17(2):89–98.
Berwick, Robert C., Kazuo Okanoya, Gabriel J. L. Beckers, and Johan J. Bolhus. 2011. Songs to syntax: The linguistics of birdsong. *Trends in Cognitive Sciences* 15(3):113–121.
Bickerton, Drek. 2009. *Adam's tongue: How humans made language, how language made humans.* New York: Hill and Wang.
Boeckx, Cedric. 2013. Biolinguistics: Forays into human cognitive biology. *Journal of Anthropological Sciences* 91:1–28.
Boeckx, Cedric, and Koji Fujita. 2014. Syntax, action, comparative cognitive science and Darwinian thinking. *Frontiers in Psychology* 5: Article 627.
Bolhuis, Johan J., Ian Tattersall, Noam Chomsky, and Robert C. Berwick. 2014. How could language have evolved? *PLoS Biology* 12(8): e1001934.
Bolhuis, Johan J., Ian Tattersall, Noam Chomsky, and Robert Berwick. 2015. Language: UG or not to be, that is the question. *PLoS Biology* 13(2): e1002063.
Buckner, Cameron. 2013. Morgan's canon, meet Hume's dictum: Avoiding anthropofabulation in cross-species comparisons. *Biology and Philosophy* 28:853–871.
Chomsky, Noam. 2005. Three factors in language design. *Linguistic Inquiry* 36:1–22.
Chomsky, Noam. 2007. Of minds and language. *Biolinguistics* 1:9–27.
Chomsky, Noam. 2013. Problems of projection. *Lingua* 130:33–49.
Clarke, Julia. 2013. Feathers before flight. *Science* 340:690–692.
Corballis, Michael C. 2002. *From hand to mouth: The origins of language.* Princeton: Princeton University Press.
Fitch, W. Tecumseh. 2005. The evolution of language: A comparative review. *Biology and Philosophy* 20:193–230.
Fujita, Koji. 2014. Recursive merge and human language evolution. In *Recursion: Complexity in cognition*, ed. Tom Roeper and Margaret Speas, 243–264. New York: Springer.
Greenfield, Patricia M. 1991. Language, tools, and brain: The ontogeny and phylogeny of hierarchically organized sequential behavior. *Behavioral and Brain Sciences* 14(4):531–595.
Greenfield, Patricia M. 1998. Language, tools, and brain revisited. *Behavioral and Brain Sciences* 21(1):159–163.
Grodzinsky, Yosef. 2004. Variation in Broca's region: Preliminary cross-methodological comparisons. In *Variation and universals in biolinguistics*, ed. Lyle Jenkins, 171–193. New York: Elsevier.
Hauser, Marc. D., Noam Chomsky, and W. Tecumseh Fitch. 2002. The faculty of language: What is it, who has it, and how did it evolve? *Science* 298:1569–1579.
Herman, Louis M. 2010. What laboratory research has told us about dolphin cognition. *International Journal of Comparative Psychology* 23:310–330.
Jackendoff, Ray. 2010. Your theory of language evolution depends on your theory of language. In *The evolution of human language: Biolinguistic perspectives*, ed. Richard K. Larson, Viviane Déprez, and Hiroko Yamakido, 63–72. Cambridge, UK: Cambridge University Press.
Marantz, Alec. 1997. No escape from syntax: Don't try morphological analysis in the privacy of your own lexicon. *University of Pennsylvania Working Papers in Linguistics* 4(2):201–225.

Miyagawa, Shigeru, Shiro Ojima, Robert C. Berwick, and Kazuo Okanoya. 2014. The integration hypothesis of human language evolution and the nature of contemporary languages. *Frontiers in Psychology* 5: Article 564.

Pigliucci, Massimo, and Gerd B. Müller. 2010. Elements of an extended evolutionary synthesis. In *Evolution – the extended synthesis*, ed. Massimo Pigliucci and Gerd B. Müller, 3–17. Cambridge, MA: MIT Press.

Stoop, Ruedi, Patrick Nüesch, Ralph Lukas Stoop, and Leonid A. Bunimovich. 2013. At grammatical faculty of language, flies outsmart men. *PLoS One* 8(8): e70284.

Stout, Dietrich. 2010. Possible relations between language and technology in human evolution. In *Stone tools and the evolution of human cognition*, ed. April Nowell and Iain Davidson, 159–184. Boulder, Colorado: University Press of Colorado.

Taylor, Alex H., Douglas Elliffe, Gavin R. Hunt, and Russell D. Gray. 2010. Complex cognition and behavioural innovation in New Caledonian crows. *Proceedings of the Royal Society B* 277:2637–2643.

Taylor, Alex H., Brenna Knaebe, and Russell D. Gray. 2012. An end to insight? New Caledonian crows can spontaneously solve problems without planning their actions. *Proceedings of the Royal Society B* 279:4977–4981.

# 10 Biological pluralism in service of biolinguistics*

*Pedro Tiago Martins, Evelina Leivada,*
*Antonio Benítez-Burraco and Cedric Boeckx*

## 1 Introduction

The first sign of a biological orientation for the study of language was the work of Noam Chomsky and Eric Lenneberg, among just a few others, who in the 1950s and 1960s rejected the structuralist linguistics of the time, believing instead that languages, although meticulously described, were not explained as a natural phenomenon. The overarching assumption of their work is that languages are not learned in the conventional sense of the term (i.e. the way one would learn a craft or how to play a musical instrument), but rather a product of a biologically determined and biologically constrained capacity of humans, located in the brain, which must be innate. This biologically determined capacity is considered to be the main focus of Generative Linguistics, and all efforts carried out within this approach since its inception are said to ultimately contribute to its study. Later, mainly in the 1970s, various interdisciplinary meetings were held, and the term "biolinguistics" was ultimately chosen as the name of the enterprise that arose in those discussions (Piattelli-Palmarini, 1974). Many linguists have indeed adhered to this conception of the field, which is something that, one would expect, must result in important insights, after around five decades of intensive research. However, a brief, random survey of the thousands of papers, chapters and books on generative linguistics will reveal a pattern: it is customary to start with a mention of the biological character of language, attributable to the genetic endowment of humans, and to convey the intention of approaching it as such, but the biological jargon is soon diluted in formal linguistic analyses as the sections unfold, with a possible reprise in the conclusion. Upon close inspection, one concludes that the larger part of the issues that most generative linguistics work covers are philological in character, albeit through the use of sophisticated tools and notation.[1] Thus, a very important (and unfortunate) realization when looking at the generative linguistics literature is that its main premise (that language is a biological property of humans) does not entail, guide or constrain linguistic research in any meaningful way. In other words, if that premise were not held, the import of most linguistic work would remain largely unaffected, which is quite odd, for that premise is, again, one of the main tenets of generative linguistics. This apparent lack of interest in the biological half of biolinguistics is one of two problems regarding how investigations into the nature of the language faculty have been carried out. The second, related problem is the conception of biology itself that has served as the (rhetorical) backbone of some of these investigations. We will discuss the treatment that notions like novelty and variation have received in the linguistic literature, sometimes under the "biolinguistics" rubric, and offer some insights and counter-evidence from evolutionary biology, in favor of a biologically informed study of language.

## 2 Novelty: the case of language

Circumstancial evidence uncontroversially suggests that language is unique to humans. While other – if not most – species display some system of computation and/or communication, it is observed that none of these systems come close to human language as a whole, and that language has a lot to do with why humans have thrived.[2] All serious linguistic traditions have held this assumption, and the Chomskyan tradition has taken it to be not only an assumption but also its main focus of inquiry (which we, as stated above, do not think has resulted in new insights, despite the good intentions.)

One of the issues frequently discussed when it comes to language is domain specificity. Apart from references to the notion of "genetic blueprint", which are based on a naive and superficial understanding of biology, upon closer inspection of the "biolinguistic" literature of the last decade or so, one also finds references to evolutionary biology and Evo-Devo. This is rather odd, as linguists seem to claim, on the one hand, that language is acquired virtually instantaneously, yet, on the other, they often claim to assume a biology in which development is important. The discussion of the contributions of both genes and developmental processes to the linguistic phenotype is a very interesting endeavor, but, alas, it is rarely pursued in linguistics. That linguists make references or are even aware of Evo-Devo literature is a virtue, but most times it is hard to see exactly which Evo-Devo linguists have in mind. The Evo-Devo literature teaches us a lesson which is more consonant with the ideas of Lenneberg, but which is actually quite incompatible with what linguists seem to need to back up their claims about the nature of language, biologically speaking. The illustration we choose to make this general point comes from a famous and influential paper published by Hauser et al. (2002).

One of the most – if not the most – well-known notions that have emerged from that paper is the distinction between the Faculty of Language in the Narrow Sense (FLN) and Faculty of Language in the Broad Sense (FLB).

This distinction has actually been taken as foundational in the field of biolinguistics, or at least in the field of linguistics when the biological foundations of language are the concern. We will, however, try to show that from a biological perspective this distinction is not useful. This should already be apparent, given the contrast between the great deal of attention it has received and the lack of results that have come out of it.

Evo-Devo is concerned with "organismal form, shape, morphological structure and the generative mechanisms underlying their evolution" (Müller, 2005). These are the main topics that an Evo-Devo researcher ultimately wants to address and understand. Regarding the generative mechanisms underlying the evolution of structures, Evo-Devo practitioners have stressed the importance of understanding the origins of novelty, both in the context of development, ontogeny and evolution, phylogeny. One of the central issues of current Evo-Devo is thus what explains the emergence of radically novel structures. As central as it might be, however, this issue is largely unresolved:

> [W]hile biologists have made great progress over the past century and a half in understanding how existing traits diversify, we have made relatively little progress in understanding how novel traits come into being in the first place.
> (Moczek, 2008: 432)

In fact, the issue of novelty was disregarded in the context of the Modern Synthesis. Echoing the general sentiment of the biologists of time, Ernst Mayr said the following:

> The problem of the emergence of evolutionary novelties consists in having to explain how a sufficient number of small gene mutations can be accumulated until the new structure becomes sufficiently large to have selective value.
> (Mayr, 1960: 357)

Current Evo-devo tries to bring back this issue into the central fold of biology. It is one thing to address the diversification of something already present, and it's another thing to find out about how that something got there in the first place. One of the things Evo-Devo quickly learned is that the extraordinary morphological diversity that exists at the level of organisms and their parts is not paralleled by corresponding diversity in genetic and developmental mechanisms. This has led to the idea that the mechanisms of development and also genetic circuits involved are highly conserved across species. Nonetheless, the phenotypes at the end product are quite different. At the lower level of genetics or maybe even development one finds something that's highly conserved and not specialized.

Müller and Wagner (1991) made an important contribution to Evo-Devo by defining a novelty as a structure as follows: "[a] morphological novelty is a structure that is neither homologous to any structure in the ancestral species nor homonomous to any other structure of the same organism" (p. 243). This definition is quite significant, and aware readers should recognize it as a generalization of what Hauser et al. said in their famous paper about language. Even though they did not mention the Evo-Devo literature, FLN was defined by Hauser et al. (2002) as something specific to language and specific to humans, something which echoes precisely the definition of Müller and Wagner (1991) of a novel trait, applied to language. In this sense, the component(s) of language they want to highlight are a very clear case of a novel trait.

However, we must point out, biologists nowadays find it extremely difficult to regard traits as novel if Müller and Wagner's (1991) distinction is taken seriously. From a reasonably representative set of examples, Moczek (2008) is left with two that conform to it: butterfly wing patterns which are unique to *Lepditoptera* (Nijhout, 1991), and firefly lanterns and their lighting patterns, which are not present in any other insect or arthropod (Lloyd 1983). It seems that the more we fine-tune our criteria for novelty on the basis of our knowledge of modern evolutionary biology, the shorter the list of possible candidates becomes.

Hauser et al. (2002) continued a trend that Chomsky has been famous for, namely that there is something highly specific to language and humans. In biological terms, the claim that a trait is unique to a species amounts to a claim for its novelty, especially if the trait in question is thought to be unprecedented in nature. What's interesting to our discussion is that, according to the authors, the real motivation for the FLN/FLB distinction was an attempt to make research advance:

> Linguists and biologists, along with researchers in the relevant branches of psychology and anthropology, can move beyond un-productive theoretical debate to a more collaborative, empirically focused and comparative research program aimed at uncovering both shared (homologous or analogous) and unique components of the faculty of language.
> 
> (Hauser et al. 2002: 1578)

We contend that this distinction actually leads to unproductive theoretical debates. Hauser et al. (2002) are correct in that it seems that some subset of the mechanisms of FLB are both unique to humans and language itself: that's the novelty aspect. The exact characterization of what constitutes FLN, however, has not been very precise. In line with the Chomskyan tradition, FLN has received the most attention in the linguistics literature, despite the vagueness of the characterization provided in the paper, as illustrated by the following passages, all taken from neighboring pages of Hauser et al. (2002):

> We hypothesize that FLN only includes recursion and is the only uniquely human component of the faculty of language.
> 
> (p. 1569)

> We assume [ . . . ] that a key component of FLN is a computational system that generates internal representations and maps them into the sensory-motor interface by the phonological system, and into the conceptual-intentional interface by the (formal) semantic system. [ . . . ] All approaches agree that the core property of FLN is recursion.
> 
> (p. 1571)

> In fact, we propose in this hypotehsis that FLN comprises only the core computational mechanisms of recursion as they appear in narrow syntax and the mappings to the interfaces.
> 
> (p. 1573)

> At minimum, then, FLN includes the capacity of recursion.
>
> (p. 1571)

The exact content of FLN is not of immediate interest for the aims of this paper, but still some brief considerations can be made about Hauser et al.'s putting their money on recursion and what that means for their distinction. If recursion is the content of FLN, this means that it is uniquely human and recently evolved. The contents of FLB, on the other hand, are expected to have a long, cross-species, traceable evolutionary history. Hauser et al. (2002) offer a list of possible experimental studies which might shed some light on what FLB and FLN are and where they came from, the replication and extension of which ultimately proves their hypothesis right or wrong. This means that, in principle, it can turn out that recursion is present in other species, and thus their hypothesis is wrong. This would be satisfactory, were it not for the fact that FLN is equated with recursion virtually as part of the hypothesis. FLN meaning recursion is in principle something that can be studied under a comparative method, assuming that recursion is a well understood and stable notion (cf. Fitch, 2010b); one must look at other species and look for recursion, experimentally proving or disproving it as what constitutes FLN. However, FLN as the set of what enters into the faculty of language and is specific to humans and to language, independently of what constitutes it – which is how Hauser et al.'s rhetoric goes – leaves no room for comparative inquiry and is not conceptually sound. It seems that FLN makes sense only insofar as it is defined after its content is determined. FLB does not present such a problem, since its definition does not clash with the comparative method like FLN's does. FLB as the set of what enters into the Faculty of Language which is neither specific to language or to humans, unlike FLN, readily opens way for inquiry in other species and domains. It seems, then, that, conceptually, it would make more sense to assume that recursion is part of FLB (being something that enters into language) and look for it in other species like the FLB component it is. Under a biological perspective, with no assumptions about what components enter into each of the senses of FL, it could be said that FLN and FLB make sense only as labels that pertain to the organization of the putative components of language, with no reference to their exact content.

In one of their later articles, the same authors (Fitch et al. 2005) make similar claims:

> It seems likely that some subset of the mechanisms of FLB is both unique to humans, and to language itself. We dub[bed] this subset of mechanisms the faculty of language in the narrow sense (FLN).
>
> (p. 1571)

Witness, however, the following passage from Fitch (2010a, 384):

> What all of these examples make clear is that the distinction between general and linguistically specialized mechanisms is hard to draw, even in those cases

where the mechanisms seem fairly clearly defined. Most areas of language are not, and will not soon be, so clearly defined, and thus the distinction itself is of little use in furthering our understanding of the mechanisms.

Fitch was one of the authors of Hauser et al. (2002), and his updated take on the issue presents us with two possible scenarios: either he is contradicting himself, or he has abandoned the distinction, which goes to show that the main motivation for its being proposed in the first place has proved unproductive with time.

Fitch et al. (2005: 182) had already hinted at this possibility in a passage whose importance has not been given the attention it deserves:

> Something about the faculty of language must be unique in order to explain the differences between humans and the other animals – if only the particular combination of mechanisms in FLB.

Actually, Chomsky himself had voiced this idea before the Hauser et al. (2002) paper, the most recent instance of it being perhaps the following passage:

> Now a question that could be asked is whether whatever is innate about language is specific to the language faculty or whether it is just some combination of the other aspects of the mind. That is an empirical question and there is no reason to be dogmatic about it; you look and you see. What we seem to find is that it is specific.
>
> (Chomsky 2000)

These rare nods in the biolinguistic literature seem to confirm the conclusion of Bloomfield et al. (2011), who say that "perhaps this is a good time to reconsider whether attempting to distinguish between qualitative and quantitative differences is helpful if the quantitative advantage is vast" (p. 948). Here the authors allude to a point that Lenneberg (1967) had already made: perhaps we should recognize that general mechanisms could be at the heart of something highly specific in terms of behavior. This doesn't seem to be a point that Hauser, Chomsky and (at least initially) Fitch appreciated. The debate between the authors and Ray Jackendoff and Steven Pinker shows as well that the distinction isn't particularly productive. Jackendoff and Pinker (2005) were right to point out that "the narrow/broad dichotomy [ . . . ] makes space only for completely novel capacities and for capacities taken intact from non-linguistic and nonhuman capacities, omitting capacities that may have been substantially modified in the course of human evolution" (p. 224). More generally, we believe Jackendoff and Pinker (2005) are right in demurring "from some of [Hauser, Chomsky and Fitch's] dichotomies, which prejudge the issues by making some hypotheses – in our view the most plausible ones – impossible to state" (p. 224). These are the hypotheses that Lenneberg, incidently, would have favored. For Hauser et al., it is crucial that FLN be a subset of FLB structures, and not a superset. In other words, Hauser et al. can't assume Lenneberg's suggestion that when there are various general mechanisms becoming integrated with one another, a novel structure can actually arise, simply as the result of interactions of shared mechanisms. Their hypothesis is simply not possible to state in light of the FLN/FLB distinction, as it would require FLN to be a superset, rather than a subset, of FLB. Thus, we conclude, much like we think Fitch has, although it is not clear that the (bio)linguistic community has followed, that the FLN/FLB distinction isn't particularly productive.

When Chomsky (2000) acknowledges that we must "look and see", perhaps he is being too simplistic, but certainly he has been consistent in that we seem to find that whatever is innate about language is specific, that is, that it is *sui generis*. It is indeed an empirical question, but biology strongly suggests that the only way to get novelty is through combinations of other aspects of the traits in question. In this case, the mental organ we call language. Here we find an important aspect of current biolinguistic thought, one that's been preserved for 40 years: we seem to want something highly specific right from the beginning, for example in the genes, but that's just not how novelty arises. In the words of West-Eberhard (2005), "phenotypic novelty is largely reorganizational rather than a product of, say, innovative genes." This seems to be a lesson that linguists haven't assimilated from biology. So even though we find a lot of references to the Evo-Devo literature, one of the central issues, namely novelty, has yet to be assimilated. This is one aspect of how biologically illiterate linguists have been for 40 or 50 years. If language researchers are serious about having a biological object of inquiry, they should pay more than lip service to biology.

## 3 Rethinking the I-language/E-language distinction

More foundational than the FLN/FLB dihotomy, the distinction between Internal language (I-language) and External language (E-language) is probably the most famous in modern linguistics. It was explicitly introduced by Chomsky (1986), and since then it has become a rule of thumb of generativism. I-language refers to the speaker-internal linguistic knowledge that reflects competence in a given language, while E-language refers to the socio-cultural construct, that is, one of the possible materializations of an I-language. Chomskyan linguistics is said to be focused exclusively on I-language; this has been useful in the sense that it ignores the messy concept of development that arises once one goes beyond the genes, and it fits nicely with Chomsky's (1955/1975) idea that acquisition is instantaneous. The problem is that not only linguists hold a naive version of genetics and biology in general, but biology is much more than genes. The fact that linguists have equated I-language with genetic endowment, on the one hand, and E-language with all things non-genetic, on the other, had the goal of circumscribing their object of study, but it has resulted only in dividing something that cannot be divided: both genes and development are crucial for the characterization of language, and the interaction of all kinds of factors is what creates the phenotypes that linguists attribute to genes (and to make matters worse, they do so in a simplistic and implausible fashion). There have been some recent attempts at reconciling internalism and externalism, suggesting that the two are mutually reinforced (Lassiter 2008, Mondal 2012), but the I/E distinction is undoubtedly a sharp one.

Contrary to what one observes in linguistics, in biology the link between the genetic makeup of an organism and the environmental factors that affect its development is exploited through the study of its phenotype. This should be true also in the case of language. In the words of Lewontin (2000: 28, emphasis added), "human beings can speak because they have the right genes *and* the right environment." This is the message of the Extended Synthesis: genes determine the capacities of organisms, yet the limits of these capacities may never be explored, depending on how adequate the environment factor eventually proves to be. An important principle in biology is Reciprocal Causation (Mayr 1961), by which an action is simultaneously cause and effect. In the case of language, reciprocal causation refers to the area of intersection behind the terms I-Language and E-language, which reflects the point at which the development of biological traits (I-properties) is affected by environmental, external triggers (E-factors). This kind of interaction ties in with the evolutionary process of niche construction, whereby organisms partly determine the selective pressures they undergo and, in a certain sense, build their own environment (Oyama et al. 2001, Robert 2004).

Linguistic data are insufficient to account for or facilitate a clear-cut distinction between I-language and E-language, which linguistis base their theories on. This insufficiency is due to the fact that both I-language and E-language are brought *together* behind the data.

In the case of language, a biolinguistic interpretation of some characteristics of instances of recent language emergence can be used to illustrate the complex dynamics between internal and external factors. There are indeed cases of languages which are still at an early stage of development, and as such have not undergone all the developmental stages which result in the end product one would expect. One such case is Al-Sayyid Bedouin Sign language (ABSL), a language that emerged in the last 70–75 years in an isolated community in Isreal, now in its third generation of signers. The presence of a gene for non-syndromic, genetically recessive, profound pre-lingual neurosensory deafness (Scott et al. 1995), along with consaguineous marriage patterns have resulted in the birth of a relatively large population of deaf individuals in a short time frame (Sandler et al. 2011). Fieldwork on ABSL suggests that even properties traditionally treated as design characteristics of language (cf. Hockett 1960) emerge as a response to environmental, externalization-related factors. These *ab initio* absent properties include:

- Signifier-signified consistency (the sound/sign and meaning pairings are synchronically stable within E-languages, disallowing inter- or intraspecific variation in general): Studies on ABSL and other languages show that there is an absence in signifier-signified consistency. Meir et al. (2010) for

ABSL, Senghas (1997) for Nicaraguan Sign Language and Washabaugh (1986) for Providence Island Sign Language all give similar reports on how consistency improves over new generations of speakers, reflecting environment needs.

- Grammaticalization (process whereby lexical items lose some of their phonological substance and/or semantic specificity and instead develop finer morpho-syntactic functions): ABSL first-generation signers have the tendency to break an event that requires two arguments into two clauses, each with its own verb sign, which predicates a different argument. Languages take time to develop fine-grained grammatical markers, such as the ones that facilitate distinguishing between the subject and the object phrases in a clause.
- Complexity: grammaticalization is one of the ways of enhancing grammatical complexity in language. Studies of ABSL report a gradual emergence of complexity in prosodic and syntactic structures (Sandler et al. 2005, 2011). The differences observed with respect to the time it takes for more fine-grained grammatical markers to develop in different communities suggest that the time factors should be viewed not just on their own, but as part of a cluster of factors which trigger adaptation depending on environmental needs.

If we take environmental factors into account, there are some lessons to derive. If it is correct that grammatical markers are of an emergent nature, this amounts to emergent 'parametric' variation, in the Chomskyan sense. If the goal is to reduce the role of genetic endowment (Chomsky 2007), analogies have to be drawn with the right kind Evo-Devo. Instead of a genocentric perspective that seems to dominate linguistics, attention should be paid to more permissive and wide-ranging frameworks, such as Developmental System Theory (DST) (Oyama 1985; Griffiths and Gray 1994). In the words of Benítez-Burraco and Longa (2010: 318):

> Development does not entail any kind of pre-existing genetic program; genes are not the source of the form. Quite the opposite: genes are just one of many developmental sources. Therefore, DST rejects the idea that genes are endowed with any special directive power. The main notion of DST is that of 'developmental system', which is to be understood as the overall collection of heterogenous influences on development.

It's becoming apparent within Evo-Devo that it is not possible to distinguish relevantly between the influence of the genes and the influence of the environment in development, since the end product is the result of the interaction of the information from both levels. In light of Evo-Devo, few dychotomies (e.g. I-Language/E-Language, Nature/Nurture, FLN/FLB, gradualism/saltationism and even adaptation/exaptation) make perfect sense.

## 4 Biological insights

In this section we shall discuss some of the ways in which one can have a fresh look at language, by learning from and applying some insights from modern evolutionary biology. In doing so, one expects to do away with conceptions of biology and language itself of the kind discussed above that have plagued linguistics and prevented a real, biologically informed study of language. We will take as the start of our discussion a well-known example of an attempt to bioligize language that presents some of these problems, and from there offer examples of work that has taken an extra step and incorporated important lessons from evolutionary biology and allied disciplines.

Anderson and Lightfoot (2000) have a conception of environment as only "linguistic environment", that is, experience from which generalizations about a particular language are extracted, while the biological part of the equation is left to the genes, whose job will be to somehow encode language:

> [ . . . ] language emerges through an interaction between our genetic inheritance and the linguistic environment to which we happen to be exposed. English-speaking children learn from their environment that the verb *is* may be pronounced [iz] or [z], and native principles prevent the reduced form from ocurring in the wrong place. Children learn from their environment that *he*, *his*, etc. are pronouns, while native principles entail where pronouns may not refer to a preceding noun. The interaction of the environmental information and the native principles accounts for how the relevant properties emerge in an English-speaking child.
>
> (p. 6)

From this passage one also gathers a secondary but related claim, namely that genes are solely responsible for all that is innate. Thus, we have two problems that characterize the treatment of biology in linguistics: that nativism and geneticism are one and the same, and that genes encode the principles that constrain final linguistic structures.

If, however, we look at the modern biological literature – which we ought to – we will find quite a different picture, one in which genes are but one part of the evolution of traits and organisms.

There is no clear bridge between genotypes and phenotypes. Here the notion of phenotypic plasticity is key, whereby organisms adapt to changes in the environment (here understood in a very broad sense), in such a way that their behavior, morphology and physiology effectively change as well (Price et al. 2003, West-Eberhard 2003). This is a very important factor for the successful adaptation of organisms to variations in the environment, and it's modernly understood as including any type of change induced by it (Kelly et al. 2012). During the heyday of the Modern Synthesis in biology, the study of phenotypic plasticity was seen as more of an obstacle than anything else, which is not something very suprising at a time when genes seemed to have all the answers, and any hint of influence of the environment was considered a problem (Falconer 1952). It seems that linguists have kept to this conception of the genes/environment interplay, but now we know better and should abandon it. As documented by Pigliucci (2001), interest in the virtues of phenotypic plasticity has increased greatly, and it seems that it is getting progressively more difficult to look at evolution without taking the environment and its relation with the genotype into account. Taking as examples virtually any species, from plants to mammals, the consensus seems to be that phenotypic plasticity can be seen as both something that can itself evolve and something that can guide evolution. This guidance may even take precedence over genetic change (West-Eberhard 2003, 2005, Lenski et al. 2006)

Moreover, even assuming this, genes themselves cannot serve as a diagnostic for what kind of lingusitic phenotype a subject ends up with. For example, mutated versions of related genes can result in different disorders or lack thereof in different populations (Slate 2011, Benítez-Burraco 2012).

Another way in which linguists often rely on well-established yet rarely scrutinized ideas (in the linguistic literature) is in their treatment of the brain, namely, the so-called language areas. Witness Anderson and Lightfoot's (2000) claim that "[ . . . ] even if it were to become clear that there is no clear segregation between language-related and non-language-related brain tissue, it would still be useful and important to treat the language capacity as a discrete and specifiable human biological system in functional if not anatomical terms [ . . . ]" (p. 19). We find this quite untenable, for there is no reason to believe that language is *located* or is processed in dedicated areas of the brain. As Poeppel (2008) has argued, there is no logical entailment between the localization of any one brain area and explanation of its function. Lenneberg (1967) already had some intuitions of this sort: "as biologists, we cannot discern meaning or purpose of specific anatomical developments" (p. 33). Moreover, the mapping of the brain is based on cognitive tasks which are understood on the basis of units that are not always compatible with what is known about the kind of units the brain operates, and this is very much apparent in the case of speech and language, where the cognitive units proposed are of a circumscribed theoretical nature and have no bearing on the kind of computations that, as far as research has shown, the brain performs. In this regard, it is also important to take into account the fact that idealized, well-defined brain areas do not reflect reality; instead, brain areas vary across the population, and even within the same subject as growth unfolds (Prat and Just 2011). This is an instance of Poeppel and Embick's (2005) Granularity Mismatch Problem, a problem which until resolved will render any map of the brain only partially informative (Poeppel 2012). Thus it seems that any attempt at making progress in relation to Lenneberg's time must include the decomposing of cognitive functions, such as language, into much wider-ranging principles, which can be generalized across cognitive domains (and species). Cross-modularity seems to be what best describes language in the brain, as a result of interacting, non-specific brain structures (Griffiths 2007).

A further lesson one can take from biology is that, much like brain areas, there is no real indication that the faculty of language, the linguistic phenotype, is uniform. Even though the notion of a fixed, shared faculty of language has allowed linguistics to surpass some conceptual barriers and isolate their object of study, we are now in a position to assess this claim and qualify it. The fact that different modalities may co-exist in the same subject (Emmorey and McCullogh 2009), and that the timing of the steps that lead to the acquisition of language is not universal (Bates et al. 1988, Dehaene et al. 1997), shows that the faculty does not unfold in a deterministic manner, unaffected by several other factors. The conceptual opposite also seems to be true: comparable phenotypes can arise out of different brain structures (Karmiloff-Smith 2010), which plausibly means that, the same way a specific brain architecture cannot be a diagnostic for a specific linguistic phenotype, a specific linguistic phenotype is also not a diagnostic for a specific brain architecture.

This largely cross-modular, reorganizational notion of the linguistic phenotype has recently been explored by Balari and Lorenzo (2013), who, inspired by Alberch's (1989) notion of phenotypic space, describe what they call the computational phenotype. According to these authors, a preexisting, non-specific computational device was recruited for language, which was then associated with a lexicon (a dictionary of symbolic units) and some means of externalization, all of which are plausibly traceable in evolutionary history. The latter – the externalization component – of language is an especially fruitful object of research these days (see, for example, Fitch 2010a). This reorganization would be a consequence of the increase in brain size of humans.

With a small adaptation, the general idea in Balari and Lorenzo (2013) is in line with Boeckx and Benítez-Burraco (2014), who put forward the hypothesis that not necessarily brain size, but rather the human, globular brain-case allowed for an expanded neuronal workspace, opening way for improved (more so than new) connections, namely those that plausibly take an existing computation device and project it across other modules. Both hypotheses challenge the notion of the language faculty as a case of novelty, which Hauser et al. (2002) defended with their notion of FLN, already discussed above.

## 5 Conclusion

Biolinguistics as a field is not easy to define. Different people disagree on some very foundational issues, and yet they can still consider themselves to be biolinguists. Many do biolinguistics without ever having heard of it. Other people even work on what we would like to call biolinguistics, yet they would rather avoid the term. A textbook on biolinguistics – the same way we have textbooks on physics or linguistics proper – would be impossible to put together. There is no consensual body of work that could be passed along as the canon. Still, we believe we can start looking ahead in search of a biolinguistics that does away with ideas that have in the meantime proven inadequate and bring in new ideas that we feel are on the right track. Thus, what we propose is a re-hauled biolinguistics, with the same spirit of what Chomsky and Lenneberg began, but with related fields as real allies, rather than just neighbors. We believe that actually integrating insights and results from the biological sciences broadly speaking, of the kind we have reviewed in the previous sections, will result not only in new discoveries about language, but also in the definition as a field that biolinguistics currently lacks.

For this goal to be achieved – and we believe that the change that might lead to it is already, albeit slowly, underway – we must start by building our linguistic theories on top of biology. The lack of biological constraining of linguistics we allude to in the first section is a real problem, and one that is not always recognized. Linguists often take shelter behind logical soundness alone, and find it satisfactory that theories and data fit together. This practice is fine as far as language(s) description is concerned, but if we believe that what we are after are the biological properties of the *language faculty*, then biology must have a very important regulatory role.[3] Thus, the notion of plausibility must first and foremost be understood as biological plausibility, and, if this is the case, typological plausibility will naturally follow. This notion must rest on an understanding of genes. Contrary to what has been claimed, a concrete, deterministic linguistic genotype, which for some is the meaning of Universal Grammar (e.g. Anderson and Lightfoot 2000), is not plausible.

Genes alone do not define a trait or how it is used. Again, the notion of phenotypic plasticity is key: the degree to which environmental choices affect the way that genes are expressed depends on the specific genotype-environment interaction in each case, for each trait. If we pay attention to the literature, we find that even in fairly "linguistic" studies, the idea that the environment is also a big player has been recently exploited, which goes to show the environment need not be a taboo for those concerned with the biological properties of language. For example, Lupyan and Dale (2010) put forward what they call the Linguistic Niche Hypothesis. Having conducted a statistical analysis of over 2,000 languages, their results suggest that language structures are influenced by the environment, just as biological organisms are shaped by ecological niches. Similarly, Wray and Grace (2007) argue that the nature of the communicative context affects the structure of language. According to their proposal, esoteric communication allows for grammatical and semantic complexity, whereas exoteric communication leads language towards rule-based regularity and semantic transparency.

More often than not, a big concern of linguists is whether a particular language will display (or be analyzed in such a way that it seems to display) a property which their theory of choice does not allow or cannot account for. When that happens, the goal then becomes to offer an alternative analysis of that property or to reformulate the theory so as to account for it or disprove it. Much of the literature of modern linguistics has resulted from this practice, and in most cases allusions to biology do little more than perpetuate either simplistic claims that we now know are false, or the idea that we don't really know how it works, and that as such we should indeed focus on proving and disproving linguistic phenomena. This back-and-forth might seem to bear winners and losers, but we contend that the battle we should pick is a different one: the quest for the biological underpinnings of the language faculty. Linguistics – be it innocently or negligently – has not yet embraced biology, and in some ways it has even hindered real progress in the study of the biological foundations of the language faculty. Shifting away from all-or-nothing, genocentric, reductionist approaches to biology of language, and instead embracing its multi-dimensional character, which encompasses genetic, developmental and environmental factors, will allow us to purse the core questions that started the bioloquistic enterprise more than 50 years ago and that we ought to be addressing if we are to unveil what's behind the uniqueness of our species.

## Notes

* This chapter is based on separate presentations by the authors in 2013, at the International Conference on Evolutionary Patterns in Lisbon. We thank the audience there for comments and discussion.
1 We are aware that ascribing a philological character to work in generative linguistics causes discomfort and even confusion among its practitioners for historical reasons, but we lack a better word for characterizing work on the particularities of specific languages, which is not what generative linguistics is supposed to be concerned with.
2 Of course, statements of this kind should not be taken as qualitative, that is, the sophistication attributed to human language does not imply a higher degree of biological sophistication in comparison with other species and traits. Rather, it just so happens that language – whatever it is – has given humans behavioral advantages as individuals and as groups which have allowed them to become a privileged, sophisticated species.
3 This is not to say that biologists have fully assimilated all the important insights from the linguistics literature and that linguists have not returned the favor; very often, one finds in the biology literature conflations of notions that linguists have learned to tell apart unequivocally since decades ago (say, communication and language).

## References

Alberch, Pere. 1989. The logic of monsters: Evidence for internal constraint in development and evolution. *Geobios* 22. 21–57.
Anderson, Stephen R & David W Lightfoot. 2000. The human language faculty as an organ. *Annual Review of Physiology* 62(1). 697–722.
Balari, Sergio & Guillermo Lorenzo. 2013. *Computational phenotypes: Towards an evolutionary developmental biolinguistics*, vol. 3. Oxford: Oxford University Press.
Bates, E, I Bretherton & L Snyder. 1988. *From first words to grammar: Individual differences and dissociable mechanisms*. Cambridge, UK: Cambridge University Press.
Benítez-Burraco, Antonio. 2012. Problematic aspects of the genetic analysis of the specific disorders of the language: Foxp2 as a paradigm. *Neurologia* 27. 225–233.
Benítez-Burraco, Antonio & Víctor M Longa. 2010. Evo-devo – of course, but which one? some comments on chomsky's analogies between the biolinguistic approach and evo-devo. *Biolinguistics* 4. 308–323.

Bloomfield, Tiffany C, Timothy Q Gentner & Daniel Margoliash. 2011. What birds have to say about language. *Nature Neuroscience* 14(8). 947.
Boeckx, Cedric & Antonio Benítez-Burraco. 2014. The shape of the human language-ready brain. *Frontiers in Psychology* 5. 282.
Chomsky, Noam. 1955/1975. *The logical structure of linguistic theory.* New York: Plenum Press.
Chomsky, Noam. 1986. *Knowledge of language.* New York: Praeger.
Chomsky, Noam. 2000. *The architecture of language.* Oxford: Oxford University Press.
Chomsky, Noam. 2007. Approaching UG from Below. In U. Sauerland and H.-M. Gärtner (eds.), *Interfaces + Recursion = Language?*, 1–31. Berlin: Mouton de Gruyter.
Dehaene, S, E Dupoux, J Mehler, L Cohen, E Paulesu, D Perani, P F van de Moortele, S Leh´ericy & D Le Bihan. 1997. Anatomical variability in the cortical representation of first and second language. *Neuroreport* 8. 3809–3815.
Emmorey, K & S McCullogh. 2009. The bimodal bilingual brain: Effects of sign language experience. *Brain Language* 109. 124–132.
Falconer, Douglas S. 1952. The problem of environment and selection. *American Naturalist 86 (830)*, 293–298.
Fitch, W Tecumseh. 2010a. *The evolution of language.* Cambridge, UK: Cambridge University Press.
Fitch, W Tecumseh. 2010b. Three meanings of "recursion": Key distinctions for biolinguistics. In R K Larson, Viviane D´eprez & Hiroko Yamakido (eds.), *The evolution of human language: Biolinguistic perspectives*, 73–90. Cambridge, UK: Cambridge University Press.
Fitch, W Tecumseh, Marc D Hauser & Noam Chomsky. 2005. The evolution of the language faculty: Clarifications and implications. *Cognition* 97. 179–210. doi:10.1016/j.cognition.2005.02.005.
Griffiths, Paul E. 2007. Evo-devo meets the mind: Towards a developmental evolutionary psychology. In R Brandon & R Sansom (eds.), *Integrating evolution and development: From theory to practice*, 195–225. Cambridge, MA: MIT Press.
Griffiths, Paul E & Russell D Gray. 1994. Developmental systems and evolutionary explanation. *The Journal of Philosophy* 91(6). 277–304.
Hauser, Marc D, Noam Chomsky & W Tecumseh Fitch. 2002. The faculty of language: What is it, who has it and how did it evolve? *Science* 298. 1569–1579.
Hockett, Charles Francis. 1960. The origin of speech. *Scientific American* 203. 88–96.
Jackendoff, Ray & Steven Pinker. 2005. The nature of the language faculty and its implications for the evolution of language. *Cognition* 97, 211–225.
Karmiloff-Smith, Annette. 2010. A developmental perspective on modularity. In Britt Glatzeder, Vinod Goel & Albrecht Mu¨ller (eds.), *Towards a theory of thinking*, 179–187. Berlin: Springer.
Kelly, Scott A., Panhuis, Tami M., Stoehr, Andrew M. (2012). Phenotypic Plasticity: Molecular Mechanisms and Adaptive Significance. *Comprehensive Physiology* 2: 1417–1439.
Lassiter, Daniel. 2008. Semantic externalism, language variation, and sociolinguistic accommodation. *Mind & Language* 23(5). 607–633.
Lenneberg, Eric H. 1967. *Biological foundations of language.* New York: Wiley.
Lenski, Richard E, Jeffrey E Barrick & Charles Ofria. 2006. Balancing robustness and evolvability. *PLoS Biology* 4(12). e428.

Lewontin, R. 2000. *The triple helix: Gene, organism, and environment.* Cambridge, MA: Harvard University Press.

Lloyd, James E. 1983. Bioluminescence and communication in insects. *Annual Review of Entomolgy* 28. 131–160.

Lupyan, G & R Dale. 2010. Language structure is partly determined by social structure. *PLoS ONE* 5(1). e8559.

Mayr, E. 1960. The emergence of evolutionary novelties. In S Tax (ed.), *Evolution after Darwin (vol. i – the evolution of life)*, 349–380. Cambridge, MA: Harvard University Press.

Mayr, Ernst. 1961. Cause and effect in biology kinds of causes, predictability, and teleology are viewed by a practicing biologist. *Science* 134(3489). 1501–1506.

Meir, Irit, Wendy Sandler, Carol Padden & Mark Aronoff. 2010. Emerging sign languages. In Marc Marschark & Patricia Spencer (eds.), *Oxford handbook of deaf studies, language, and education*, 267–280. Oxford: Oxford University Press.

Moczek, Armin P. 2008. On the origins of novelty in development and evolution. *BioEssays* 30(5). 432–447.

Mondal, Prakash. 2012. Can internalism and externalism be reconciled in a biological epistemology of language? *Biosemiotics* 5(1). 61–82.

Müller, Gerd B. 2005. Evolutionary developmental biology. In F M Wuketits & F J Ayala (eds.), *Handbook of evolution, vol. 2: The evolution of living systems (including hominids)*, 87–115. Weinheim: Wiley-VCH.

Müller, Gerd B & Gunter P Wagner. 1991. Novelty in evolution: Restructuring the concept. *Annual Review of Ecology and Systematics* 22, 229–256.

Nijhout, H Frederik. 1991. *The development and evolution of butterfly wing patterns.* Washington, DC: Smithsonian Institution Press.

Oyama, Susan. 1985. *The ontogeny of information: Developmental systems and evolution.* Cambridge, UK: Cambridge University Press.

Oyama, S, P Griffiths & R D Gray. 2001. Introduction: What is developmental system theory? In *Cycles of contingency: Developmental systems and evolution*, Cambridge, MA: MIT press.

Piattelli-Palmarini, Massimo. 1974. *A debate on bio-linguistics, endicott house, dedham, ma (May 20–21, 1974).* Paris: Centre Royaumont pour une Science de l'Homme.

Pigliucci, Massimo. 2001. *Phenotypic plasticity: beyond nature and nurture.* Baltimore: Johns Hopkins University Press.

Poeppel, D. 2008. The cartographic imperative: Confusing localization and explanation in human brain mapping. *Bildwelten des Wissens* 6. 13.

Poeppel, D. 2012. The maps problem and the mapping problem: Two challenges for a cognitive neuroscience of speech and language. *Cognitive Neuropsychology* 29(1–2). 34–55.

Poeppel, David & David Embick. 2005. Defining the relation between linguistics and neuroscience. In Anne Cutler (ed.), *Twenty-first century psycholinguistics: Four Cornerstones*, 103–118. Mahwah, NJ/London: Lawrence Erlbaum.

Prat, C S & M A Just. 2011. Exploring the neural dynamics underpinning individual differences in sentence comprehension. *Cerebral Cortex* 21. 1747–1760.

Price, Trevor D, Anna Qvarnstr̈om & Darren E Irwin. 2003. The role of phenotypic plasticity in driving genetic evolution. *Proceedings of the Royal Society of London. Series B: Biological Sciences* 270(1523). 1433–1440.

Robert, J T. 2004. *Embryology, epigenesis and evolution: Taking development seriously.* West Nyack, NY: Cambridge University Press.

Sandler, Wendy, Mark Aronoff, Irit Meir & Carol Padden. 2011. The gradual emergence of phonological form in a new language. *Natural Language & Linguistic Theory* 29(2). 503–543.

Sandler, Wendy, Irit Meir, Carol Padden & Mark Aronoff. 2005. The emergence of grammar: Systematic structure in a new language. *Proceedings of the National Academy of Sciences of the United States of America* 102(7). 2661–2665.

Scott, D A, R Carmi, K Elbedour, G M Duyk, E M Stone & V C Sheffield. 1995. Nonsyndromic autosomal recessive deafness is linked to the dfnb1 locus in a large inbred bedouin family from israel. *American Journal of Human Genetics* 57. 965–968.

Senghas, Richard Joseph. 1997. *An" unspeakable, unwriteable" language: Deaf identity, language & personhood among the first cohorts of nicaraguan signers*. University of Rochester. Rochester, N.Y. Department of Anthropology dissertation.

Slate, M W. 2011. The erosion of the phenotypic specificity in psychiatric genetics: Emerging lessons from cntnap2. *Biology Psychiatry* 69. 816–817.

Washabaugh, William. 1986. *Five fingers for survival*. Ann Arbor: Karoma.

West-Eberhard, Mary Jane. 2003. *Developmental plasticity and evolution*. Oxford: Oxford University Press.

West-Eberhard, Mary Jane. 2005. Developmental plasticity and the origin of species differences. *Proceedings of the National Academy of Sciences of the United States of America* 102(Suppl 1). 6543–6549.

Wray, Alison & George W Grace. 2007. The consequences of talking to strangers: Evolutionary corollaries of socio-cultural influences on linguistic form. *Lingua* 117(3). 543–578.

# 11 On the current status of biolinguistics as a biological science*

*Masanobu Ueda*

## 1 Introduction

Generative grammar, currently known as biolinguistics, and henceforth referred to as biolinguistics, emerged as the embodiment of a naturalistic approach to the study of language in the early 1950s. Chomsky (2000: 106) contends that the naturalistic approach to the study of language and mind "considers language and similar phenomena to be elements of the natural world to be studied by ordinary methods of empirical inquiry." It is based on the assumption that mind and language are aspects of the world on a par with "chemical," "optical" or "electrical" aspects of the world which have long been the objects of study in natural sciences, and that there is "no metaphysical divide" between these aspects of the world. Thus, it is quite natural that biolinguistics has been claimed since its inception to be a science of language.

There are, however, discrepancies even among generative linguists in their understanding of the nature of biolinguistics as a science.[1] Why is it that there are such discrepancies concerning the nature of biolinguistics, whose methodological legitimacy Chomsky (2000:76) considers to be based on truisms of science? This seems to be at least partly a result of the tacit assumption of an ahistorical notion of science that the methodological or conceptual discussions of biolinguistics so far tend to presuppose. This attitude allows for a degree of arbitrariness in characterizing and assessing the current status of biolinguistics as a science.

In view of these facts, it is quite interesting to note Friedman's (1993:37) observation that "Thomas Kuhn's *The Structure of Scientific Revolutions* (1962) forever changed our appreciation of the philosophical importance of the history of science," so that "careful and sensitive attention to the history of science must remain absolutely central in any serious philosophical consideration of science."

Furthermore, Friedman (1993: 38) claims that it is legitimate to extend this "lesson of Kuhn's" to the history of philosophy:

> Now I do not suppose that this claim about the relevance of conceptual revolutions in science to the history of philosophy is a controversial one. Indeed, it is now becoming more and more common for historians of

philosophy, especially those concerned with the modern period, to emphasize precisely this relevance. Thus it is now a commonplace that the articulation of characteristically modern philosophy by Descartes and his successors must be viewed against the background of the scientific revolution of the sixteenth and seventeenth centuries. By emphasizing Descartes's concern to replace the Aristotelian-Scholastic natural philosophy with the "mechanical natural philosophy" of the new science, we can achieve a fuller and deeper understanding of such characteristically modern preoccupations as, for example, the distinction between primary and secondary qualities, the "veil of perception," the mind-body problem, and so on. Viewed against the background of the scientific revolution that created modern "mechanistic" natural science itself, these modern philosophical preoccupations no longer appear arbitrary and capricious, as stemming, perhaps, from otherwise unaccountable obsessions with certainty or with "mirroring reality." Instead, they can be understood as natural attempts to come to grips with a profound reorganization of the very terms in which we conceptualize ourselves and our world.

This passage suggests that seemingly "arbitrary and capricious" modern preoccupations, such as "the distinction between primary and secondary qualities," can be more naturally appreciated when viewed against the historical background.

This chapter is an attempt to extend Friedman's approach to the philosophical consideration of biolinguistics and its status as a science. I will show that this approach makes it possible to unify at least some of the methodological characteristics of biolinguistics noted so far by Chomsky and other generative linguists into a coherent picture.

The chapter is organized as follows. In section 2, I will briefly show how modern science was formed in the Scientific Revolution of the seventeenth century, explicating some of the relevant characteristics in the formation of the first conceptual framework of modern science, namely Newtonian dynamics. In section 3, I will attempt to show that, viewed from the perspective of the Scientific Revolution, at least some of the methodological characteristics of biolinguistics, occasionally mentioned separately in the literature, can be accounted for in a unified way.

## 2 The Scientific Revolution and the formation of modern science

A tremendous body of literature has been published on the Scientific Revolution. In particular, Henry (2008) presents a fairly balanced and accessible updated account. In the following discussion, I will assume Henry's (2008) general picture of the Scientific Revolution, adding further details when necessary.

Henry (2008: 1) characterizes the Scientific Revolution as "the name given by historians of science to the period in European history when, arguably, the conceptual, methodological and institutional foundations of modern science were first established." Although "the precise period in question varies from historian to historian," as Henry (2008:1) notes, it is often identified symbolically with the 144-year period from the publication of Copernicus's *De Revolutionibus Orbium Coelestium* (On the Revolutions of the Heavenly Sphere) in 1543 to that of Newton's seminal *Principia Mathematica Philosophiae Naturalis* (*Mathematical Principles of Natural Philosophy*) in 1687.[2]

In the rest of this section, I will show some of the characteristics of the formation of modern science in the Scientific Revolution which are directly relevant for the exploration of methodological characteristics of biolinguistics in section 3.

## 2.1 How was modern science formed in the Scientific Revolution?

What exactly happened in the Scientific Revolution? Henry (2008: 5–6) succinctly summarizes an answer to this question as follows:

> A simple but essentially accurate way of summing up what took place in the Scientific Revolution, then, is to say that the natural philosophy of the Middle Ages, which had tended to remain aloof from mathematical and more pragmatic or experiential arts and sciences, became amalgamated with these other approaches to the analysis of nature, to give rise to something much closer to our notion of science. The Scientific Revolution should not be seen as a revolution in science, because there was nothing like our notion of science until it began to be forged in the Scientific Revolution out of previously distinct elements.

This passage indicates that modern science was formed through the amalgamation of two independent approaches to nature: what Henry calls "mathematical and more pragmatic or experiential arts and sciences" and natural philosophy.[3] According to Henry (2008:5), while mathematical and pragmatic or experiential arts and sciences were such "technically developed disciplinary traditions" as astronomy, optics, and mechanics, natural philosophy "aimed to describe and explain the entire system of the world."

## 2.2 Mathematical and pragmatic or experiential arts and sciences

Henry (2008: 18–19) states that before they came to be amalgamated with natural philosophy, the mathematical and pragmatic or experiential arts and sciences underwent an important change in attitude toward mathematical analysis:

> To put it simply, the Scientific Revolution saw the replacement of a predominantly instrumentalist attitude to mathematical analysis with a more realist outlook. Instrumentalists believe that mathematically derived theories are put forward merely hypothetically, in order to facilitate mathematical calculations and predictions. Realists, by contrast, insisted that mathematical

analysis reveals how things must be; if the calculations work, it must be because the proposed theory is true, or very nearly so.

Henry (2008: 19) adds that "the new realism can be seen at work" in the astronomy of Copernicus, Brahe and Kepler. As Henry (2008: 24–25) notes, Kepler, for example, proclaimed that the mathematical astronomy of his *New Astronomy* has explanatory power:

> Here, then, Kepler was clearly announcing that this astronomy was not merely abstract mathematics for use in practical calculations, but was presenting a physical account of the way the world system really worked.

The above passage suggests that in astronomy, mathematics has come to be used "to explain, not just to describe, the workings of the physical world" (Henry 2008: 25).

## 2.3 Galileo's science of motion

Henry (2008: 26) further observes that the use of mathematics was also found in "the mathematical science of (terrestrial) mechanics." Henry (2008: 27), recognizing that "the greatest figure in this movement is Galileo Galilei," describes the nature of Galileo's work as follows:

> Although Galileo is most famous for his defense of Copernican theory, his initial interest was in terrestrial mechanics and, in particular, kinematics. Like many of his contemporaries, he was dissatisfied with the Aristotelian account of motion, and struggled to arrive at a better theory. During the course of his career, his account of free fall, for example, took him from a mere refinement of Aristotle's belief that bodies fall with speeds proportional to their weight, to the realization that acceleration in free fall is a constant (in a vacuum) for all bodies.

Galileo's science of motion includes at least three interesting methodological characteristics relevant to a discussion of the nature of biolinguistics. Takahashi (2006: 422–423) summarizes the first two of them. First, Galileo's science of motion laid the foundations for the method of modern science. It is the method which includes the idealization and abstraction of phenomena, the use of mathematical models, and the empirical examination of predictions of mathematical models by experiments. Second, Galileo stopped asking why motion occurs, but instead studied how motion is mathematically described. Galilei (1638/1954: 166–167) expressed an opinion against Aristotle's view about "the cause of the acceleration of natural motion" in Salviati's (Galileo's own spokesman) words as follows:[4]

> Salv. The present does not seem to be the proper time to investigate the cause of acceleration of natural motion concerning which various opinions

> have been expressed by various philosophers, some explaining it by attraction to the center, others to repulsion between the very small parts of the body, while still others to attribute it to a certain stress in the surrounding medium which closes in behind the falling body and derives it from one of its positions to another. Now, all these fantasies, and others too, ought to be examined; but it is not really worth while. At present it is the purpose of our Author merely to investigate and to demonstrate some of the properties of accelerated motion (whatever the cause of this acceleration may be). . . .

Finally, Wisan (1978: 3) argues that Galileo uses mathematics "in order to achieve the logical certainty of mathematics" in reasoning:

> Also, while much has been said concerning Galileo and mathematics (and Archimedes), it has been insufficiently noticed that to an important degree his 'mathematicism' consists in the attempt to reduce natural science to the Greek mathematical model in order to achieve the logical certainty of mathematics. To do this meant substituting the kind of reasoning used by mathematicians for that of traditional logicians, for to Galileo the greater certainty of mathematics comes not from contemplation of ideal objects (pace Koyré) but from use of a superior technique of reasoning.

This passage is interesting in that it clearly shows one of the roles that mathematics plays in modern science. In section 3.3, I will show how these methodological characteristics of Galileo's science of motion are connected to those of biolinguistics.

### 2.4 Natural philosophy

During the Scientific Revolution, natural philosophy was also transformed before it was amalgamated with mathematical sciences. Henry (2008: 69) states as follows:

> By the end of the century [the seventeenth century], the mechanical philosophy had effectively replaced scholastic Aristotelianism as the new key to understanding all aspects of the physical world, from the propagation of light to the generation of animals, from pneumatics to respiration, from chemistry to astronomy. The mechanical philosophy marks a definite break with the past and sets the seal upon the Scientific Revolution.

Westfall (1971: 1) makes two interesting remarks about the mechanical philosophy. First, the mechanical philosophy "conceived of nature as a huge machine and sought to explain the hidden mechanisms behind phenomena" and "concerned itself with the causation of individual phenomena." Second, Westfall (1971:1) observes that "the explication of mechanical causation frequently stood athwart the path that led toward exact description, and the full fruition of the scientific revolution required a resolution of the tension between the two dominant trends."

It is interesting to note that while Galileo's science of motion refrained from explaining the cause of motion, the mechanical philosophy was closely connected with the notion of "causation" as a part of explanation. From this perspective, the essence of modern science might be interpreted to lie at least partly in the integration of mathematical analysis with mechanical philosophy through the introduction of some proper notion of "causality." Viewed against this background, Galileo's science of motion without causality, or kinematics, was a significant first step toward modern science of motion, although it was not until the development of Newtonian dynamics that the fully developed conceptual framework of modern science emerged.

## 2.5 Synthesis in Newtonian dynamics

Henry (2008: 31–32) considers the publication of Newton's *Principia* as "the completion of mathematization of natural philosophy:

> The publication of Newton's *Principia* marks the completion of the trend towards the mathematization of natural philosophy which began in the sixteenth century. But perhaps it is true to say that we make that judgement about the *Principia* because Newton, unlike Galileo or Descartes, succeeded in getting the mathematics and the physics substantially correct.

Westfall (1971: 159) explains how the tradition of mathematical description and the tradition of mechanical philosophy were reconciled by the concept of "force" in Newtonian dynamics as follows: [5]

> Newton believed that nature is ultimately opaque to human understanding. Science cannot hope to obtain certain knowledge about the essences of things. Such had been the program of the mechanical philosophy in the 17th century, and the constant urge to imagine invisible mechanisms sprang from the conviction that a scientific explanation is only valid when it traces phenomena to ultimate entities. To Newton, in contrast, nature was a given, aspects of which might never be intelligible. When they learned to accept the same limitation, other sciences such as optics, chemistry, and biology, likewise ceased to play with imaginary mechanisms and, describing instead of explaining, they formulated a set of conceptions adequate to their phenomena. Newton believed that the aim of physics is an exact description of phenomena of motion in quantitative terms. Thus the concept of force could be admitted into scientific demonstrations even if the ultimate reality of force were not comprehended. In Newton's work, it made possible the reconciliation of the tradition of mathematical description, represented by Galileo, with the tradition of mechanical philosophy, represented by Descartes. By uniting the two, Newton carried the scientific work of the 17th

century to that plane of achievement which has led historians to speak of a scientific revolution. And modern science continues to pursue its effective course within the framework thus established.

Westfall's explanation indicates that, by introducing a notion of causality called "force," Newton integrated mathematical description and mechanical philosophy, which were not in harmony with each other, into Newtonian dynamics, the first conceptual framework of modern science, although "the ultimate reality of force" remains to be explored.

In the next section, I will show that, viewed from the perspective of the Scientific Revolution briefly reviewed in this section, some of the methodological characteristics of biolinguistics can be accounted for in a unified way.

## 3 Methodological characteristics of biolinguistics

Chomsky repeatedly states that the cognitive revolution of the 1950s constitutes the reshaping of the achievements and insights of what he calls "the first cognitive revolution," which took place as a part of the Scientific Revolution of the seventeenth century.[6] Chomsky (2000: 6), for example, expresses this view as follows:

> The "cognitive revolution" renewed and reshaped many of the insights, achievements, and quandaries of what we might call "the first cognitive revolution" of the seventeenth and eighteenth century, which was part of the scientific revolution that so radically modified our understanding of the world.

In fact, Henry (2008: 80) notes that leading mechanical philosophers, such as Descartes and Hobbes, extended their approach to "the explanation of vital phenomena and animal (including human) behavior," beyond mechanics, kinetics and dynamics, with which it was originally connected.

Given this, Chomsky's view simply suggests that biolinguistics has been developing toward a modern science of language virtually along the lines analogous to those along which modern science developed in the Scientific Revolution of the seventeenth century. I will show that, viewed in this light, at least a few other seemingly arbitrary methodological characteristics of biolinguistics discussed so far are naturally accounted for in a unified way.

### 3.1 Synthesis of two traditions

Chomsky (1968/2006: 20–21, 57) states that there have been "two major traditions that have enriched the study of language in their separate and very different ways." One is "the tradition of philosophical grammar that flourished from the seventeenth century through romanticism." Chomsky (1968/2006: 14) characterizes philosophical grammar as "a kind of 'natural philosophy' or, in modern terms, 'natural science'," which seeks explanation rather than "the accumulation of data." The other is structural linguistics, "which has dominated research for the past century, at least until the early 1950s." Chomsky (1968/2006: 19–20) acknowledges its contribution as follows:

> Structural linguistics has enormously broadened the scope of information available to us and has extended immeasurably the reliability of such data. It has shown that there are structural relations in language that can be studied abstractly. It has raised the precision of discourse about language to entirely new levels.

On the other hand, however, Chomsky clearly notes that its attempt to construct "discovery procedures," which are "techniques of segmentation and classification," are a failure, "because such techniques are at best limited to the phenomena of surface structure and cannot, therefore, reveal the mechanisms that underlie the creative aspect of language use and the expression of semantic content."

At the same time, despite its failure, Chomsky underscores the importance of the attempt in the sense that it was directed at the problem of "specifying the mechanisms that operate on the data of sense and produce knowledge of language – linguistic competence." Chomsky (1968/2006: 57) also states that "a kind of synthesis of philosophical grammar and structural linguistics" began "to take shape" in the late 1960s.

Viewed in the light of the Scientific Revolution, Chomsky's account indicates again that biolinguistics has been developing along the lines similar to modern science during the Scientific Revolution, through the synthesis of structural linguistics, which corresponds to "mathematical and more pragmatic or experiential sciences", and a kind of "natural philosophy", the philosophical grammar. In the next section, I will show how structural linguistics has been transformed before it could be amalgamated with what counts as "natural philosophy" in linguistics.

### 3.2 Methodological parallelism

Now let us turn to the method of biolinguistics. There are at least three methodological parallels between modern science and biolinguistics in their formation.

First, as we saw in section 2.1, just as mathematicians espoused an instrumentalist attitude toward mathematical analysis, structural linguists espoused an instrumentalist attitude toward linguistic analysis. As Freidin (1994:653–654) notes, Zellig Harris, a leading structural linguist, for example, states in his *Methods in Structural Linguistics* (1951) that "methods of research used in descriptive, or more exactly, structural, linguistics" are "the operations which the linguist may carry out in the course of his investigations, rather than theory of the structural analyses which result from these investigations." This statement suggests that Harris takes an instrumentalist attitude toward linguistic analysis, just like mathematicians in mathematical sciences, such as astronomy, before the Scientific Revolution. It is quite interesting to note that Freidin (1994:660) also argues that Harris "does not adopt the realist position," whereas "from the very outset Chomsky adopts a realist interpretation of grammar." So, it is obvious that in the cognitive revolution of the 1950s, the instrumentalist attitude to linguistic analysis was replaced with a realist attitude, as in new astronomy during the Scientific Revolution.

Second, like Galileo, Chomsky also stopped seeking the causation of verbal behavior, as indicated in the passage from Chomsky (1959: 55):

> The questions to which Skinner has addressed his speculations are hopelessly premature. It is futile to inquire into the causation of verbal behavior until much more is known about the specific character of this behavior; and there is little point in speculating about the process of acquisition without much better understanding of what is acquired.

I will discuss consequences of this shift in the goal for the method of biolinguistics in section 3.3.

Third, Chomsky (1980: 218) states that "it is common to adopt what has sometimes been called 'the Galilean Style' – that is, to construct "abstract mathematical models of the universe to which at least the physicists give a higher degree of reality than they accord the ordinary world of sensations.'" The Galilean Style – the characterization of the method of physics by the theoretical physicist Stephen Weinberg – elegantly captures modern science's way of understanding the physical world. However, it is also interesting to note that although it explicitly states that the inner reality of the universe is better captured in terms of mathematical models than sensation, their causal structure is not specifically mentioned, or at best remains implicit in it. I will turn to this point in section 3.3.

Although the Galilean Style is characterized on the basis of physics, Chomsky (1980: 118), like mechanical philosophers of the Scientific Revolution, such as Descartes, claims as follows:

> A comparable approach is particularly appropriate in the study of an organism whose behavior, we have every reason to believe, is determined by the interaction of numerous internal systems operating under conditions of great variety and complexity.

Chomsky (1980: 223) further notes that, under this approach, "the grammar of a language, conceived as a system of rules that weakly generates the sentences of a language and strongly generates their structures, has a claim to that 'higher degree of reality' that the physicist ascribes to his mathematical models of the universe."

Finally, let us briefly discuss the role that mathematics plays in biolinguistics. In the interview included in Chomsky (2004), when asked why mathematics does "not seem to work so extensively for human language," Chomsky (2004:174) answered as follows:

> Well, it doesn't work for anything. There's no mathematics in biology. Not even in chemistry. Mathematics is of course used, but not in the sense we are talking about here, at least as far as I know.

As Chomsky suggests, if mathematics does not play a genuine role in biolinguistics, as it does in physics, in what sense does mathematics work for biolinguistics, as stated in the Galilean Style? It is important to note that, as we saw in section 2.3, Wisan (1978: 3) argues that Galileo uses it "in order to achieve the logical certainty of mathematics" in reasoning. Thus, in Galileo's science of motion, mathematics was used at least partly to attain "rigorous demonstrations from evident principles."

As Chomsky (1979: 125) states, similarly in biolinguistics, "a certain quasi-mathematical mode of expression is presupposed in the overall program" and a formal system, such as "some variety of recursive function theory," is used "to formulate precise principles and precise rules within a formalized system," although "it would not be correct to consider that as mathematics." Thus, the role of mathematics in the Galilean Style should be interpreted in a broader sense, taking into consideration the role of mathematics in early modern science.

## 3.3 Shift to mechanisms

Chomsky (2000: 6) states that in the cognitive revolution of the 1950s, "there was an important change of perspective: from the study of behavior and its products (such as texts) to the inner mechanisms that enter into thought and action." In light of the Scientific Revolution, it is obvious that this "change of perspective" corresponds to the replacement of Aristotelianism by mechanical philosophy in the Scientific Revolution.[7]

Freidin (2012: 893) characterizes this change of perspective as "the first step towards the current perspective" involving "a shift in focus from the external forms of a language to the grammatical system that generates it, a system that is assumed to exist in the mind of the speaker, constituting the core part of a speaker's knowledge of his or her language."

Under this perspective, as Freidin (2012: 893) states, "in its initial conception, the grammar consists of two distinct formal mechanisms for generating the sentences of a language: phrase structure rules and transformations. The former constructs initial syntactic structures which the latter could modify in certain ways, including the combination of pairs of initial structures – yielding complex and compound sentences."

Now, it is interesting to note that Chomsky (1965:9) states that the model of grammar consisting of these formal mechanisms is a neutral characterization of "the knowledge of the language that provides the basis for actual use of language by a speaker-hearer" as follows:

> . . . a generative grammar is not a model for a speaker or a hearer. It attempts to characterize in the most neutral possible terms the knowledge of the language that provides the basis for actual use of language by a speaker-hearer. When we say that a sentence has a certain derivation with respect to a particular generative grammar, we say nothing about how the speaker or hearer might proceed, in some practical or efficient way, to construct a derivation.

As Chomsky (2007: 6) clearly states, the formal nature of the model of grammar remains the same in the Minimalist Program. (Here "the system" corresponds to "the Merge-based system".)

> But the system implies no temporal dimension. In this respect, generation of expressions is similar to other recursive processes such as construction of formal proof. Intuitively, the proof "begins" with axioms and each line is added to earlier lines by rules of inference or additional axioms. But this implies no temporal ordering. It is simply a description of the structural properties of the geometrical object "proof."

From the perspective of the Scientific Revolution, the formal nature of mechanisms of biolinguistics is rather expected, since, like Galileo, Chomsky stopped seeking the "causation of behavior" when he started developing biolinguistics.

On the other hand, as philosophers of science, Craver and Darden (2013: 5), for example, states that "whatever the origins of this mechanistic perspective, and however it is related to the austere forms of mechanism that developed in the sixteenth and seventeenth centuries, it is now so thoroughly woven into the fabric of contemporary biology one might easily forget that biology could have taken a different form."

Furthermore, as Glennan (2009:315) states, in biological sciences, such as neurobiology and molecular biology, "mechanism is undoubtedly a causal concept, in the sense that ordinary definitions and philosophical analyses explicate the concept in terms of other causal concepts such as production and interaction."

This difference in the ontic status of mechanism between biolinguistics and other biological sciences seems to suggest that biolinguistics as a biological science is still at a stage roughly equivalent to modern science prior to Newtonian dynamics, while such biological sciences as neurobiology and molecular biology are at a more advanced stage of modern science.

### 3.4 Basic questions in ethology and biolinguistics

In order to further clarify why mechanisms of biolinguistics are different from those of other biological sciences, let us compare the basic questions of ethology and those of biolinguistics.

According to Bolhuis and Giraldeau (2005: 2), ethology, currently often called classical ethology, is the study of animal behavior, which "became an independent scientific discipline," "in the middle of the twentieth century," "mainly through the efforts of two biologists, the Austrian Konrad Lorenz (1903–89) and the Dutchman Niko Tinbergen (1907–88)."

Tinbergen (1963/2009: 2) characterizes ethology in methodological terms as follows:

> In the course of thirty years devoted to ethological studies I have become increasingly convinced that the fairest characterisation of Ethology is "the

biological study of behaviour". By this I mean that the science is characterised by an observable phenomenon (behaviour, or movement), and by a type of approach, a method of study (the biological method). The first means that the starting point of our work has been and remains inductive, for which description of observable phenomena is required. The biological method is characterised by the general scientific method, and in addition by the kind of questions we ask, which are the same throughout Biology and some of which are peculiar to it. Huxley likes to speak of "the three major problems of Biology": that of *causation*, that of *survival value*, and that of *evolution* – to which I should like to add a fourth, that of *ontogeny*.

Viewed in light of the Scientific Revolution, Tinbergen's characterization of ethology has two noteworthy features. First, its method, called "the biological method," consists of two parts: the general scientific method and the four questions. This seems to be a result of the fact that modern science developed through the amalgamation of mathematical sciences and mechanical philosophy. The two parts correspond to the two aspects of modern science. Namely, the general scientific method corresponds to mathematical description, and the four basic questions to mechanical natural philosophy.

Second, why does ethology have the three further questions, i.e., survival value, evolution and ontogeny, in addition to causation? In order to answer this question, it is necessary to understand that ethology has developed as a biological study of animal behavior by extending the conceptual framework of modern science from physics to biology.

Given this, it is interesting to note that Mayr (2004: 30) observes that there is one respect that fundamentally distinguishes all biological processes from all inanimate processes, i.e., dual causation, as follows:

> Furthermore, all biological processes differ in one respect fundamentally from all processes in the inanimate world; they are subject to dual causation. In contrast to purely physical processes, these biological ones are controlled not only by natural laws but also by genetic programs. This duality fully provides a clear demarcation between inanimate and living processes.
>
> The dual causality, however, which is perhaps the most important diagnostic characteristic of biology, is a property of both branches of biology.[8] When I speak of dual causality I am of course not referring to Descartes' distinction of body and soul but rather to the remarkable fact that all living processes obey two causalities. One of them is the natural laws that, together with chance, control completely everything that happens in the world of the exact sciences. The other causality consists of the genetic programs that characterize the living world so uniquely. There is not a single phenomenon or a single process in the living world that is not in part controlled by a genetic program contained in the genome. There is not a single activity of any organism that is not affected by such a program. There is nothing comparable to this in the inanimate world.

Thus, it is Tinbergen's insight to include the questions of survival value, evolution, and ontogeny, which all pertain directly to genetic programs, but only indirectly to natural laws, in addition to the question of causation. Therefore, the four questions in (1) all have dual causation, being controlled primarily by genetic programs and ultimately by natural laws. (Tinbergen's questions in (1) are reordered to match Chomsky's questions in (2).)

(1) a. causation
 b. ontogeny
 c. evolution
 d. survival value

Turning to biolinguistics, we have seen that it has also developed by extending the conceptual framework of modern science. In addition, as Chomsky (1980: 185) notes, it has incorporated into its "natural philosophy" a view that it is possible to "regard the language capacity virtually as we would a physical organ of the body" and to "investigate the principles of its organization, functioning, and development in the individual and the species."

Therefore, it is natural that biolinguistics have basic questions comparable to those of ethology. Actually Chomsky (1995: 17–18) proposes the following five questions:

(2) a. What does Jones know when he has a particular language?
 b. How did Jones acquire this knowledge?
 c. How does Jones put this knowledge to use?
 d. How did these properties of the mind/brain evolve in the species?
 e. How are these properties realized in mechanisms of the brain?

Boeckx (2010: 23), among others, suggests that Tinbergen's four questions (1) "correspond point by point to" Chomsky's five questions (2).[9] When (1) and (2) are closely examined, however, it turns out that there are three discrepancies between them. First, Chomsky divides Tinbergen's first question (1a) causation into three separate questions, i.e., (2a) what does Jones know? (2c) how does Jones put this knowledge to use? and (2e) how are these properties realized in mechanisms of the brain? Second, once (1a) is divided into three questions, its first part (2a) does not include the notion of causality, as we see when we discuss the nature of mechanisms in biolinguistics in section 3.2. Third, Tinbergen's question (1d) of survival value is not included in Chomsky's set of basic questions (2). Putting aside the third discrepancy, let us consider why there are these discrepancies between (1) and (2), focusing on the first and the second.

The first and the second discrepancies are related to each other and are the result of the nature of idealization and abstraction adopted in biolinguistics. Chomsky (1980: 188) explains the idealization involved in biolinguistics as follows:

> Continuing to think of the system of grammatical rules as a kind of "mental organ", interacting with other mental organs with other functions and

properties, we face a rather typical problem of natural science, namely, the problem of appropriate idealization and abstraction. In an effort to determine the nature of one of these interacting systems, we must abstract away from the contribution of others to the actual performance that can be observed.

This clearly indicates that, by abstracting away from other cognitive systems interacting with the system of grammatical rules (knowledge of language), it has become virtually impossible to conduct any causal analysis of language production and understanding, reducing the first problem of the knowledge of language in (2a) to formal analysis, while transferring causal analysis of linguistic processes to (2c) and (2d).

This analysis of the discrepancies suggests that "formal mechanisms for generating the sentences of a language" assumed in linguistic analysis, such as phrase structure rules or Merge, do not contain the notion of causality, while physical mechanisms assumed in (2d) (and probably also in (2c)) do, as those assumed in other biological sciences, such as neurobiology and molecular biology, do.

Finally, viewed from this perspective, the Minimalist Program is an attempt to attain an amalgamation of non-causal formal mechanisms in linguistic analysis with causal physical mechanisms of linguistic and other biological processes so that biolinguistics will be a full-fledged modern science of language.

## 4 Conclusion

In this chapter, I have attempted to show that a perspective from the history and philosophy of science, particularly that of the Scientific Revolution, makes it possible to unify at least some seemingly unrelated or even curious methodological characteristics of biolinguistics occasionally mentioned in the literature into a coherent picture so that it could serve to elucidate the current status of biolinguistics as a biological science.

## Notes

* I am grateful to Chuck Brown, Cindy Brown, Koji Fujita, Jeffry Gayman, Yasuhiko Kato, and John Matthews for invaluable comments and stylistic suggestions. My thanks also go to Yurie Murayama for clerical assistance.
1 Recently, for example, Al-Mutairi (2014: 2–3) observed that it is "necessary to first take a broad view of the general development of Chomskyan linguistics, with the primary aim of clarifying some of the misconceptions that have been expressed by (ironically enough) some well-known popularisers of Chomsky's work." In fact, Al-Mutairi takes issue with the account of the development of generative grammar by Boeckx and Hornstein (2010). See Boeckx and Hornstein (2010) and Al-Mutairi (2014) for discussion.
2 Schuster (1990: 231) notes, for example, that Copernicus should not be simply considered to represent the first stage of the Scientific Revolution, but should be interpreted carefully in the historical context. Cf. Koyré (1957) and Henry (2008).
3 Westfall (1971) expressed a similar view about the Scientific Revolution nearly 40 years earlier. Westfall recognized two major themes dominating the Scientific

Revolution of the seventeenth century: the Platonic-Pythagorean tradition and the mechanical philosophy. These two themes correspond to Henry's "mathematical and more pragmatic or experiential arts and sciences" and "natural philosophy," respectively. Cohen (2010a: 9) suggests that in addition to the two approaches mentioned above, there is another mode of "nature-knowledge," which Cohen refers to as "an empiricist and practice-oriented form of experimental science." Although it is important to determine how this factor might affect the development of biolinguistics, I will leave this problem open here. See Cohen (2010b) for a more detailed recent historiographical characterization of the Scientific Revolution. Cf. Henry (2008)

4 See also Pitt (1978) for discussion.
5 Chomsky (2000: 84) presents a different interpretation of the synthesis of mathematical description and mechanical philosophy Newton established:

Just as the mechanical philosophy appeared to be triumphant, it was demolished by Newton, who reintroduced a kind of "occult" cause and quality, much to the dismay of leading scientists of the day, and of Newton himself. The Cartesian theory of mind (such as it was) was unaffected by his discoveries, but the theory of body was demonstrated to be untenable. To put it differently, Newton eliminated the problem of "the ghost in the machine" by exorcising the machine; the ghost was unaffected.

Although this discrepancy points to an important issue concerning the notion of causality, which is worth careful discussion, I will leave it for future study.

6 See Gardner (1985) for the notion of the Cognitive Revolution of the 1950s.
7 Bechtel (2008/2009: 2) notes as follows:

Here modern science has taken a different path, outdoing Descartes at his own endeavor by finding mechanistic explanations for mental as well as bodily phenomena. Cognitive scientists, and their predecessors and colleagues in such fields as psychology and neuroscience, assume that the mind is a complex of mechanisms that produce those phenomena we call "mental" or "psychological."

8 Mayr (2004: 25) comments upon the establishment of modern biology as follows:

The two-hundred-year period from about 1730 to 1930 witnessed a radical change in the conceptual framework of biology. The period from 1828 to 1866 was particularly innovative. Within these 38 years, both branches of modern biology – functional and evolutionary biology – were established. See Mayr (2004) for further discussion.

9 Boeckx and Grohmann (2007) and Jenkins (2011) express a similar view.

# References

Al-Mutairi, Fahad Rashed. 2014. *The minimalist program: The nature and plausibility of Chomsky's biolinguistics*. Cambridge, UK: Cambridge University Press.
Bechtel, William. 2008/2009. *Mental mechanisms: philosophical perspectives on cognitive neuroscience*. New York and London: Psychology Press Taylor & Francis Group.

Boeckx, Cedric. 2010. *Language in cognition: Uncovering mental structures and the rules behind them*. West Sussex, UK: Wiley-Blackwell.
Boeckx, Cedric, and Kleanthes K. Grohmann. 2007. The biolinguistics manifesto. *Biolinguistics* 1:1–8.
Boeckx, Cedric, and Norbert Hornstein. 2010. The varying aims of linguistic theory. In *Chomsky notebook*, ed. Jean Bricmont and Julie Franck, 115–141. New York: Columbia University Press.
Bolhuis, Johan J., and Luc-Alain Giraldeau. 2005. The study of animal behavior. In *The behavior of animals: mechanisms, functions, and evolutions*, ed. Johan J. Bolhuis and Luc-Alain Giraldeau, 1–10. Oxford: Blackwell.
Chomsky, Noam. 1959. A review of B. F. Skinner's *Verbal Behavior*. *Language* 35:26–58.
Chomsky, Noam. 1965. *Aspects of the theory of syntax*. Cambridge, MA: MIT Press.
Chomsky, Noam. 1968/2006. *Language and mind*. Cambridge, UK: Cambridge University Press.
Chomsky, Noam. 1979. *Language and responsibility: Based on conversations with Mitsou Ronat*. Hassocks, UK: Harvester Press.
Chomsky, Noam. 1980. *Rules and representations*. New York: Columbia University Press.
Chomsky, Noam. 2000. *New horizons in the study of language and mind*. Cambridge, UK: Cambridge University Press.
Chomsky, Noam. 2004. *The generative enterprise revisited*. Berlin/New York: Mouton de Gruyter.
Chomsky, Noam. 2005. Three factors in language design. *Linguistic Inquiry* 36.1:1–22.
Chomsky, Noam. 2007. Approaching UG from below. In *Interface + recursion = language?: Chomsky's minimalism and the view from syntax-semantics*, ed. Uli Sauerland and Hans-Martin Gärtner, 1–29. Berlin/New York: Mouton de Gruyter.
Cohen, H. Floris. 2010a. The onset of the scientific revolution: Three near-simultaneous transformations. In *The science of nature in the seventeenth century: Patterns of change in early modern natural philosophy*, ed. Peter R. Anstey and John A. Schuster, 9–34. Dordrecht: Springer.
Cohen, H. Floris. 2010b. *How modern science came into the world: Four civilizations, one 17th-century breakthrough*. Amsterdam: Amsterdam University Press.
Craver, Carl F., and Lindley Darden. 2013. *In search of mechanisms: Discoveries across the life sciences*. Chicago: University of Chicago Press.
Freidin, Robert. 1994. Conceptual shifts in the science of grammar: 1951–92. In *Noam Chomsky critical assessments volume I: Linguistics: Tome II*, ed. Carlos P. Otero, 653–690. New York: Routledge.
Freidin, Robert. 2012. A brief history of generative grammar. In *The Routledge companion to philosophy of language*, ed. Gillian Russel and Delia Graff Fara, 895–916. New York: Routledge.
Friedman, Michael. 1993. Remarks on the history of science and the history of philosophy. In *World changes: Thomas Kuhn and the nature of science*, ed. Paul Horwich, 37–54. Cambridge, MA: MIT Press.
Galilei, Galileo (1638/1954) *Dialogues concerning two new sciences*. Translated by Henry Crew and Alfonso de Salvio. New York: Dover Publications, Inc.
Gardner, Howard. 1985. *The mind's new science: A history of the cognitive revolution*. New York: Basic Books.

Glennan, Stuart. 2009. Mechanisms. In *the Oxford Handbook of Causation,* ed. Helen BeeBee, Chirstopher Hichcoch, and Peter Menzies, 315–325. Oxford: Oxford University Press.

Harris, Zellig. 1951. *Methods in structural linguistics.* Chicago: University of Chicago Press.

Henry, John. 2008. *The scientific revolution and the origins of modern science.* New York: Palgrave Macmillan.

Jenkins, Lyle. 2011. The three design factors in evolution and variation. In *The biolinguistic enterprise: New perspectives on the evolution and nature of the human language faculty,* ed. Anna Maria Di Sciullo and Cedric Boeckx, 169–179. Oxford: Oxford University Press.

Koyré, Alexandre. 1957. *From the closed world to the infinite universe.* New York: Johns Hopkins Press.

Mayr, Ernst. 2004. *What makes biology unique?* Cambridge, UK: Cambridge University Press.

Pitt, Joseph C. 1978. Galileo: causation and the use of geometry. In *New perspectives on Galileo,* ed. Robert E. Butts and Joseph C. Pitt, 181–196. Dordrecht: D. Reidel.

Schuster, John A. 1990. The scientific revolution. In *Companion to the history of modern science,* ed. Robert C. Olby, Geoffrey N. Cantor, John R. R. Christie, and Jonathan S. Hodge, 217–242. London: Routledge.

Takahashi, Kenichi. 2006. *Garireo no meikyuu – shizenwa suugakuno gengode kakareteiruka?* [*Galileo's labyrinth – Is nature written in mathematical language?*]. Tokyo: Kyooritsushuppan.

Tinbergen, Niko. 1963/2009. On aims and methods of ethology. In *Tinbergen's legacy: function and mechanism in behavioral biology,* ed. Johan J. Bolhuis and Simon Verhulst, 1–24. Cambridge, UK: Cambridge University Press.

Ueda, Masanobu. 2013. Tinbergenno 4tsuno mondaino seibutsugengogakuniokeru ichizuke to sono hoohoorontekigani [On the status of Tinbergen's four questions in biolinguistics and its methodological implications]. *Sophia Linguistica* 61:85–96.

Westfall, Richard S. 1971. *The construction of modern science: Mechanisms and mechanics.* New York: John Wiley & Sons, Inc.

Wisan, Winfred L. 1978. Galileo's scientific method: A reexamination. In *New perspectives on Galileo,* ed. Robert E. Butts and Joseph C. Pitt, 1–57. Dordrecht: D. Reidel.

# Part IV
# Evolutionary considerations

# 12 Proposing the hypothesis of an earlier emergence of the human language faculty*

*Masayuki Ike-uchi*

## 1 Introduction

This short chapter develops a hypothesis that claims that the human language faculty (Merge/UG) as a genetic endowment emerged much earlier than has often been assumed: at the latest, 130–150 thousand years ago (kya). It is argued that the convergence of archaeological/paleoanthropological and genetic evidence supports this claim (Ike-uchi 2012, 2014).

We use the following two basic assumptions/frameworks to address the problems of the origins and evolution of human language.

First, we assume the single origin hypothesis of *Homo sapiens* and assume that anatomically modern humans first appeared in East Africa around 200 kya. We are all descendants of these early humans. This hypothesis, I assume, is a widely accepted view.

Second, our hypothesis will be developed within the framework of the recent minimalist program of generative grammar. Based on recent discussions (e.g., Hauser et al. 2002, Chomsky 2013, 2015, and Fujita 2013), as a kind of possible "idealization," I assume that human UG consists only of Merge, i.e., the so-called Merge-only hypothesis. However, I do not necessarily exclude the possibility that there are other operations of core syntax in UG. One strong candidate is Inheritance. With regard to the two other operations, i.e., Transfer and Labeling, I assume that they are not core syntactic operations (but see Bošković 2015 on the latter), essentially following Chomsky (2013, 2015). I assume that Agree is not a biologically evolved product.

As has often been noted, the more UG devices are posited, the more complicated their evolutionary processes are. Hence, if Inheritance is an indispensable core UG operation, then the question of how and when UG emerged becomes more difficult and produces more ramifications. For instance, we will need to consider the problem of the timing of the emergence of at least two operations, Merge and Inheritance. I believe that the solution to this problem is unreachable at the current stage of inquiry of evolutionary linguistics. Hence, in this chapter, as a type of idealization, I assume the Merge-only hypothesis. In other words, we would say that if Merge and Inheritance are both in UG, then they simultaneously emerged in human evolution.

Now the following question is considered: when did Merge/UG emerge in the history of human evolution?[1] I approach this question in this way. First, based on the Merge-only hypothesis, we assume that all currently existing human languages have one common property: that is, all of them have recursive Merge as the structure-building operation.[2] Then the simplest (almost null, I suspect) hypothesis is that genetically endowed Merge/UG first emerged and existed in the language of *Homo sapiens* in East Africa at the origin of human language. Hence, Merge/UG certainly already existed (at the latest) prior to the Out-of-Africa dispersal of *Homo sapiens* (Renfrew 2007: 78).[3] Consequently, determining when this move occurred can assist us in discovering when Merge/UG emerged, at least as a first approximation.

Suppose that human language did not emerge in this way. That is, suppose that every current human language had individually emerged in each and every place after modern humans migrated out of Africa and dispersed worldwide. Then, one would need to explain why it was recursive Merge rather than, for example, a non-structural linear concatenation operation that emerged in a language in every place in the world. Thus, the burden of proof is on the proponents of the other hypotheses, including those based on, for example, convergent evolution.[4]

As noted, the question then concerns when the Out-of-Africa of *Homo sapiens* occurred in human history. The claim that has been traditionally and widely accepted is that *Homo sapiens* migrated out of Africa around 50–80 kya (at the earliest 100 kya) (e.g., Klein 2009a, 2009b, Ash and Robinson 2010, Chomsky 2013, Berwick et al. 2013, Brahic 2014).[5] However, recent archaeological/paleoanthropological and genetic evidence reveals the possibility of a much earlier Out-of-Africa movement – that their dispersal occurred around 130 kya. I will present five pieces of archaeological/paleoanthropological evidence in favor of this claim and then present and discuss two pieces of genetic evidence that support it.

## 2  Archaeological/paleoanthropological evidence

The archaeological/paleoanthropological arguments introduced in this section essentially take the pattern of either (i) or (ii). (i) Stonewares or tools excavated in the stratum/place in question are quite similar to those produced by *Homo sapiens* in (North) East Africa. Technically, one can speculate as to when those layers and tools were formed and made. Thus it can be concluded that at the time in question a group of *Homo sapiens* was already there. (ii) The hominin fossils discovered in the layer formed some hundred thousand years ago are those of *Homo sapiens*. Hence, it can be assumed that they were there at that time.

First, we go to Oman in the Arabian Peninsula (Rose et al. 2011). The stone tools found at the Nubian Complex in the Dhofar region of Oman are from 106 kya. These tools thus provide the archaeological evidence of the earlier existence of a northeast African Middle Stone Age technocomplex in southern Arabia (particularly, the use of the Levallois method). Hence, *Homo sapiens* was already in the Arabian Peninsula around 106 kya and thus left East Africa through a southern route much earlier than has been assumed thus far, as early as 110 kya.[6]

Second, we turn to the United Arab Emerites. According to Armitage et al. (2011), assemblages (i.e., their Levallois assemblage C) from Jebel Faya, UAE, have crucial affinities to those of the late Middle Stone Age in (North) East Africa. Armitage and his colleagues determined that they were created approximately 125 kya. Hence, this evidence demonstrates the presence of *Homo sapiens* in eastern Arabia by approximately 125 kya. In order for *Homo sapiens* to be in eastern Arabia by 125 kya, they would have needed to leave Africa some time earlier than 125 kya. That is, they migrated from East Africa into south Arabia much earlier than is usually assumed, likely around 130 kya at the latest.

Third, turning to China, fragmentary human remains, two molars and an anterior mandible, were discovered in 2007 at Zhirendong in South China. Their date was confirmed by U-series on an overlying flowstone layer and an associated faunal sample to be from the initial Late Pleistocene, i.e., approximately 100–113 kya. These remains provide evidence of a modern human emergence in East Asia at least 100 kya (Liu et al. 2010b, Dennell 2010). This finding in turn suggests that the Out-of-Africa exodus occurred earlier than at least 100 kya. *Homo sapiens* reached southern China by 100 kya and thus had to have left Africa much earlier.[7]

Fourth, the Lunadong (LND) hominin fossil assemblage found at Luna Cave, Guangxi, southern China, comprises one left upper second molar (M2) and one right lower second molar (m2), both of which are permanent teeth. According to Bae et al. (2014), at least M2 is clearly assigned to modern *Homo sapiens*, while m2 is less clear, but is likely to be affiliated with modern *Homo sapiens*, in terms of metric and geometric morphometric analysis. Notably, the teeth are securely dated between 127 and 70 kya according to uranium-series dating of associated flowstones. The same logic can apply to the former date.

Fifth, turning to central China, the seven hominin teeth discovered in Huanglong Cave in Hubei Province are assumed to be from modern *Homo sapiens*. The U-series on thin flowstone formations dates them to 81–101 kya (Shen et al. 2013).[8] Shen et al. (2013: 166) argue that this evidence "would . . . be compatible with a scenario in which fully modern human morphology first evolved in East Africa after 200 ka, and diffused to the rest of the world quickly thereafter, arriving in eastern Asia 100 ka ago (not 50 ka ago), or earlier." The logic is the same as in the above cases: to arrive in central China, eastern Asia, as early as 100 kya, modern *Homo sapiens* would have left Africa earlier than 100 kya.[9]

Taken together, these pieces of recent archaeological/paleoanthropological evidence lead us to conclude that anatomically modern *Homo sapiens* left Africa through the southern route, not as recently as 50–80 kya but at the latest around 130 kya.[10,11]

## 3 Genetic evidence

We now consider some recent genetic evidence.

The first piece of evidence is based on the idea of a "molecular clock."[12] Mutations accumulate along time. Hence, roughly speaking, when we multiply the number of genetic mutations in DNA sequences by the average human mutation rate, we can reasonably determine, for instance, when the two hominins split from their common ancestor or when they migrated from Africa. This determination is the gist of our argumentation. For the past 15 years, the speed of the molecular clock has been estimated in terms of the number of mutational differences in matching segments of DNA between humans and primates based on the fossil records. The speed has been assumed to be rather high. Thus, according to this phylogenetic mutation rate, it has been supposed that the Out-of-Africa dispersal of *Homo sapiens* occurred 70 kya at the earliest (Oppenheimer 2012, Gibbons 2012). A new method of obtaining the mutation rate that has recently been developed calculates the mutation rate of the entire genome of present-day humans by counting the number of new mutations in the nuclear DNA of a newborn baby compared with its parents (Gibbons 2013). The value that Scally and Durbin (2012) cite is $0.5 \times 10^{-9}$ bp$^{-1}$year$^{-1}$, which they claim is half of the previous fossil-calibrated phylogenetic rate. In other words, the molecular clock ticks more slowly than previously assumed. The calculation based on this new rate shifts the Out-of-Africa migration of modern humans to 90–130 (rather than less than 70) kya (Scally and Durbin 2012, Gibbons 2012).[13]

Second, based on genetic polymorphisms and cranial shape variables of modern human populations from Africa and Asia, Reyes-Centeno et al. (2014) claim support for a multiple-dispersal model in which only Australo-Melanesian populations are isolated descendants of an early migration, whereas other Asian populations are from members of subsequent dispersal events. The researchers note that their results support an initial exodus out of Africa into Asia by a coastal route along the Arabian Peninsula beginning as early as 130 kya and a later migration wave into northern Eurasia starting by 50 kya.

Interestingly, the number resulting from recent genetic studies (i.e., 130 kya) nicely matches the number obtained from the arguments developed in the preceding section based on archaeological/paleoanthropological evidence from stone tools and fossils.[14]

In sum, recent archaeological/paleoanthropological evidence and genetic evidence converge to reasonably conclude that the Out-of-Africa dispersal of *Homo sapiens* occurred much earlier than previously hypothesized, that is, as early as or at the latest 130 kya.[15] Hence, based on the simplest hypothesis noted above, I propose that human Merge/UG emerged (at the latest) around 130–150 kya in East Africa.[16] This is the hypothesis of an earlier emergence of the human language faculty (see also Ike-uchi 2012, 2014).

## 4 A possible scenario

According to our hypothesis of an earlier emergence of the human language faculty, the scenario of the emergence of Merge/UG and the Out-of-Africa dispersal is as follows (Ike-uchi 2014). First, due to genetic mutation, Merge/UG emerged in the brain of one individual member of a certain group of *Homo sapiens* in (North) East Africa (at the latest) approximately 130–150 kya. Merge/UG subsequently spread among a certain number of people in that group (but not necessarily among all the people in the group at that time) as a result of natural selection because this human language trait was clearly adaptive. Then, around 130 kya, a group of modern humans, some of whom had Merge/UG and were leading the group, migrated out of Africa through the southern route, crossing the Bab al-Mandab Straits, moving to Oman and Jebel Faya of the Arabian Peninsula, and finally arriving in southern China. Among each group in each place, language genes spread among members, and as a result Merge/UG became fixed among modern humans. These humans replaced indigenous archaic humans in each area (if any were present). Subsequently, around 50–100 kya, some groups of humans, all of whom had Merge/UG, left Africa through the northern route via the Nile Valley, reaching Eurasia. I suppose that they coexisted and interbred with original linguistic *Homo sapiens* with Merge/UG in each area. In this way, *Homo sapiens* with Merge/UG spread throughout the world. I assume that after Merge/UG spread and became fixed among a certain large number of modern humans and groups and thus a certain large number of people obtained Merge/UG, the change and evolution of culture, behavior and tools occurred.

Note that this proposal is consistent with the gradual cultural evolution hypothesis (e.g., McBrearty and Brooks 2000). It is thus incompatible with the so-called "Great Leap Forward," which claims that the emergence of human language occurred around 50 kya and then triggered vast cultural and behavioral change/evolution over a short period (Klein 2009b).

## 5 Conclusion

In this chapter, which provides ample archaeological/paleoanthropological and genetic evidence from recent studies, I have proposed that the Early Exodus of humans out of Africa occurred around 130,000 years ago. Based on the simplest hypothesis on the relationship between the Out-of-Africa migration and the emergence of human Merge/UG, I have then suggested the hypothesis of an earlier emergence of the human language faculty, which claims that human Merge/UG emerged as early as 130,000–150,000 years ago. It is much earlier than has been widely assumed that modern humans became "truly linguistic humans."[17]

## Notes

* This is a revised and expanded version of section 4 of Ike-uchi (2014). I would like to thank Shun-ichiro Inada and Koji Fujita for allowing me to gain access to some of the papers relevant to this article.
1 Of course, the question is "when did UG first emerge in the brain of one *Homo sapiens* in evolutionary history?" One reviewer for Evolang X asserts that this formulation of the question is "preposterous." Fundamentally, language is a matter of the human brain as the internalized knowledge of language. How could the origins and evolution of human language/UG be otherwise formulated?
2 Here, I assume that Pirahã also has the operation of recursive Merge, although it apparently does not have embedded clausal structures. The question of whether language has a generative procedure such as Merge is completely different from the question of whether it has embedded structures.

3  Also, for that matter, prior to the intra-Africa migration from (North) East Africa, which is not explored in this chapter.
4  I do not deny that so-called convergence would be one possibility. However, it would be extremely difficult (in fact, almost impossible) to discover sufficient niche conditions that caused one particular abstract common linguistic property to emerge in every place in the world.
5  For some recent general remarks on "Out-of-Africa," see Klein (2009) and Tattersall (2009).
6  See Adler et al. (2014) for the presence of Levallois technology in the Southern Caucasus and the possibility of a different interpretation.
7  However, see the critiques by Kaifu and Fujita (2012) and Kaifu (2013), and Curnoe et al. (2012).
8  See also Liu et al. (2010a).
9  Furthermore, according to Shen et al. (2002), the Liujiang (southern China) well-preserved fossils of *Homo sapiens* are dated by U-series dating to at least 68 kya, but more likely to 111–139 kya (or older than 153 kya). The same logic can also apply here. See also Liu et al. (2013) and Xiao et al. (2014) for some unclear cases of hominin fossils.
10  According to Wang et al. (2014), Middle Pleistocene bifaces from the Fengshudao site located in the Bose Basin, Guangxi, southern China, are around 803 kya. They indicate that "although Fengshudao may be a case of western Acheulean hominins dispersing into the Bose Basin from nearby South Asia, it is quite possible that the Fengshudao bifaces may be considered an example of convergent evolution." That is, it is fairly certain that they are not the stone tools directly associated with those originally from East Africa.
11  Hominin fossils found in early twentieth century in Qafseh and Skhul caves in the Levant, Israel, Middle East, are viewed as evidence which shows that *Homo sapiens* was there 92–115/119 kya (Shea (2003, 2008)). Humans were considered to have migrated out of Africa through the northern route via the Nile valley. In this chapter, I essentially follow Shea (2008) and Tattersall (2009), accepting the hypothesis that humans could not reach other areas because they became extinct 75 kya because of the advent of the Ice Age. Note that the timing of this dispersal is not incompatible with the hypothesis of an earlier human UG emergence. However, if we consider the human bones found in Misliya Cave, Israel, which are believed to be 150,000 years old (Brahic (2014)), the timing of the emergence of human UG may need to be slightly modified. See note 16.
12  The concept of a molecular clock is "[T]he idea that nucleotide substitutions accumulate at a constant rate over time and that this rate can therefore be used to estimate divergence times between sequences" (Scally and Durbin (2012: 752)).
13  Recently, Fu et al. (2013) have proposed a rate of $2.67 \times 10^{-8}$ substitutions per site per year for the whole molecule, which is higher than that proposed by Scally and Durbin (2012) and is "approximately 1.6 fold higher than the fossil-calibrated rate . . ." (Fu et al. (2013: 556)). Furthermore, the authors note that the calculation based on this rate implies that Out of-Africa occurred less than 62.4–94.9 kya (p. 556). A tentative solution for this inconsistency that I could suggest at this point is that Scally and Durbin (2012) used the nuclear genome with a lower mutation rate, while Fu et al. (2013) used ancient mitochondrial DNA, which is not a typical genome, in their investigation (Gibbons (2013)).
14  However, see Fagundes et al. (2007). Based on a Bayesian analysis of the genomic data under a simple African replacement model with exponential population growth, these authors suggest that "[T]he Out-of-Africa migration, initially involving only ≈450 effective individuals would have occurred some 51 Kya, . . ." (p. 17615). See also Mellars et al. (2013).

15 I admit that this view of an early exit from Africa currently remains a minority view (Brahic (2014)).
16 The number 150 kya is tentatively calculated based on the assumption that it must have taken UG some time to spread among a certain number of *Homo sapiens*.
17 Of course, much further investigation is necessary. For instance, as is correctly pointed out by Dennell (2010: 513), it is expected that *Homo sapiens* fossils and/or MSA stone tools will be discovered in the areas between Arabia and (southern) China.

## References

Adler, D. S., K. N. Wilkinson, S. Blockley, D. F. Mark, R. Pinhasi, B. A. Schmidt-Magee, S. Nahapetyan, C. Mallol, F. Berna, P. J. Glauberman, Y. Raczynski-Henk, N. Wales, E. Frahm, O. Jöris, A. MacLeod, V. C. Smith, V. L. Cullen, and B. Gasparian. 2014. Early Levallois technology and the lower to middle Paleolithic transition in the Southern Caucasus. *Science* 345:1609–1613.

Armitage, Simon, Sabah Jasim, Anthony Marks, Adrian Parker, Vitaly Usik, and Hans-Peter Uerpmann. 2011. The southern route "Out of Africa": Evidence for an early expansion of modern humans into Arabia. *Science* 331:453–456.

Ash, Patricia, and David Robinson. 2010. *The emergence of humans*. Chichester, UK: Wiley-Blackwell.

Bae, Christopher, Wei Wang, Jianxin Zhao, Shengming Huang, Feng Tian, and Guanjun Shen. 2014. Modern human teeth from Late Pleistocene Luna Cave (Guangxi, China). *Quaternary International* 354:169–183.

Berwick, Robert, Angela Friederici, Noam Chomsky, and Johan Bolhuis. 2013. Evolution, brain, and the nature of language. *Trends in Cognitive Sciences* 17:89–98.

Bošković, Željko. 2015 From the complex NP constraint to everything: On deep extraction across categories. Lingbuzz/002278.

Brahic, Catherine. 2014. Human exodus may have reached China 100,000 years ago. *New Scientist*, 08 August 2014.

Chomsky, Noam. 2013. Problems of projection. *Lingua* 130:33–49.

Chomsky, Noam. 2015. Problems of projection: Extensions. In *Structures, strategies and beyond: Studies in honour of Adriana Belletti*, ed. Elisa Di Domenico, Cornelia Harmann, and Simona Matteini, 3–16. Amsterdam: John Benjamins.

Curnoe, Darren, Ji Xueping, Andy Herries, Bai Kanning, Paul Taçon, Bao Zhende, David Fink, Zhu Yunsheng, John Hellstrom, Luo Yun, Gerasimos Cassis, Su Bing, Stephen Wroe, Hong Shi, William Parr, Huang Shengmin, and Natalie Rogers. 2012. Human remains from the Pleistocene-Holocene transition of Southwest China suggest a complex evolutionary history for East Asians. *PLoS ONE* 7:e31918.

Dennell, Robin. 2010. Early *Homo sapiens* in China. *Nature* 468:512–513.

Fagundes, Nelson, Nicolas Ray, Mark Beaumont, Samuel Neuenschwander, Francisco Salzano, Sandro Bonatto, and Laurent Excoffier. 2007. Statistical evaluation of alternative models of human evolution. *Proceedings of the National Academy of Sciences* 104:17614–17619.

Fu, Qiaomei, Alissa Mittnik, Philip L. F. Johnson, Kirsten Bos, Martina Lari, Ruth Bollongino, Chengkai Sun, Liane Giemsch, Ralf Schmitz, Joachim Burger, Anna Maria Ronchitelli, Fabio Martini, Renata G. Cremonesi, Jiří Svoboda, Peter Bauer,

David Caramelli, Sergi Castellano, David Reich, Svante Pääbo, and Johannes Krause. 2013. A revised timescale for human evolution based on ancient mitochondrial genomes. *Current Biology* 23:553–559.

Fujita, Koji. 2013. From generative grammar to evolutionary linguistics: A new enterprise of generative grammar. In *The state-of-the-art of generative linguistic studies*, ed. Masayuki Ike-uchi and Takuya Goro, 95–123. Tokyo: Hituji-shobo.

Gibbons, Ann. 2012. Turning back the clock: Slowing the pace of prehistory. *Science* 338:189–191.

Gibbons, Ann. 2013. Clocking the human exodus Out of Africa. *ScienceNOW*:21 March.

Hauser, Marc, Noam Chomsky, and Tecumseh Fitch. 2002. The faculty of language: What is it, who has it, and how did it evolve? *Science* 298:1569–1579.

Ike-uchi, Masayuki. 2012. Recent archaeological evidence suggests much earlier emergence of human UG. In *Proceedings of the 9th international conference on the evolution of language*, ed. Thomas Scott-Phillips, Mónica Tamariz, Erica Cartmill, and James Hurford, 454–455. Singapore: World Scientific.

Ike-uchi, Masayuki. 2014. Notes on the emergence of FLN and FLB: An MP-based approach. In *The design, development and evolution of language: Exploration in biolinguistics*, ed. Koji Fujita, Naoki Fukui, Noriaki Yusa, and Masayuki Ike-uchi, 214–237. Tokyo: Kaitaku-sha.

Kaifu, Yousuke. 2013. *Homo sapiens*'s expansion into Eurasia: Recent development. In Homo sapiens *and paleoanthropic man: Replacement of Neanderthals by modern humans from the viewpoint of Paleolithic archaeology*, ed. Yoshihiro Nishiaki, 3–17. Tokyo: Rokuichi-shobo.

Kaifu, Yousuke, and Masaki Fujita. 2012. Fossil record of early modern humans in East Asia. *Quaternary International* 248:2–11.

Klein, Richard. 2009a. Darwin and the recent African origin of modern humans. *Proceedings of the National Academy of Sciences* 106:16007–16009.

Klein, Richard. 2009b. *The human career: Human biological and cultural origins*. Chicago: University of Chicago Press.

Liu, Wu, Chang-Zhu Jin, Ying-Qi Zhang, Yan-Jun Cai, Song Xing, Xiu-Jie Wu, Hai Cheng, R. Lawrence Edwards, Wen-Shi Pan, Da-Gong Qin, Zhi-Sheng An, Erik Trinkaus, and Xin-Zhi Wu. 2010b. Human remains from Zhirendong, South China, and modern human emergence in East Asia. *Proceedings of the National Academy of Sciences* 107:19201–19206.

Liu, Wu, Lynne Schepartz, Song Xing, Sari Miller-Antonio, Xiujie Wu, Erik Trinkaus, and María Martinón-Torres. 2013. Late Middle Pleistocene hominin teeth from Panxian Dadong, South China. *Journal of Human Evolution* 64:337–355.

Liu, Wu, Xianzhu Wu, Shuwen Pei, Xiujie Wu, and Christopher Norton. 2010a. Huanglong Cave: A Late Pleistocene human fossil site in Hubei Province, China. *Quarternary International* 211:29–41.

McBrearty, Sally, and Alison Brooks. 2000. The revolution that wasn't: A new interpretation of the origins of modern human behavior. *Journal of Human Evolution* 39:453–563.

Mellars, Paul, Kevin Gori, Martin Carr, Pedro Soares, and Martin Richards. 2013. Genetic and archaeological perspectives on the initial modern human colonization of southern Asia. *Proceedings of the National Academy of Sciences* 110:10699–10704.

Oppenheimer, Stephen. 2012. Out-of-Africa, the peopling of continents and islands: Tracing uniparental gene trees across the map. *Philosophical Transactions of the Royal Society B: Biological Sciences* 367:770–784.

Renfrew, Colin. 2007. *Prehistory: The making of the human mind*. New York: Modern Library.
Reyes-Centeno, Hugo, Silvia Ghirotto, Florent Détroit, Dominique Grimaud-Hervé, Guido Barbujani, and Katerina Harvati. 2014. Genomic and cranial phenotype data support multiple modern human dispersals from Africa and a southern route into Asia. *Proceedings of the National Academy of Sciences* 111:7248–7253.
Rose, Jeffrey, Vitaly Usik, Anthony Marks, Yamandu Hilbert, Christopher Galletti, Ash Parton, Jean Marie Geiling, Viktor Černý, Mike Worley, and Richard Roberts. 2011. The Nubian Complex of Dhofar, Oman: An African Middle Stone Age industry in southern Arabia. *PLoS ONE* 6.11:e28239.
Scally, Aylwyn, and Richard Durbin. 2012. Revising the human mutation rate: Implications for understanding human evolution. *Nature Reviews Genetics* 13:745–753.
Shea, John. 2003. Neandertals, competition, and the origin of modern human behavior in the Levant. *Evolutionary Anthropology* 12:173–187.
Shea, John. 2008. Transitions or turnovers? Climatically-forced extinctions of *Homo sapiens* and Neanderthals in the east Mediterranean Levant. *Quaternary Science Reviews* 27:2253–2270.
Shen, Guanjun, Wei Wang, Qian Wang, Jianxin Zhao & Kenneth Collerson, Chunlin Zhou, and Phillip Tobias. 2002. U-Series dating of Liujiang hominid in Guangxi, Southern China. *Journal Human Evolution* 43:817–829.
Shen, Guanjun, Xianzhu Wu, Qian Wang, Hua Tu, Yue-xing Feng, and Jian-xin Zhao. 2013. Mass spectrometric U-series dating of Huanglong Cave in Hubei Province, central China: Evidence for early presence of modern humans in eastern Asia. *Journal of Human Evolution* 65:162–167.
Tattersall, Ian. 2009. Human origins: Out of Africa. *Proceedings of the National Academy of Sciences* 106:16018–16021.
Wang, Wei, Christopher Bae, Shengmin Huang, Xin Huang, Feng Tian, Jinyou Mo, Zhitao Huang, Chaolin Huang, Shaowen Xie, and Dawei Li. 2014. Middle Pleistocene bifaces from Fengshudao (Bose Basin, Guangxi, China). *Journal of Human Evolution* 69:110–122.
Xiao, Dongfang, Christopher Bae, Guanjun Shen, Eric Delson, Jennie Jin, Nicole Webb, and Licheng Qiu. 2014. Metric and geometric morphometric analysis of new hominin fossils from Maba (Guangdong, China). *Journal of Human Evolution* 74:1–20.

# 13 Two aspects of syntactic evolution[1]

*Michio Hosaka*

## 1 Introduction

There has been growing interest in language evolution among non-linguists as well as linguists. Even in the dawn of evolutionary theory, Darwin (1859, 1871) himself mentioned the adaptive emergence of language as the extension of natural selection on biological evolution, but the dispute about it has neither reached nor gotten close to a valid conclusion until recently. Pinker and Bloom (1990) posit that the human language faculty is a complex biological adaptation that evolved by natural selection for communication. On the contrary, Chomsky refuses their claim and strongly insists that language is not the production of natural selection and did not evolve adaptively solely for communication. Although both points of view seem to be quite contrary, their proposals are not incompatible if viewed from a different angle. It can be thought that they see opposite sides of the same coin. This chapter proposes that there are both adaptive and non-adaptive aspects in the evolutionary process of language. Specifically, on the basis of empirical evidence of case phenomena attested in Icelandic and the historical change of the English language, it is argued that language should be divided into Language of Thought (LoT) and Language of Communication (LoC) on the basis of the assumption that case plays an important role in identifying thematic roles of nouns exclusively at LoC. Then it will be proved that the former has a non-adaptive process of evolution, whereas the latter is vulnerable to selection pressure.

## 2 Issues concerning case/Case

Within the standard framework of GB theory, NP lacking case must move to a position where it is assigned case because of the Case Filter in (1).

(1) *NP if NP has phonetic content but no Case.

For example, as illustrated by (2b), *the book*, which does not receive case in its base position, must move to Spec, IP in order to receive nominative case.

(2) a. The book was bought by John.
    b. [$_{IP}$ the book$_i$ [$_{I'}$ was+I [$_{VP}$ [$_{VP}$ bought $t_i$ ] by John]]]

The early stage of the Minimalist Approach tries to explain the role of case under the Visibility Condition in (3).

(3) Visibility Condition
A chain is visible for θ-marking only if it has a Case position.
(Chomsky (1995: 119))

It means that in (2b) *the book*$_i$ and $t_i$ form a chain, in which case is assigned to its left end and a thematic role is assigned to its right end. Then the chain becomes visible so that the noun can be interpretable.

However, Icelandic quirky subjects as exemplified in (4) cast doubt on the validity of the Visibility Condition.

(4) Honum    var    hjálpað
    him(D)   was    helped (Zaenen, Maling and Thráinsson (1985: 442))

*Honum*, which is already assigned dative, does not need to move to the subject position for the purpose of receiving nominative case. Therefore, within the present minimalist framework, case theory is integrated into a feature checking system with the assumption that the movement of NP to the subject position is driven by another factor such as the EPP feature checking. In addition, under the present feature checking mechanism, it is assumed that case features such as uNom and uAcc do not have any corresponding features to agree with, unlike a φ feature and an EPP feature, as illustrated in (5).

(5) [$_{TP}$ he$_{i<EPP>}$ [$_{T'}$ T$_{<EPP, u\phi>}$ [$_{vP}$ $t_{i<EPP, i\phi, uNom>}$ sells$_j$+v$_{<u\phi>}$ [$_{VP}$ $t_j$ books $_{[i\phi, uAcc]}$ ]]]]

This assumption makes it difficult to recognize the raison d'être of case itself and incurs various other proposals such as Pesetsky and Torrego's (2011), who claim that case is closely related to tense, as in (6), and Marantz's (1991) and Haeberli's (2001), who argue against abstract case as in (7).[2]

(6) The well-known correlation between tense and nominative case suggests that case might in fact be an uninterpretable instance of *tense* (T).
(Pesetsky and Torrego (2011: 68))

(7) a. Giving content to the theory of morphological case allows for the elimination of abstract Case theory from the theory of syntax.
(Marantz (1991: 234))

  b. First, it is argued that the UG concept of abstract Case, which has played a central role for the analysis of nominal constituents in the generative literature, can be eliminated from the grammar. . . .
(Haeberli (2001: 279))

In this paper, on the basis of the new dynamic model of language presented in the following section, it is argued that close attention should be paid to visibility in order to clarify the role of case, and that the diachronic view of the English language can provide new clues to solving the quirky subject issues.

## 3 Dynamic model of language

The standard derivational model of the Minimalist syntax is roughly depicted in Figure 13.1.

The above model is still controversial regarding where structural case is assigned and realized. Early versions of the minimalist program assumed that realized case forms already exist in the Lexicon and are structurally licensed during the derivational process. On the contrary, the recent versions of the minimalist program presume that the realization of morphological case exists in the externalization process required by the SM interface, while case features are structurally licensed before SPELL-OUT. As mentioned before, such licensing of case features is considered to be epiphenomenal and simply exists for the rule that case features must be deleted before SPELL-OUT because they are uninterpretable at the CI Interface. This assumption seems insufficient to give adequate explanations for case phenomena which have long been widely disputed from both synchronic and diachronic points of view. Therefore, I propose that the language model in Figure 13.2 be adopted to explain why case exists.

In this model, the selection of semantic features and their merge is the first step, which leads to the rise of *Language of Thought* (LoT) directly connected to the Conceptual Interface which produces SEM1. On the contrary, *Language of Communication* (LoC) required by the Communicative Interface brings forth SEM2, which contains an interpretation different from that of SEM1.[3,4]

The study of language can be divided into the study of linguistic competence and the study of language use. The target of generative grammar is linguistic competence whereas language use is mainly dealt with in cognitive grammar, functional linguistics and sociolinguistics. Competence is closely related to the universality of languages while use connects to language diversity. It is natural that linguists on both sides of the issue conflict with one another as they are

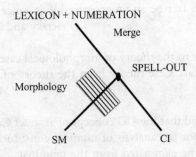

*Figure 13.1* The Standard Derivation Model.

*Two aspects of syntactic evolution* 201

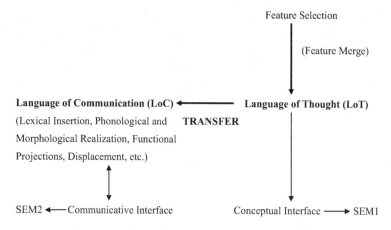

*Figure 13.2* The Dynamic Model of Language.

looking into the different aspects of language. Nevertheless, competence and use are equally worth deep study, and uniting each approach can lead us to a comprehensive understanding of language. In this model, it is assumed that universal language competence lies in LoT, whereas languages, which are actually used and various, are formed in LoC.[5] Chomsky repeatedly insists that there has been no relevant evolution of the language faculty since the trek from Africa some 50,000 to 100,000 years ago.[6] This unchangeable faculty exclusively lies in LoT, whereas various languages which emerged as the result of the historical change are in LoC. Therefore, this paper assumes that the syntactic derivation before Transfer is limited to the external merge of semantic features and that Functional Projections (FPs), where movement and agreement operate, and morpho-phonological realization exist in LoC after Transfer.

For example, (8b, c) can be assumed to be the derivational process of thinking that *Jane hit Tom*.

(8) a. Jane hit Tom.
 b. [$_{\sqrt{hit}}$ √Tom √hit ][7]
 c. [$_{\sqrt{hit}}$ √Jane [$_{\sqrt{hit}}$ √Tom √hit ]]

*Hit* accompanies the theme (*Tom*) and the agent (*Jane*). First the theme (*Tom*) is merged to *hit* as in (8b) and subsequently the agent (*Jane*) is merged as in (8c). In other words, nothing but a simple merge operation (i.e. external merge) is postulated in the process of the derivation, with no case, no agreement and no movement presumed. It is conceivable that this simple ability to merge semantic features marks a watershed between human beings and other primates. In addition, LoT is related exclusively to a person producing a language and is never interfered with by the Communicative Interface. In other words, LoT is literally a perfect language in that SEM1 does not contain any ambiguity.

However, language in use is full of imperfections, which are the properties of LoC required by the Communicative Interface, and SEM2 is solely related to the interpretation of the output language. The semantic ambiguity occurs there, and the duality of semantics that Chomsky refers to is attributed to the difference between SEM1 and SEM2.[8] (9a) and (9b), for example, seem to be synonymous, but (9a) is an appropriate answer to *What did John do?* whereas (9b) is to *What did John do with my book?*

(9)  a. John gave your book to Mary.
     b. Your book, John gave _____ to Mary.

(Rizzi (2011: 2))

Therefore, there is attested semantic difference between (9a) and (9b). If such topicalization occurs after Transfer, it is impossible to yield the correct output to the CI interface, which requires us to presume different semantic interfaces at CI and SM.[9]

Then, what exists in the process of LoC? There are assumed to be morphophonological realizations, movement and functional projections, which are required by the Communicative Interface and, susceptible to historical change, yield language variation. In addition, the mechanism to produce such diversity is related to not only language itself but also to the universal factors to produce principles and constraints in common, which can be regarded as self-organization.[10]

## 4 Quirky subject revisited

The problem of case in Icelandic is a daunting challenge to the generativists. Notice that not a nominative noun but a dative noun occupies the subject position of each sentence, as exemplified in (10).

(10) a. Jóni      líkuðu     þessir    sokkar
        Jon.DAT  like.PL   these    socks.NOM
        'Jon likes these socks.'
     b. Þeim     var       hjálpað
        them.DAT was.SG   helped
        'They were helped.'
     c. Um veturinn    voru      konunginum     gefnar    ambáttir
        In the.winter  were.PL  the.king.DAT   given     slaves.NOM
        'In the winter, the king was given (female) slaves.'

(Bobaljik 2008: 298)

In addition, it is proved through several syntactic tests that these are not topic and focus positions but subject positions.[11] Then it is obvious that the movement of the dative to the subject position in (10) is not driven by case. Therefore, in the present framework of the Minimalist approach it is assumed that the dative moves there in order to check the EPP feature.

There exists something similar to Icelandic quirky subjects in Old English (OE) and Middle English (ME). Examples in (11) represent the quirky subject phenomena in OE.

(11) a. licade   us        efencuman
        liked    us.DAT    come
        'it pleased us to come' (Ælfred, Bede, 4.5.276.12; Visser (1963: I 28))
    b. Me        hingrode.
        me.DAT   hungered
        'I was hungry' (Mt 25.35, WSCp)
    c. and him        sy    geþancod
        and him.DAT   be    thanked
        'and may he be thanked' (LS 14(Margaret Ass 15) 255; Allen (1995: 353))

It is well known that these lead to nominative subject constructions in Present-Day English (PDE) through the process from impersonal constructions to personal constructions.[12] Although both quirky subjects in Icelandic and OE seem to be the same, an example from German as shown in (12) casts doubt on their similarity.

(12) Mir       wurde   geholfen.
     me.DAT    was     helped
     'I was helped (by somebody).' (Sigurðsson (2004: 143))

The dative (*mir*) in (12) is proved to not be a subject on the basis of several syntactic properties by Sigurðsson (2004) among others. (13) is a syntactic test based on Conjunction Reduction. It shows that a non-nominative subject coreferential with a preceding nominative subject is acceptable as the missing argument in a conjunction in Icelandic, as illustrated in (13a), whereas they are inacceptable in German as seen in (13b).

(13) a. *Icelandic*
        Ég        hafði mikið að gera og (mér)      var samt           ekki hjálpað
        I.NOM     had   much  to do  and (me.DAT)   was nonetheless    not  helped
        'I had a lot to do and (I) was nonetheless not helped.'
    b. *German*
        Ich       hatte viel zu tun, und *(mir)     wurde trotzdem     nicht geholfen
        I.NOM     had   much to do   and (me.DAT)   was   nonetheless  not   helped
                                                   (Sigurðsson (2004: 144))

Though it is impossible to carry out a syntactic test in OE and ME, Allen (1995) offers some valuable observations as to the subjecthood. (14) is an ME counterpart of Icelandic quirky subjects.

(14) Mildheorted beð þe man   þe      reouð his nehgebures  unselðe
     merciful    is  the man  who  regret his neighbors'  unfortune
     and likeð   here alre selðe
     and pleases their all fortune
     'merciful is the man who feels sorry about his neighbours' unhappiness
     and is pleased by the good fortune of them all'
                              (M.OEH II 95.33; adapted from Allen (1995: 244))

It follows that the gap before *likeð* can be regarded as a dative subject just like in Icelandic. Then Allen insists that such examples are found more abundantly in Early ME.[13] In addition, Allen's investigation of the verb *lyst* indicates that dative subjects were changing into nominative subjects during the ME period. Table 13.1 shows that dative subjects were predominant before the fourteenth century, whereas nominative subjects began to be used after the fifteenth century and became abundant in the sixteenth century.

If her observation is on the right track, it is assumed that nominative subjects became obligatory in the course of the historical change of English, which began at a stage of the German type (with non-dative subjects) and proceeded through a stage of the Icelandic type (with dative subjects).

Then, how can we place such a process in the Dynamic Model of Language? First, as (15) shows, LoT has the same representation in OE, ME, ModE and PDE respectively.

(15) Language of Thought
     [$_{\sqrt{like}}$ $\sqrt{John}_{<exp>}$ [$_{\sqrt{like}}$ $\sqrt{these\ socks}_{<th>}$ $\sqrt{like}$]]] (OE, ME, ModE, PDE)

What we should notice is that there is a thematic hierarchy when each semantic feature is merged. (16) is a thematic hierarchy postulated based on Allen (1995), in which the merging process works from the lowest to the highest. Then it produces the structure of (15).

(16) Agent > Experiencer > Recipient > Theme > Patient > Goal[14]

*Table 13.1 Lyst* in the NO PROP construction from OE to the sixteenth century (Allen (1995:288))

|       | AMB V | NOM V | DAT/ACC V | it V (to) EXP | it (to EXP) V |
|-------|-------|-------|-----------|---------------|---------------|
| OE    | 1     | 0     | 7         | 0             | 0             |
| 13thC | 0     | 0     | 3         | 0             | 0             |
| 14thC | 9     | 0     | 201       | 1             | 2             |
| 15thC | 0     | 23    | 8         | 0             | 0             |
| 16thC | 11    | 117   | 12        | 0             | 0             |

Next, the derivational process at LoC is mainly controlled by the need to make the thematic roles assigned at LoT visible to hearers, which can be called the Visibility Requirement.

(17) Visibility Requirement
Thematic roles of NP must be visible at the Communicative Interface.

The former Visibility Condition has been considered to be imposed at the CI interface.[15] In the model presented here, however, SEM1 is visible intrinsically because it is a direct output from LoT and what matters is SEM2 at the SM interface. In other words, as (18) shows, thematic roles at LoC are invisible to hearers and several systems such as case inflections, functional projections and prosody help to make them visible.

(18) a. [$_{\sqrt{like}}$ $\sqrt{John}_{<exp>}$ [$_{\sqrt{like}}$ $\sqrt{these\ socks}_{<th>}$ $\sqrt{like}$]]] → CI interface → SEM1
b. [$_{\sqrt{like}}$ $\sqrt{John}_{<exp \leftarrow necessary\ to\ identify>}$ [$_{\sqrt{like}}$ $\sqrt{these\ socks}_{<th \leftarrow necessary\ to\ identify>}$ $\sqrt{like}$]]]
→ SM interface → SEM2

In this paper we focus on morphological case and functional projections. (19) is an example of the morphological case realization.[16]

(19) Thematic roles and morphological cases
   a. *Agent* is made visible by *Nom.*
   b. *Experiencer* is made visible by *Dat.*
   c. *Recipient* is made visible by *Dat.*
   d. *Theme* is made visible by *Nom* or *Acc.*[17]

In (20) functional projections help to make thematic roles visible. For example, $F_{<ag>}$ makes an agent role visible through the Spec-Head relation.

(20) Thematic roles and FPs
   a. *Agent* is made visible by $F_{<ag>}$.
   b. *Experiencer* is made visible by $F_{<exp>}$.
   c. *Recipient* is made visible by $F_{<rec>}$.
   d. *Theme* is made visible by $F_{<th>}$.
   e. *Patient* is made visible by $F_{<pat>}$.
   f. *Goal* is made visible by $F_{<gl>}$.

(21) indicates the change of the ways to keep the visibility in the history of the English language.

(21) morphological case > morphological case and FP > FP

That is, as (22) illustrates, thematic roles were made visible by morphological case in early English, whereas in PDE, FP plays a role in identifying them as in (23).

(22) [$_{FP}$ eall ðis $_{<th\leftarrow acc,\ top>}$ [$_{F'}$ aredað+F$_{<top>}$ [$_{VP}$ se reccere$_{<ag\leftarrow nom>}$ [$_{V'}$ swiðe ryhte
     all this             arranges        the ruler         very rightly
[$_{V'}$ ~~eall ðis aredað~~]]]]]
~~all this arranges~~
'the ruler arranges all this very rightly' (CP 168.3; van Kemenade (1987: 17))

(23) [$_{FP}$ John$_{<ag>}$ [$_{F'}$ gave+F$_{<ag>}$ [$_{FP}$ Mary$_{<rec>}$ [$_{F'}$ gave+F$_{<rec>}$ [$_{FP}$ a ring$_{<th>}$ [$_{F'}$ ~~gave~~ +F$_{<th>}$
[$_{VP}$ ~~John~~ [$_{V'}$ ~~Mary [a ring gave~~]]]]]]]]][20]

This means that the rise of FP brought about a new device to identify thematic roles on the basis of its hierarchical structure. It is presumed that such a development of FP is considered as a kind of exaptation,[19] which expanded FP as in (24).

(24) F$_{<top>}$ > F$_{<pred>}$, F$_{<\theta>}$

That is, FP, which emerged to identify discourse features such as topic and focus, expanded itself to identify thematic roles and predicational relation.[20]

The structural change as illustrated in (25) is assumed to lie in the process from impersonal constructions to personal constructions discussed in Jespersen (1927).

(25) a. [$_{FP}$ þam cynge$_{<exp\leftarrow dat>,\ <top>}$ [$_{F'}$ licodon+F$_{<top>}$ [$_{VP}$ ~~þam cynge~~$_{<exp\leftarrow dat>}$
    [$_{V'}$ peran$_{<th\leftarrow nom>}$ ~~licodon~~]]]
b. [$_{FP}$ the king$_{<exp\leftarrow dat>,\ <pred>}$ [$_{F'}$ likeden+F$_{<pred>}$ [$_{VP}$ ~~the king~~ [$_{V'}$ peares$_{<th\leftarrow nom\ /}$
    $_{acc>}$ ~~likeden~~]]
c. [$_{FP}$ the king$_{<pred>,<exp>}$ [$_{F'}$ liked+F$_{<pred,\ exp>}$ [$_{FP}$ pears$_{<th>}$ [$_{F'}$ ~~liked~~+F$_{<th>}$ [$_{VP}$ ~~the king~~
    [$_{V'}$ ~~pears liked~~]]

As shown in (25a), F$_{<top>}$ was used as a device to identify topic and focus in an early stage, and in (25b) F$_{<pred>}$ emerged to identify predication, and then, at the same time as the loss of inflection, F$_{<\theta>}$ emerged to make thematic roles visible. This can be also regarded as the change from Topic Prominent Language to Subject Prominent Language.[21]

Let us go back to the problem of quirky subjects in Icelandic. As the loss of morphological inflection is not attested there, Icelandic still remains at stage b. It means that, as illustrated in (26), Icelandic keeps the morphological device to identify thematic roles and has developed PredP to clarify the relation between a subject and a predicate so that the dative behaving as a subject emerged.

(26) [$_{FP}$ Jóhi$_{<exp\leftarrow dat>,<pred>}$ [$_{F'}$ likuðu+F$_{<pred>}$ [$_{VP}$ ~~Jóhi~~$_{<exp\leftarrow dat>}$ [$_{V'}$ þessir sokkar$_{<th\leftarrow nom>}$
~~likuðu~~]]]

In addition, it enables us to explain the word order difference between Icelandic and German. In the subordinate clause of Icelandic, the V2 order

Table 13.2 Typological difference on the basis of morphological case and functional projections

|  | m-Case | $F_{<pred>}$ | $F_{<\theta>}$ |
| --- | --- | --- | --- |
| German | + | − | − |
| Icelandic | + | + | − |
| Old English | + | − | − |
| Middle English | ± | + | ± |
| ModE & PDE | − | + | + |

is maintained as in (27a), whereas the V-final order is used in German as in (27b).

(27) a. að   henni/stelpunum      *líkuðu*     hestarnir       *Icelandic*
       that her-DAT/girls-the-DAT liked-3.PL horses-the-NOM
       'that she/the girls liked the horses'
   b. dass ihr/den Mädchen       die Pferde *gefielen*      *German*
       that her-DAT/the-DAT girls the horses pleased
       'that the horses pleased her/the girls'

(Haider (2010: 23))

That is because, as shown in (28), the verb movement is presumed because of $F_{<pred>}$ in Icelandic while the verb is not moved in German because $F_{<pred>}$ is not assumed.[22]

(28) a. að [$_{FP}$ henni/stelpunum [$_{F'}$ líkuðu+$F_{<pred>}$ [$_{VP}$ ~~henni/stelpunum~~ [$_{V'}$ hestarnir ~~líkuðu~~ ]]]]
   b. dass [$_{VP}$ ihr/den Mädchen [$_{V'}$ die Pferde gefielen]]

These differences of case and FP are summarized in Table 13.2.

This shows that the morphosyntactic change in English is considered to proceed from the German type to the PDE type via the Icelandic type.

## 5 Adaptive and non-adaptive evolution of language

Lastly, we will argue about the raison d'être of case again and how (non-) adaptively language changed on the basis of the dynamic model repeated below. Chomsky often remarks that the initial evolution of language is more likely connected to a tool for thought.[23] If he is on the right track, it is not words but semantic features that are merged first. The merged structure is realized at LoT, which is connected to SEM1. As there are no words at this stage, no word order and no case can be attested. What exists are only features and the external merge operation putting them together. In this sense, LoT can be considered as a perfect language unchanged through time.

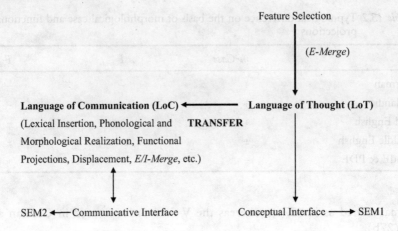

Figure 13.3 The Dynamic Model of Language

As (29) illustrates, at LoT a structure is built up by the external merge of semantic features in accordance with the thematic hierarchy and sent out to the Conceptual Interface.

(29) Language of Thought
 a. √hit <ag, pat>
 b. Thematic Hierarchy: Agent > Patient
 c. [√hit √Jane<ag> [√hit √Tom<pat> √hit]]

On the contrary, at LoC in (30) semantic features are directly sent out to the Communicative Interface, but as the thematic roles are not interpreted by hearers, some kinds of devices to identify thematic roles such as morphological case and functional projections are indispensable. Then, for example, FPs and syntactic movements (external merge and internal merge) are obligatory in PDE as in (30a), while postpositions play a significant role in identifying thematic roles in Japanese as in (30b).[24]

(30) Language of Communication
 a. [FP Jane<ag> [F' hit+F<ag> [FP Tom<pat> [F' hit+F<pat> [VP Jane [V' Tom hit ]]]]]]
 b. [VP Jane ga<ag←ga> [V' Tom wo<pat←wo> tataita ]]]

To put it briefly, it is considered that case exists for the purpose of the visibility of thematic roles at the Communicative Interface. The ways to make them visible are selected in the course of derivation at LoC and are synchronically and diachronically various depending on languages. In other words, languages keep changing adaptively at LoC. In addition, the explanation above suggests the possibility that language phenomena can be explained without mentioning controversial grammatical relations such as a subject and an object.[25]

Many minimalist theorists including Chomsky repeatedly insist that language did not evolve as a communication system. However, languages in the world are actually varied, and the diversity is closely connected to the Communicative Interface. The model presented in this chapter tries to explain both sides of language. One is LoT, which is non-adaptive and unchangeable, while the other is LoC, which is adaptive and diverse. The language which emerged some 50,000 to 100,000 years ago is no doubt LoT, and the diversity of language in both synchronic and diachronic aspects lies in LoC, which is still constantly changing.

## Notes

1 This chapter is a revised version in English of Hosaka (2014b), which was originally written in Japanese. I'd like to thank the following people for their valuable comments on early versions of the chapter: Heizo Nakajima, Masayuki Ikeuchi, Koji Fujita, Masanobu Ueda, Satoshi Oku, Yoshiki Ogawa.
2 Baker (2013) also mentions the role of case as quoted in the following:

> But given Chomsky's (2000b) promotion of Agree to primitive status, . . ., the role of case theory is now often back-grounded, and case is seen as an additional side effect of the primary relation of Agree, . . . . (Baker 2013: 632, emphasis added)

3 This model seems to be inconsistent with the Single-Cycle Hypothesis (Chomsky 2005: 18) and appears to be an older assumption with multiple representations. However, LoT and LoC are based on "virtual conceptual necessity" unique to thought and communication, respectively. Presupposing both representational levels is highly adequate from the viewpoint of language evolution discussed later. In addition, note that this assumption of LoT is not incompatible with the following remarks by Chomsky because LoT and LoC are considered to be continuum in this model.

> One suggestion that Jerry [Fodor] proposes which seems to me to require more evidence is that there is a language of thought. And the question is whether the language of thought is any different from whatever our universal, internal language is. As far as I can see, we can't tell anything about the language of thought other than it's reflection of whatever our language is. And if it's true – as it is likely – that the existing and, indeed the attainable languages are only superficially different, then the core that they share has a good claim to be the language of thought – so far as I can see. (Chomsky 2012a: 71–72)

4 SEM2 can be regarded as "interpretive components of thought"(Chomsky 2007: 13).
5 It makes us assume both syntax with only external merge at LoT and syntax with external and internal merge at LoC. The former can be called the primary syntax (Narrow Syntax) and the latter the secondary syntax (Broad Syntax). It entails that phenomena such as movement and agreement are postulated only in the secondary syntax at LoC.
6 See Berwick and Chomsky (2011: 19) among others.
7 √ means a bundle of semantic features.
8 It is claimed that this duality of semantics is related to Internal Merge in Chomsky (2007, 2008, 2012b).

9  It can be argued that there are appropriate discourse features before Transfer, but it is more adequate as the model of thought to exclude any interference from the Communicative Interface.
10 In van Trijp (2012), the rise of case is regarded as one of the examples of self-organization, but he does not mention the loss of case inflection. In this chapter, the rise and loss of case inflection are explained in terms of the balance between computational efficiency and communicative efficiency at LoC.
11 Freidin and Sprouse (1991) argue about subjecthood on the basis of the empirical evidence such as anaphoric bindings, inversion in interrogative sentences and subject raising.
12 See Jespersen (1927: 11.2) for the traditional assumption.
13 Allen also points out some examples of conjunction reduction in OE, which leads us to the possibility that PredP mentioned later had already emerged in Late OE.
14 Other proposals about a thematic hierarchy include the following:

Agent > Theme > Goal > Obliques (manner, location, time, . . .) (Larson (1988))
Agent > Goal > Theme (Jackendoff (1972), Grimshaw (1990))
Agent > Theme > Location (Baker (1997))

15 Sigurðsson (2012:193) states, "Event licensing entails θ-licensing. While thus being syntactically licensed, θ-roles are interpreted or read off at the C-I interface, on the basis of structural and lexical information transferred from syntax."
16 As the relation between semantic features and morphological cases is not universal but idiosyncratic, (19) should be regarded as an instance of Accusative languages.
17 Assuming the hierarchy of Nom > Acc, we can give a theoretical explanation to Burzio's generalization (Burzio 1986: 178–186) that case is assigned to the object iff a θ-role is assigned to the subject. As for the unaccusativity in (i), theme is made visible by Nom in (ia) whereas theme must be made visible by Acc in (ib) because Nom has a priority to make the agent visible.

(i) a. dass der Fehler gefunden wurde/war
       that the mistake-NOM found was/was
       'that the mistake was found'
    b. dass er den Fehler gefunden hat
       that he-NOM the mistake-ACC found has
       'that he has found the mistake' (adapted from Haider (2010: 276))

Moreover, as for a nominative object in Icelandic in (ii), as there is no agent in the sentence, the theme is made visible by Nom.

(ii) Jóni líkuðu þessir sokkar
     Jon.DAT like.PL these socks.NOM
     'Jon likes these socks.'

18 See Hosaka (2013) for the derivation of the double object construction.
19 The concept of exaptation can be attributed to evolutionary biology by Gould and Vrba (1982). Owing to Lass (1990), the explanation by exaptation prevailed even in the field of historical linguistics. The proposal here is based on the assumption that linguistic structures as well as lexical and morphological phenomena are vulnerable to exaptation.
20 In Tanaka (2010) the rise of PredP in English infinitival constructions is discussed and Bowers (2002) argues about the existence of PredP in Icelandic.
21 As explained in Li and Thompson (1976), from the typological point of view, languages in the world are divided into Topic Prominent Languages such as

Japanese and Subject Prominent Languages such as PDE. Interestingly enough, Proto-Indo European is supposed to be a Topic Prominent Language. It entails that English changed from a Topic Prominent Language to a Subject Prominent Language in the course of its history. In this chapter, such change is recaptured as a structural change (the rise of PredP).
22 See Hosaka (2014a) for the dynamic feature of the FP structure.
23 See Chomsky (2008, 2012a, b, 2013) among others.
24 Baker's following remarks are worth mentioning:

> However, it is also true that if a functor takes more than one argument, it must have some way to tell which argument is which; this is necessary in order to distinguish restaurant reviews ("Man eats shark") from suspense movies ("Shark eats man"). The UTAH performs this function of distinguishing the different arguments of the verb by way of virtually the only method available in Chomsky's very spare system: it "merges" the arguments into the representation at systematically different points. Therefore, there does seem to be a place for the UTAH within the limits of "virtual conceptual necessity."
> (Baker 1997: 122)

In this chapter, case and FPs at LoC are assumed to identify thematic roles projected at LoT on the basis of UTAH.
25 There have been long-standing debates concerning the correspondence between meanings, case and grammatical relations in the field of generative grammar as well as relational grammar. In this chapter, assuming that case which is realized through morphology or structure exists for the purpose of identifying thematic roles at LoC, it is possible to explain language structures without referring to grammatical relations.

# References

Allen, Cynthia L. 1995. *Case Marking and Reanalysis*. Oxford: Oxford University Press.
Baker, Mark 1997. Thematic Roles and Syntactic Structure. In *Elements of Grammar*, ed. L. Haegeman, 73–137. Dordrecht: Kluwer.
Baker, Mark C. 2013. Agreement and Case. In *The Cambridge Handbook of Generative Syntax*, ed. Marcel den Dikken, 607–654. Cambridge, UK: Cambridge University Press.
Berwick, Robert C., and Noam Chomsky. 2011. The Biolinguistic Program: The Current State of its Evolution and Development. In *Biolinguistic Enterprise: New Perspectives on the Evolution and Nature of the Human Language Faculty*, ed. Anna Maria Di Sciullo and Cedric Boeckx, 19–41. Oxford: Oxford University Press.
Bobaljik, Jonathan David. 2008. Where's φ? Agreement as a Post-syntactic Operation. In *Phi-Theory: Phi-features Across Interfaces and Modules*, ed. Daniel Harbour, David Adger, and Susan Béjor, 295–328. Oxford: Oxford University Press.
Bowers, John. 2002. Transitivity. *Linguistic Inquiry* 33:183–224.
Burzio, Luigi (1986) *Italian Syntax*. Dordrecht: Kluwer.
Chomsky, Noam. 1995. *The Minimalist Program*. Cambridge, MA: MIT Press.
Chomsky, Noam. 2005. Three Factors in Language Design. *Linguistic Inquiry* 36:1–22.
Chomsky, Noam. 2007. Approaching UG from Below. In *Interfaces + Recursion = Language? Chomsky's Minimalism and the View from Syntax-Semantics*, ed. Uli Sauerland and Hans-Martin Gärtner. 1–29. Berlin: Mouton de Gruyter.

Chomsky, Noam. 2008. On Phases. In *Foundational Issues in Linguistic Theory: Essays in Honor of Jean-Roger Vergnaud*, ed. Robert Freidin, Carlos P. Oterro, and Maria Luisa Zubizarreta, 133–166. Cambridge, MA: MIT Press.

Chomsky, Noam. 2012a. *The Science of Language: Interviews with James McGilvray*. Cambridge, UK: Cambridge University Press.

Chomsky, Noam. 2012b. Poverty of the Stimulus: Willingness to be Puzzled. In *Rich Language From Poor Inputs*, ed. Massimo Piattelli-Palmarini and Roberts C. Berwick, 61–67. Oxford: Oxford University Press.

Chomsky, Noam. 2013. Problems of Projection. *Lingua* 130:33–49.

Darwin, Charles. 1859. *On the Origin of Species*. London: John Murray.

Darwin, Charles. 1871. *The Descent of Man, and Selection in Relation to Sex*. London: John Murray.

Freidin, Robert, and Rex A. Sprouse. 1991. Lexical Case Phenomena. In *Principles and Parameters in Comparative Grammar*, ed. Robert Freidin, 392–416. Cambridge, MA: MIT Press.

Gould, Stephen Jay, and Elisabeth S. Vrba. 1982. Exaptation-A Missing Term in the Science of Form. *Paleobiology* 8.1:4–15.

Grimshaw, Jane. 1990. *Argument Structure*. Cambridge, MA: MIT Press.

Haeberli, Eric. 2001. Deriving Syntactic Effects of Morphological Case by Eliminating Abstract Case. *Lingua* 111:279–313.

Haider, Hubert. 2010. *The Syntax of German*. Cambridge, UK: Cambridge University Press.

Hauser, Marc D., Noam Chomsky and W. Tecumseh Fitch. 2002 The Faculty of Language: What Is It, Who Has It, and How Did It Evolve? *Science* 298(22): 1569–1579.

Hosaka, Michio. 2013. Niju mokutekigo kobun saikou. In *Seiseigengogaku no ima*, ed. Masayuki Ikeuchi and Takuya Goro, 67–93. Tokyo: Histuzi Shobo.

Hosaka, Michio. 2014a. Dynamic Change of Language Structure: Focusing on Word Order Change in the History of English. In *Studies in Modern English: The Thirtieth Anniversary Publication of the Modern English Association*, ed. Ken Nakagawa, 135–151. Tokyo: Eihosha.

Hosaka, Michio. 2014b. Kaku no sonzaiigi to tougo henka (The Raison d'être of Case and Syntactic Change). In *Gengo no sekkei hattatsu shinka*, ed. Fujita, Koji, Naoki Fukui, Noriaki Yusa, and Masayuki Ikeuchi, 257–278. Tokyo: Kaitakusha.

Jackendoff, Ray. 1972. *Semantic Interpretation in Generative Grammar*. Cambridge, MA: MIT Press.

Jackendoff, Ray. 1999. Possible Stages in the Evolution of the Language Capacity. *Trends in Cognitive Sciences* 3(7): 272–279.

Jespersen, Otto. 1927. *A Modern English Grammar on Historical Principles* III. Heidelberg: Carl Winters Universitätsbuchhandlung.

Kemenade, Ans van. 1987. *Syntactic Case and Morphological Case in the History of English*. Dordrecht: Foris Publications.

Larson, Richard. 1988. On the Double Object Construction. *Linguistic Inquiry* 19:335–392.

Lass, Roger. 1990. How to Do Things with Junk: Exaptation in Language Evolution. *Journal of Linguistics* 26:79–102.

Li, Charles N., and Sandra A. Thompson. 1976. Subject and Topic: A New Typology of Language. In *Subject and Topic*, ed. Charles N. Li, 459–489. New York: Academic Press.

Marantz, Alec. 1991. Case and Licensing. *ESCOL '91: Proceedings of the Eighth Eastern States Conference on Linguistics*, 234–253. Columbus: The Ohio State University.

Pesetsky, David, and Esther Torrego. 2011. Case. In *The Oxford Handbook of Linguistic Minimalism*, ed. Cedric Boeckx, 52–72. Oxford: Oxford University Press.

Pinker, Steve, and Paul Bloom. 1990. Natural Language and Natural Selection. *Behavioral and Brain Sciences* 13, 707–784.

Pinker, Steve, and Ray Jackendoff. 2005. The Faculty of Language: what's special about it? *Cognition* 95(2): 201–236.

Rizzi, Luigi. 2011. Syntactic Cartography and the Syntacticisation of Scope-Discourse Semantics. In *Philosophical Papers Dedicated to Kevin Mulligan*, ed. Anne Reoul, http://www.philosophie.ch/kevin/festschrift/

Sigurðsson, Halldór Ármann. 2004. Icelandic Non-nominative Subjects: Facts and Implications. In *Non-nominative Subjects*, ed. Pri Bhaskararao and K. V. Subbarao, Vol. 2, 137–159. Amsterdam and Philadelphia: John Benjamins.

Sigurðsson, Halldór Ármann. 2012. Minimalist C/case. *Linguistic Inquiry* 43: 191–227.

Tanaka, Tomoyuki. 2010. Agreement, Predication, and the Rise of Functional Categories in Nonfinite Clauses. *English Linguistics* 27.2:374–398.

Trijp, Remi van. 2012. The Evolution of Case Systems for Marking Event Structure. In *Experiments in Cultural Language Evolution*, ed. Luc Steels, 169–205. Amsterdam: John Benjamins.

Visser, F. Th. 1963. *An Historical Syntax of the English Language*, 4 vols. Leiden: E. J. Brill.

Zaenen, Annie, Joan Maling, and Höskuldur Thráinsson. 1985. Case and Grammatical Functions: The Icelandic Passive. *Natural Language and Linguistic Theory* 3:441–483.

# Part V
# Topics in neurobiology

# 14 Syntax in the brain*

*Noriaki Yusa*

## 1 Introduction

Humans share with other species the ability to associate sounds and meanings in a particular way (Kaminki, Call and Fisher 2004). However, the unbounded nature of the association is uniquely designated to humans. While other species convey a limited, fixed array of meanings by using a rigid order of body motions, humans convey an unbounded set of different meanings by using language. This unique property of unboundedness is generated by the internalized mechanisms (internalized language, *I*-language) in the brain. The basic property of I-language is to provide "an unbounded array of hierarchically structured expressions that receive interpretations at two interfaces, sensorimotor for externalization and conceptual-intentional for mental processes"(Chomsky 2013: 647). Linear order is a part of externalization required by properties of the sensorimotor system, during the process of which hierarchical structure is mapped onto linear sequences of words. Linear or sequential information is not specific to humans. In this sense, hierarchical structures are a defining feature of human language, forming a division between human and nonhuman communication systems (Anderson 2008; Moro 2011).

Frank et al. (2012), however, challenge the taken-for-granted assumption that hierarchical structure is basic to human language (Chomsky 1957, Hauser et al. 2002). Drawing from evidence related to psycholinguistics, computational linguistics and cognitive neuroscience, they claim that sequential structure is more fundamental to language *use* (i.e., production and comprehension) than hierarchical structure. Chomsky (1957) has already shown in a convincing way that sequential information based on probabilistic generalizations about surface distribution cannot characterize the essential properties of human language. In fact, many important generalizations about linguistic knowledge (e.g., knowledge regarding subject-verb agreement, reflexive pronouns, negative polarity items; knowledge regarding syntactic constituency (i.e., syntactic unit) in terms of pronominal anaphora, phrasal movement, ellipsis) have been made in terms of hierarchical structure. Frank et al. (2012), however, deny the involvement of hierarchical structure in language use or sentence processing. They claim that even if the offline (with no time limit) mental representations may be hierarchical, the real-time comprehension and production of language do not resort to such abstract hierarchical structure as has been explicated in the literature of generative grammar, because sequential information and simple extra-linguistic heuristics are sufficient enough for language use. Under Frank et al's view, which is surprisingly still prevalent in computational cognitive science (Berwick et al. 2011), the computational-representational theories of mind developed by theoretical linguists cannot be linked to language processing theories. This implies that the generative system responsible for linguistic knowledge does not underlie language comprehension and production, that is, hierarchical structure is not accessed in language processing. Similarly, language acquisition is characterized as learning word sequences or clusters of words ("chunks" or "constructions") on the basis of a surface-based probabilistic tendency. This raises a renewed research interest in the relation between hierarchical structure in linguistic knowledge and its significance in real-time language use (see Lewis and Philips 2015 for this problem). In addition, they claim that since language comprehension and production primarily rely on sequential information, hierarchical structure is problematic for evolution. The next section discusses the phenomenon called "agreement attraction," which Frank et al. (2012) cite as indicating most clearly that sequential information plays a crucial role in real-time language use.

## 2 Agreement attraction

Agreement attraction is an error in subject-verb agreement as in (1), where the copular verb *were* erroneously agrees with the linearly closer plural noun *balls* rather than the singular head noun of the subject noun phrase *coat*:

(1) *The coat with the ripped cuffs by the orange balls *were* . . . (Frank et al. 2012: 3)

The phenomenon of agreement attraction is commonly observed in production (Bock and Miller 1991) and often yields an illusion of grammaticality in comprehension (Pearlmutter, Garnsey and Bock 1999). Agreement attraction provides us with an important source of information about how linguistic knowledge interacts with language processing. In other words, agreement attraction enables us to assess how grammatical information, including hierarchical structure, is encoded in the process of building grammatical dependencies in real-time language use.

Unfortunately, Frank et al. (2012) have not considered major findings from previous research on agreement attraction, which speak against their claim that human sentence processing is insensitive to hierarchical structure and depends on linear or sequential information. There are at least five findings in agreement attraction that support hierarchical structure. First, if language processing exclusively depended on linear information, we should expect to see agreement attraction shown in (2a) and (2b) in equal proportion:

(2) a. *The keys to the door *is* missing.
    b. *The key to the doors *are* missing.

In comparison to (2b), however, agreement attraction as in (2a) is rarely induced in production (Eberhard, Cutting and Bock 2005), nor do speakers fail to notice the unacceptable subject-verb agreement in comprehension (Wagers 2014, Lewis and Phillips 2015). Only intervening plural nouns attract agreement as in (1) and (2b). Second, agreement attraction is selective (Phillips et al. 2011): agreement attraction results in illusions of grammaticality in cases like (1) but does not induce illusions of ungrammaticality in cases like (3):

(3) The key to the doors *is* missing.

If linear distance were the only relevant factor in language production, speakers would judge (3) as unacceptable since the plural noun *doors* should induce plural agreement on the verb. However, illusions of ungrammaticality do not occur in grammatical sentences like (3). Third, agreement attraction is elicited inside a "plurally headed relative clause":

(4) *The squirrels that the cat *are* methodically stalking . . . (Wagers 2014)

In (4), the plural head noun *squirrels* induces the plural number agreement on the verb, which indicates that the linear proximity of the noun with respect to the verb is not responsible for agreement attraction. Fourth, Franck, Vigliocco and Nicol (2002) show that when the subject contains two stacked PPs as in (5), a plural noun hierarchically closer to the head of the subject as in (5a) induces more agreement attraction errors than a plural noun linearly closer to the verb as in (5b):

(5) a. *The inscription on the doors of the toilet *are* . . .
    b. *The inscription on the door of the toilets *are* . . .

This again suggests a role of hierarchical structure in language processing. Fifth, Franck, Vigliocco and Nicol (2002) conduct an elicitation experiment and report that English native speakers make agreement attraction errors in interrogative sentences like (6a) at almost the same rates as in declarative sentences like (6b):

(6) a. **Are* the helicopter for the flights safe?
    b. *The helicopter for the flights *are* safe.

The polar interrogative changes the relative word order of the attractor and the verb, but elicits agreement errors at almost identical frequency. This suggests that what matters is not the surface linear proximity of the attractor to the verb but structural closeness of the attractor to the verb in the derivation of the sentence. The attraction errors reviewed so far univocally show that it is not linear distance between the attractor and the verb that elicits attraction errors, but the hierarchical structure involved. Readers are referred to Lewis and Phillips (2015) for more evidence that hierarchical structure matters in agreement attraction. In the next section, structure dependence is discussed in relation to the brain.

## 3 Structure-dependence in the brain

It has been observed that our linguistic knowledge makes use of structure-dependent rules that require an analysis of a sentence into an abstract hierarchical structure and operates on abstract grammatical constructs such as phrases or clauses, instead of structure-independent rules that just operate on the rigid-order of words (Chomsky 1957). Structure dependence is so ubiquitous in language that it is difficult to appreciate its importance in the design of language. Structure dependence is interesting for several reasons. First, the only physical aspect of syntax, i.e., linearity, does not play a role in syntactic rules (Moro 2008, Yusa 2012a, b), suggesting that core properties of language are not transparent to perceptual systems. Structure dependence is an abstract principle with no overt realization in any language, making it difficult to receive enough information of structure dependence. It is highly possible for a person to go through life without encountering any relevant examples that would help to choose between structure-dependent rules and structure-independent rules (Berwick et al. 2011). This poses challenging questions that characterize the factors underling language acquisition, since evidence bearing on structure dependence is almost non-existent (Yang 2002). Second, structure dependence is a counterintuitive principle: the structure-dependent rule based on abstract hierarchical structure is computationally more complex than the structure-independent rule relying on linear order (Chomsky 2013). Third, the principle puts severe limits on logically possible rules that human languages exploit (Moro 2008). This is a matter of the poverty of the stimulus problem in language acquisition. Structure dependence is derived from Merge, which applies to X, Y, forming Z= {X, Y} and closeness is derived from laws of nature (Chomsky 2013). Structure-dependence is a deep-rooted principle of human language that is not readily accessible from observation. Readers are referred to Berwick et al. (2011) and Chomsky (2013) for detailed arguments about structure dependence.

There are two relevant behavioral studies that tested the principle of structure dependence. Crain and Nakayama (1987) report that English-speaking kids aged 3 to 5 do not produce yes/no questions that violate the principle of structure-dependence in their mother tongue. This result suggests that the principle is a deep-rooted innate property. More interesting is the case of Christopher, an unprecedented linguistic savant, who can "read, write and communicate in any of fifteen to twenty languages" (Smith and Tsimpli 1995: 1). However, he could not acquire an invented language that involves a number of rules referring to linear order, like "suffix an emphatic particle *nog* to the third orthographic word" (Smith and Tsimpli 1995). This clearly shows that languages do not use simple computations based on linear order or fixed positions in a sentence.

Of much relevance is a recent study on the acquisition of word order in an artificial language. Culbertson and Adger (2014) show in an elegant way that language learners prefer structural knowledge to distributional knowledge of their native language when learning a new language (an artificial language based on their native language). The fact that learners' inference for the word order in a new language rests on abstract structure more than linear ordering of words in their language clearly shows that learners are biased toward abstract hierarchical structure in language acquisition, contrary to the claim in Frank et al. (2012).

Neuroimaging data also disconfirm Frank et al.'s claim. The development and availability of neuroimaging techniques have provided us with new windows into where exactly within the brain specific language processing (syntactic, semantic, or phonological) is localized. Major current models of language processing refer to anatomical regions (e.g., Broca's area), pathways (e.g., dorsal or ventral), or networks of interconnected brain regions (see Friederici 2011 and Poeppel 2012 for the review). To examine the neuroimaging evidence that Broca's area responds differently to structure-dependent rules and structure-independent rules (Musso et al. 2003), I will limit myself to Broca's region.[1] In Andrea Moro's words (2008: 161), "[in] the same way as impossible sentences have led to significant advancement in the understanding of the formal mechanisms behind linguistic competence, languages may lead to significant advancement in the understanding of the neurobiological nature of the limits of variation across languages." This motivated Andrea Moro and his colleagues to find neuroimaging data for structure dependence in the brain. They investigated the neural correlates of Germans acquiring new linguistic knowledge of two parametrically different languages (Italian or Japanese) by examining whether the brain would distinguish between structure-dependent and structure-independent operations. Universal Grammar (UG)-consistent, structure-dependent rules, for example, include a rule such as "dropping the subject pronoun in Italian," while UG-inconsistent, structure-independent rules include a rule such as "forming negation by inserting *nai* after the third of the sentence in Japanese." Participants learned the rules quite accurately possibly due to ceiling effects of learning. They carried out fMRI experiments to compare the acquisition of syntactic rules based on hierarchical structure to ones based on fixed positions in the sentence. What they found is that syntactic rules following structure dependence selectively activate Broca's area, including the pars triangularis (BA 45), while syntactic rules following structure-independent rules do not. Broca's area is involved in the acquisition of new linguistic knowledge respecting the principle of UG. Whatever property in the brain it is that leads to the use of structural information in language processing, this is one of the most exciting discoveries in cognitive neuroscience research based on linguistic theory and has far-reaching consequences for research on language faculty. To strengthen their conclusion, I would like to play the devil's advocate. Musso et al. (2003) report that during three-minute pauses between sessions in fMRI experiments, the participants learned three real rules and three unreal rules. In addition, they used the same sentence types during fMRI sessions as the ones during three-minute instruction sessions. It is not clear whether this really reflects real language acquisition. In the next section, I review the results of Yusa et al. (2011), which replicate the involvement of Broca's area in acquiring new linguistic knowledge based on structure dependence but in a way immune to the possible criticisms mentioned above.

## 4 Neural correlates of the acquisition of a new syntactic rule

Yusa et al. (2011) investigate whether adult second language learners' knowledge goes beyond the input or stimuli that they receive during classroom instruction and whether instruction induces any functional changes in the brain.[2] Japanese participants learned a new English rule (i.e., negative inversion, NI) with monoclausal negative inversion constructions as in (7) during one month of instruction. Then, they were tested on "uninstructed biclausal sentences" containing relative clauses as in (9) as well as "instructed monoclausal sentences" as in (8).

(7) a. I will never eat sushi.
    b. Never will I eat sushi.
    c. *Never I will eat sushi.

(8) a. Those students are never late for class.
    b. Never are those students late for class.
    c. *Never those students are late for class.

(9) a. Those students who will fail a test are never hardworking in class.
    b. *Never will those students who fail a test are hardworking in class.
    c. Never are those students who will fail a test hardworking in class.

Simplex sentences such as, *Those students are never late for class*, can be converted to inverted sentences such as, *Never are those students late for class*, either by a structure-independent rule (i.e. move the first auxiliary after a fronted negative adverb), or by a structure-dependent rule (i.e. move the main clause auxiliary after a fronted negative adverb). The structure-independent rule does not require analyzing the sentence in terms of its hierarchical structure. We just scan the sentence, *Those students are never late for class*, from left to right until we hit the first auxiliary *are*, then move it after the fronted negative adverb *never*.

However, the structure-independent rule fails in biclausal sentences containing relative clauses. For example, with the complex sentence, *Those students who will fail a test are never hardworking in class*, the structure-independent rule moves the first auxiliary *will* and produces the ungrammatical result, *Never will those students who fail a test are hardworking in class*. The structure-dependent rule, on the other hand, targets the main clause auxiliary *are* and correctly generates the grammatical sentence, *Never are those students who will fail a test hardworking in class.* It might be objected that with the biclausal sentence as in (9a), moving the second auxiliary would generate grammatical negative inversion. However, this strategy does not work for complex sentences, such as *Those restaurants are rarely full of people who have been there before*, where moving the second auxiliary produces the ungrammatical sentence, *Rarely have those restaurants are full of people who been there before.*

Japanese learners of English participated in our experiment, all of whom started to study English as a second language in Japan after the so-called critical or sensitive period. The participants' knowledge of negative inversion was measured by a grammaticality judgment task (GJT) as a pre-test (Test 1) before the instruction, and again as a post-test (Test 2) after the instruction. The participants also underwent fMRI scans during the GJT so that any instruction effects on the brain could be measured. They were divided into instruction and non-instruction groups on the basis of their standardized English test scores and their error rates for negative inversion in simplex sentences during the first fMRI experiment (Test 1). For one month, the instruction group received instruction about NI only with simplex sentences after Test 1. That is, the participants were not exposed to examples instantiating crucial information such as the notion of structure dependence. Note that no reference was made to structure dependence in the instruction sessions. On the other hand, the non-instruction group received no instruction about NI but took a pre-test (Test 1) as well as a post-test (Test 2). The second fMRI measurement and GJT task (Test 2) were given to both the instruction and non-instruction groups.

The behavioral results of the instruction group show that the participants became significantly more accurate on the GJT of the uninstructed biclausal sentences as well as the instructed monoclausal sentences from the pre-test (Test 1) to the post-test (Test 2). No significant change, however, was observed in the non-instruction group.

Moreover, the fMRI data after the instruction also show a significant activation in Broca's area in the instruction group. Importantly, significant activations in the left inferior frontal gyrus (Broca's area) were found for the *uninstructed* complex sentences after the instruction. More specifically, the par triangularis (F3t or Brodmann's area (BA 44/45) of the left IFG was more significantly activated when the participants processed the inversion in uninstructed complex sentences. Again, there was no significant cortical activation change between Test 1 and Test 2 in the non-instruction group.

This study provides new insight into the role of Broca's area, where nature (UG) and nurture (instruction) contribute to enabling L2 learners to be *more* aware of certain phenomena far beyond what the instruction offered them. Our study used complex sentences involving relative clauses in Test 2, which were *structurally different* from simple sentences used during the instruction sessions. This enabled us to overcome the possibility that the participants might have just memorized the relevant rules and then repeated back what they had learned, which might have happened in the case of the participants reported in Musso et al.

(2003). Moreover, contrary to Musso et al. (2003), exposure to the relevant L2 input in this study was not brief, providing the participants with situations where linguistic knowledge was engaged. Our results show that structure dependence is still operative in the acquisition of a new rule by late second language learners, who do not learn a rule violating the principle of structure dependence.

It is worth noting here that structure dependence was regarded as pure UG before the Minimalist Program because there was no serious thought of trying to extract 'third factor' elements from structure dependence (Noam Chomsky, *pc*). In the Minimalist Program, however, UG consists only of the simplest recursive operation called Merge, which applies to two syntactic objects X and Y and merges them to form a new syntactic object Z: structure principle restricts computations to order-free structures (with externalization ancillary) and relies on the simplest computational procedure. The formation of polar interrogatives involving subject-auxiliary inversion is a classic example of structure dependence. Chomsky (2013) argues that subject-auxiliary inversion is derived from the operation of minimal search (as a third-factor principle of computational efficiency) along with the predicate-internal hypothesis. The results reported in Musso et al. (2003) and Yusa et al. (2011) could be interpreted as providing neuroimaging support for the operation of minimal structural search in language processing. In the next section, we will look in more detail at Broca's area, in which Merge might be implemented.

## 5 Broca's area: BA 44 and BA 45

Broca's area, usually defined as the pars opercularis (BA 44) and the pars triangularis (BA 45) of the left inferior frontal gyrus (IFG), has been reported to be involved in linguistic and non-linguistic process (Tettamanti and Weniger 2006, Friederici et al. 2006). Note here that the posterior part (BA 44) of Broca's area extends into the premotor area (BA 6) and the anterior part borders the pars orbitalis (BA 47). Syntactic processes have been reported to be implicated in BA 44 or BA 45,[3] though these two parts are cytoarchitectonically heterogeneous: BA 45 has a granular layer IV and BA 44 is dysgranular. The agranular frontal operculum (Friederici et al. 2006) and premotor cortex (BA 6) (Friederici 2006) are active in the processing of finite state grammar. In terms of the principle of gradation (Sanides 1962), "the brain evolution proceeded from agranular to dysgranular and then to complete granular cortex" (Friederici 2011: 186). In this sense, the dysgranual BA 44 is unique in that it occupies a place between the phylogenetically older agranular BA 6 and the phylogenetically younger granular BA 45. This uniqueness of BA 44 has interesting implications for language evolution (Yusa 2012b). It is not implausible to stipulate that the functions in BA 45 are evolved from those in BA 44, which are in turn evolved from those in BA 6 (See Fujita 2014 for a discussion of the evolution of Merge). Given this, it is not implausible, either that hierarchy in general is processed in BA 44 and linguistic hierarchy is in BA 45, which will be discussed below.

Most of the evidence for the role of BA 44 in syntactic processes, be it noted, comes from experiments using invented lexical items or artificial grammar (e.g., Tettamanti et al. 2002, 2009, Friederici et al. 2006, Bahlmann et al. 2008).[4] BA 44 has also been implicated in non-linguistic tasks. BA 44 is selectively implicated in the acquisition of "non-rigid" dependencies established between visuospatial symbols (Tettamanti et al. 2009), which clearly shows that this portion of the brain is responsible for processing hierarchy in general. Processing of hierarchical object manipulation is also reported to activate BA 44, which has been proposed as the neural correlates of Greenfield's (1991) "Action Grammar" (Fazio et al. 2009, Higuchi et al. 2009).[5] A virtual lesion of BA 44 causes a problem with hierarchically encoding other people's behavior (Clerget et al. 2009). Based on a large body of literature on BA 44, Yusa (2012a, b) suggests that domain-general Merge might be implicated in BA 44, which modulates the processing of hierarchical structure across cognitive domains. Given this suggestion, it is not surprising that BA 44 is activated in both linguistic and non-linguistic tasks. It is therefore not the case that hierarchy should be a problem with language evolution, as Frank et al. (2012) claims.

On the other hand, evidence for the role of BA 45 in syntactic processes comes from the acquisition/processing of real rules in natural languages (Musso et al. 2003, Yusa et al. 2011).[6] BA 45 is also responsible for embedding of syntactic constituents (Shetreet et al. 2009). The number of embeddings affects BA 45 more than BA 44 (Pallier et al. 2011). BA 45 is also reported to be selectively activated by movement or internal-Merge operations (Santi and Grodzinsky 2010). If internal-Merge is a unique property of human language, it is not implausible to assume that BA 45 might have something to do with deep properties of human language. Based on these observations, Yusa (2012a, b) suggests that domain-specific Merge might be implemented in the pars triangularis. Recently, Fedorenko et al. (2012) claimed that traditional group-based fMRI analyses have obscured the "functional heterogeneity within Broca's area." Individual-subject fMRI analyses clearly show that the pars triangularis is robustly sensitive to linguistic stimuli but not to non-linguistic stimuli. This is in conformity to the claim that BA 45 is responsible for domain-specific Merge (see Yusa (2012b) for more discussions of BA 44 and 45 and Fujita and Yusa (submitted) for the evolution of language in terms of Merge in the brain).

With regard to BA 44, a dorsal pathway connecting BA 44 and the posterior temporal lobe via the arcuate fasciculus is hypothesized to be involved in mapping sound to articulation. In contrast, a ventral pathway connecting BA 45 and the temporal lobe via the extreme capsule is responsible for the mapping of sound to meaning (Hickok and People 2004, Weiller et al. 2009).[7] Indeed, the functional roles of these pathways continue to be discussed, but it is worth bearing in mind that a better understanding of the essential properties of these pathways converging in Broca's area might provide new insights into how unique language is to humans.

## 6 Conclusion

This chapter reviews and discusses recent research on hierarchical structure in sentence processing. First, Frank et al.'s (2012) argument against hierarchical structure in language use was not justified, supporting the "one-system view" (Lewis and Philips 2015) that the cognitive system responsible for linguistic knowledge is employed in language use. Second, evidence was presented to support the claim that structure-dependence is neurologically real, strengthening the claim that the principle is a part of language design. Finally, it was suggested that domain-general Merge and domain-specific Merge might be implicated in BA 44 and BA 45, respectively.[8] Regarding the evolution of language, what evolves is I-language of human beings. Therefore, research on the evolution of language necessarily implicates the research on the evolution of the capacity to acquire and use language. This chapter showed that how languages are processed in the brain may deepen our understanding of the evolution of language.

## Notes

\* This is a revised version of Yusa (2012b), written in Japanese, and Yusa (2012a), incorporating the recent findings of language processing, although the basic claim regarding the functional specificity of Broca's area remains the same. I would like to thank Neal Snape and Manami Sato for helpful comments on the earlier version of this paper. This study was supported in part by the JSPS Grant-in-Aid for Scientific Research (A) (23242025, Principal Investigator: Koji Fujita, Kyoto University) and Challenging Exploratory Research (25580133, Principal Investigator: Noriaki Yusa, Miyagi Gakuin Women's University).
1 See Pallier et al. (2011), who identify brain regions correlative to the number of syntactic constituents as opposed to linear distance, contrary to Frank et al's (2012) claim.
2 See Yusa (2015) for the significance of bilingual research in terms of the study on human language faculty and language evolution. Section 4 is in part an excerpt from Yusa et al. (2011).
3 BA 44 has been reported to be consistently activated in the processing of linguistic hierarchical structure-dependencies (see Tettamanti and Perani 2012 for a review), but the data do not seem to be crystal-clear. For example, Rogalsky and Hickok (2011) claim that phonological short-term memory can explain the involvement of BA 44 in sentence comprehension.
4 Makuuchi et al. (2009) show that the main effect of hierarchy involves BA 44, but see Rogalsky and Hickok (2010) for a different claim, as mentioned in footnote 3.
5 Fujita (2014) claims that Action Grammar is a precursor to Recursive Merge in faculty of language in a narrow sense (Hauser et al. 2002).
6 Initial activation in BA 45 for invented languages in Musso et al. (2003) might suggest that BA 45 was sensitive to real words used in the experiments of the acquisition of invented languages. This might provide neuroimaging data that words can be constructed in syntax as Distributed Morphology claims.
7 See Friederici (2009) for the claim that there are in fact two dorsal pathways and two ventral pathways.
8 See Boeckx et al. (2014) for relevant discussions regarding BA 44 and 45.

# References

Anderson, Stephen R. 2008. The logical structure of linguistic theory. *Language* 84:795–814.

Bahlmann, Jörg, Ricarda I. Schubotz, and Angela D. Friederici. 2008. Hierarchical artificial grammar processing engages Broca's area. *NeuroImage* 42:525–534.

Berwick, Robert C., Noam Chomsky, and Massimo Piattelli-Palmarini. 2011. Poverty of the stimulus' revisited: recent challenges reconsidered. *Cognitive Science* 35:1207–1242.

Bock, Kathryn, and Carrol L. Millter. 1991. Broken agreement. *Cognitive Psychology* 23:45–93.

Boeckx, Cedric, Anna Martínez-Alvarez, and Evelina Leivada. 2014. The functional neuroanatomy of serial order in language. *Journal of Neurolinguistics* 32:1–15.

Chomsky, Noam. 1957. *Syntactic structures*. The Hague: Mouton.

Chomsky, Noam. 2013. What kind of creatures are we? *The Journal of Philosophy* CX:645–662.

Clerget, Emeline, Aline Winderickx, Luciano Fadiga, and Etienne Olivier. 2009. Role of Broca's area in encoding sequential human actions: A virtual lesion study. *Neuroreport*, 20(16):1496–1499.

Crain, Stephen, and Mineharu Nakayama. 1987. Structure dependence in grammar formation. *Language* 63:522–543.

Culbertson, Jennifer, and David Adger. 2014. Language learners privilege structured meaning over surface frequency. *PNAS* 111(6):5842–5847.

Eberhard, Kathleen M., J. Cooper Cutting, and Kathryn Bock. 2005. Making syntax of sense: Number agreement in sentence production. *Psychological Review* 112(3):531–559.

Fazio, Patrik, Anna Cantagallo, Laila Craighero, Alessandro D'Ausilio, Alice C. Roy, Thierry Pozzo, Ferdinando Calzolari, Enrico Granieri, and Luciano Fadiga. 2009. Encoding of human action in Broca's area, *Brain*, 132(7):1980–1988.

Fedorenko, Evelina, John Duncan, and Nancy Kanwisher. 2012. Language-selective and domain-General regions lie side by side within Broca's area. *Current Biology* 22(21):2059–2062.

Franck, Julie, Gabriella Vigliocco, and Janet Nicol. 2002. Subject-verb agreement errors in French and English: The role of syntactic hierarchy. *Language and Cognitive Processes* 17:371–404.

Frank Stefan L., Bod Rens Bod, and Christiansen Morten H. 2012. How hierarchical is language use. *Proceedings of Royal Society B.* 279:4522–4531

Friederici, Angela D. 2006. Broca's area and the ventral premotor cortex in language: Functional differentiation and specificity. *Cortex* 42(4):472–475.

Friederici, Angela D. 2009. Pathways to language: Fiber tracts in the human brain. *Trends in Cognitive Sciences* 13(4):175–181.

Friederici, Angela D. 2011. The brain differentiates hierarchical and probabilistic grammars. In *of minds and language*, eds. Massimo, Piattelli-Palmarini, Juan Uriagereka and Pello Salaburu, 184–194. Oxford: Oxford University Press.

Friederici, Angela D., Jörge Bahlmann, Stefan Heim, Ricarda I. Schubotz, and Alfred Anwander. 2006. The brain differentiates human and non-human grammars: functional localization and structural connectivity. *PNAS* 103(7):2458–2463.

Fujita, Koji. 2014. Recursive merge and human language evolution. In *Recursion: complexity in cognition*, eds. Tom Roeper and Margaret Speas, 243–264. New York: Springer.

Fujita, Koji, and Noriaki Yusa (submitted).The evolution of a Merge-ready brain.
Greenfield, Patricia M. 1991. Language, tools, and brain: The ontogeny and phylogeny of hierarchically organized sequential behavior. *Behavioral and Brain Sciences* 14(4):531–595.
Hauser, Marc D., Noam Chomsky, and Tecumseh W.Fitch. 2002. The faculty of language: what is it, who has it, and how did it evolve? *Science* 298:1569–1579.
Higuchi, Satomi, Thierry Chaminade, Hiroshi Imamizu, and Mitsuo Kawato. 2009. Shared neural correlates for language and tool use in Broca's area. *NeuroReport*, 20(15):1376–1381.
Kaminski, Juliane, Joseph Call, and Julia Fisher. 2004. Word learning in a domestic dog: Evidence for "fast mapping." *Science* 304:1682–1683.
Lewis, Shevaun, and Colin Phillips 2015. Aligning grammatical theories and processing models. *Journal of Psycholinguistic Research* 44:27–46.
Makuuchi, Michiru, Jörg Bahlmann, Alfred Anwarder and Angela D. Friederici. 2009. Segregating the core computational faculty of human language from working memory. *PNAS* 106 (20):8362–8367.
Moro, Andera. 2008. *The boundaries of babel: The brain and the enigma of impossible languages*. Cambridge, MA: MIT Press.
Moro, Andrea. 2011. A closer look at the turtle's eyes. *PANAS* 108(6):2177–2178.
Musso, Mariacristina, Andrea Moro, Volkmar Glauche, Michel Rijntjes, Jürgen Reichenbach, Christian Büchel, and Cornelius Weiller. 2003. Broca's area and the language instinct. *Nature Neuroscience* 6:774–781.
Pallier, Christoph, Anne-Dominique Devauchelle, and Stanislas Dehaene. 2011. Cortical representation of the constituent structure of sentences. *PNAS* 108(6):2522–2527.
Pearlmutter, Neal J., Susan M. Garnsey, and Kathryn Bock.1999. Agreement processes in sentence comprehendson. *Journal of Memory and Language* 41:427–456.
Phillips, C., Wagers, M. W., and Lau, E. F. 2011. Grammatical illusions and selective fallibility in real-time language comprehension. In *Experiments at the interfaces*. ed. J. Runner, 153–186. Bingley, UK: Emerald.
Poeppel, David. 2012. The maps problem and the mapping problem: Two challenges for a cognitive neuroscience of speech and language. *Cognitive Neuropsychology* 29(1–2):34–55.
Rogalsky, C., and Hickok, G. 2011 The role of Broca's area in sentence comprehension. *Journal of Cognitve Neuroscience* 23(7):1664–1680.
Sanides, Friedrich. 1962. *Die Architektonik des menschlichen Strinhirns*. Berlin: Springer.
Santi, Andrea, and Yosef Grodzinsky. 2010. fMRI adaptation dissociates syntactic complexity dimensions. *NeuroImage* 51:1285–1293.
Shetreet, Einat, Naam Friedmann, and Uri Hadar. 2009. An fMRI study of syntactic layers: Sentential and lexical aspects of embedding. *Human Brain Mapping* 48(4):707–716.
Smith, Neil, and Ianthi-Maria Tsimpli. 1995. *The mind of a savant: Language, learning and modularity*. Oxford: Blackwell.
Tettamanti, Marco, Hatem Alkadhi, Andrea Moro, Daniela Perani, Sypros Kollias, and Dorothea Weniger. 2002. Neural correlates for the acquisition of natural language syntax. *NeuroImage* 17(2):700–709.
Tettamanti, Marco, and Daniela Perani. 2012. The neurobiology of structure-dependency in natural language grammar. In *The handbook of the neuropsychology of language I*. ed. Miriam Faust, 229–251. Oxford: Wiley-Blackwell.

Tettamanti, Marco, Irene Rotondi, Daniela Perani, Giuseppe Scotti, Ferruccio Fazio, Stefano F. Cappa, and Andrea Moro. 2009. Syntax without language: Neurological evidence for cross-domain syntactic computations. *Cortex* 45(7):825–838.

Tettamanti, Marco, and Dorothea Weniger. 2006 Broca's area: A supramodal hierarchical processor? *Cortex* 42(4):491–494.

Wagers, Matthew W. 2014. Syntax in forward and in reverse. In *The Routledge handbook of syntax*, eds. Andrew Carnie, Yosuke Sato and Daniel Siddipi, 409–425. New York: Routledge.

Weiller, Cornelius, Mariachristina Musso, Michel Rijntjes, and Dorothee Saur. 2012. Please don't underestimate the ventral pathway in language. *Trends in Cognitive Sciences* 13 (9):369–370.

Yang, Charles. 2002. *Knowledge and learning in natural language*. Oxford: Oxford University Press.

Yusa, Noriaki. 2012a. Srtucture dependence in the brain. In *Five approaches to language evolution. Proceedings of the workshops of the 9th International Conference on the Evolution of Language*, 25–26. Kyoto: Evolang9 Organizing Committee.

Yusa, Noriaki. 2012b. Hierarchical structure and recursive computation in Broca's area. In *Constructing evolutionary linguistics*, eds. Koji Fujita and Kazuo Okanoya, 77–94. Tokyo: Kaitakusha.

Yusa, Noriaki. 2015. Remarks on the study of language evolution. In *The design, development and evolution of language: Exploration in biolinguistics* eds. Koji Fujita, Naoki Fukui, Noriaki Yusa, and Masayuki Ike-uchi, 128–155. Tokyo: Kaitakush.

Yusa, Noriaki, Masatoshi Koizumi, Jungho Kim, Naoki Kimura, Shinya Uchida, Naoki Miura, Ryuta Kawashima, and Hiroko Hagiwara. 2011. Second-language instinct and instruction effects: Nature and nurture in second-language acquisition. *Journal of Cognitive Neuroscience* 23(10):2716–2730.

# 15 The central role of the thalamus in language and cognition*

*Constantina Theofanopoulou
and Cedric Boeckx*

## 1 Beyond the cortex

It is common to come across statements like the following when trying to characterize what makes human cognition special: "the enlargement and species-specific elaboration of the cerebral neocortex during evolution holds the secret to the mental abilities of humans" (Rakic 2009: 724). The belief that the neocortex holds the key to "humaniqueness", as Hauser (2009) calls it, is deeply rooted indeed. The same was true 50 years ago, which is why Lenneberg (1967) could write: "Traditionally, all intellectual functions including speech and language have been thought to be located in the cerebral cortex, and more speculations have been directed to this thin sheet of tissue than toward any other cerebral component" (p. 62).

But Lenneberg knew better, as he stressed that "there are many other structures that are demonstrably connected with the cortex and with each other (often only by circuitous routes). Every structure of the brain is physiologically active and at least some of the structures have been hypothesized to play a part in the same intellectual functions that are more frequently imputed to the cortex." (p. 62) As an example, he cited Campion and Elliott-Smith (1934), who proposed that "*thought* consisted of cortico-*thalamic* circulation of impulses" (emphasis ours). Lenneberg also discussed Penfield's idea of a "centrencephalic integrating center", a "central system within the brain stem which is responsible for integration of varied specific functions from different parts of the hemispheres." Penfield too was aware of the explanatory potential of thalamic involvement. "It is proposed," he wrote (Penfield and Robert, 1959, p. 207) "as a *speech hypothesis*, that the function of all three cortical speech areas (that is, Broca's, Wernicke's, and the supplementary motor speech area) in man are coordinated by projections of each to parts of the thalamus, and that by means of these circuits, the elaboration of speech is somehow carried out."

Unfortunately, Lenneberg's remarks, which came not long after Lashley's (1950) conclusion opposing localization and in favor of distribution in brain studies, have not figured prominently in studies on language and human cognition more generally. As Staudigl et al. (2012) observe: "Human cognitive neuroscience currently has a strong cortico-centric focus when it comes to exploring the neural basis of higher cognitive functions. That is, many studies ignore the role of subcortical structures." But there are signs that the role of subcortical structures is being appreciated again. Recently, Lieberman's work has done much to draw attention to the role of the basal ganglia (Lieberman 2002, 2009, 2013). As he puts it:

> The traditional theory equating the brain bases of language with Broca's and Wernicke's neocortical areas is wrong. Neural circuits linking activity in anatomically segregated populations of neurons in subcortical structures and the neocortex throughout the human brain regulate complex behaviors such as walking, talking, and comprehending the meaning of sentences.
> (Lieberman 2002: 36)

Likewise, the role of the cerebellum, another subcortical structure, is also being reappreciated (Barton 2012, Barton and Venditti 2014, Murdoch 2010, Beaton and Mariën 2010), as is the role of the hippocampus (Rubin et al. 2014, Duff and Brown-Schmidt 2012). The shift of focus towards subcortical structures is in fact part of a larger sea-change in neurolinguistics, which calls for the need to operate with an "extended language network" (Ferstl et al. 2008), one that goes much beyond the classical, Broca-Wernicke model.

The present contribution is to be seen in this light. We will focus on the thalamus, thanks to which the basal ganglia, the cerebellum, and the hippocampus interface with the cortex in a reciprocal fashion. As a matter of fact, this reciprocal linkage should, by the logic of co-evolution, lead us to adopt a more balanced view of cortical expansion, and view it as part of a cortical-subcortical network reorganization. Paraphrasing Buzsáki (2006: 179), we can say that the neocortex co-evolved with thalamic connectivity.

Of all the subcortical structures, the thalamus is perhaps the one that has figured less prominently in the literature. But, as a recent special issue in *Brain & Language* (2013, vol. 126) devoted to the role of the thalamus in language shows, this has been a mistake. The *Brain & Language* issue in part revisits the importance of the early work by Bruce Crosson and George Ojemann, who pioneered deep brain stimulation approaches to examine the impact of selective silencing or stimulation of the thalamus on performance of language tasks. Tellingly, the editor of the special issue, Daniel A. Llano, closes his editorial with the following words (p. 21):

> The idea that understanding the thalamus will pay broad scientific dividends is not new. A. Earl Walker, in his groundbreaking monograph about the thalamus in 1938 (Walker 1938), made this very point to us many years ago: "The thalamus holds the secret of much that goes on within the cerebral cortex."

We second this opinion, and will try to demonstrate in the rest of the article why the thalamus is critical to capturing the role of language in cognition.

## 2 The thalamus at the center

A series of considerations have led us to reanalyze and reemphasize the importance of the thalamus as a region-of-interest for the neurobiology of language. We will discuss several such considerations in this section, from conceptual to empirical and technical. The order in which we discuss these considerations should not be construed as reflecting the intrinsic importance we attach to each. All of them are important to making the thalamus as central as we think it is.

There is a lot of evidence from a range of fields that humans are unique – or, to put it in the context of an evolutionary continuum, far better than other species – in transcending the signature limits of core knowledge systems, going beyond modular boundaries (Boeckx 2011a, b and references therein). This ability, which has all the characteristics of a phase transition, is at the heart of cognitive novelty, and subsequently, material and cultural innovation, leading to the establishment of a new cognitive phenotype. Hauser (2009) referred to this as 'humaniqueness', which he defines as follows: the ability to "create and easily understand symbolic representations of computation and sensory input", to "apply the same rule or solution to one problem to a different and new situation", and to "combine and recombine different types of information and knowledge in order to gain new understanding".

Boeckx (2013a, b) and Theofanopoulou and Boeckx (submitted) put forth the idea that cross-modular conceptual combinations, of the sort made possible by the core combinatorial rules of the language faculty, takes the form of cross-frequency oscillation synchrony and coherence across distant, local brain networks, allowing for the binding of features distributed across core knowledge systems. We think that this linking capacity (to be refined in subsequent sections) could be performed by the thalamus. As a matter of fact, a capacity of long-distance synchrony and modulation has routinely been attributed to the thalamus in other cognitive domains, such as vision (Saalmann et al. 2012, Saalmann and Kastner 2011), so we think that it makes sense that this capacity of the thalamus would have been recruited in the context of linguistic cognition. Indeed, it can be said that the central role of the thalamus is already firmly established in cognitive functions such as consciousness, attention, working memory or the central executive role, for which a crucial role for language is often recognized. As a result, we think that a rapprochement between language and the thalamus makes sense.

The latter point is further reinforced by considerations such as Chomsky's renewed emphasis on the idea that language serves as an internal "instrument of thought" (Chomsky 2012, Berwick et al. 2013). As he writes (Chomsky 2012: 11), "probably 99.9% of [the] use [of language] is internal to the mind. You can't go a minute without talking to yourself. It takes an incredible act of will not to talk to yourself". With passages like this one, Chomsky seems to be suggesting that language is crucially involved in cognitive behaviors like mind wandering, foresight, internal planning and mentation, etc. This gains particular relevance in a neurobiological context, as many of these cognitive traits are features of the so-called "default mode" network of the brain, of which the thalamus is recognized as a key node. The default mode network refers to a set of cortical and subcortical regions that are among the highest energy-consuming brain regions and that have undergone the highest degree of expansion in our lineage. The default mode network is typically involved in tasks such as mind wandering, self-reference, recollecting one's past and imagining one's personal future, internal mentation, and autobiographical plans. Buckner et al. (2008) take the network to underlie "stimulus-independent thought", and cite older sources referring to this type of activity as "contribut[ing] a great deal to the style and flavor of being human."

Interestingly, once the combinatorial potential of language is seen as contributing to a particular mode of structuring and generating thought, capable of linking conceptual domains, the language faculty can be brought closer to standard descriptions of consciousness, working memory or the central executive role. As Miller (2013) puts it, "working memory is, essentially, what we think of as thought" (p. 411). Thanks to this, connections between the literature on language and the literature on working memory become more transparent. Indeed, the possibility that psychologists appealing to the central executive function of working memory (Coolidge and Wynn 2005, Wynn and Coolidge 2007, Coolidge and Wynn 2007, Aboitiz 1995, Aboitiz et al. 2006), or related systems (Garofoli and Haidle 2013), to account for the origins of human-specific cognitive behavior were in fact attributing this transformative role to language strikes us as very high.

An additional consideration leading to the thalamus arises in the context of evolutionary discussions (Boeckx 2013a, Boeckx and Benítez-Burraco 2014). It is now well-established that *Homo sapiens* has a more globular braincase compared to our closest relatives (both extant and extinct). Paleoneurological studies based on endocranial geometry suggested that a spatial dilation of the deep parietal areas was the major morphological difference between modern and non modern human brains (Bruner et al. 2003, Bruner 2004, 2010). In our species, the morphogenetic change associated with this parietal bulging was then localized in a very early postnatal period, in a stage which is absent in chimpanzees or in Neandertals (Gunz et al. 2012, 2010, Neubauer et al. 2010). Boeckx (2013a), and, in a much more articulated fashion, Boeckx and Benítez-Burraco (2014) suggested that the developmental changes expressed at the levels of brain morphology and neural connectivity that occurred in our species after the split from Neanderthals-Denisovans and that gave rise to a more globular braincase configuration entailed significant changes not only at the cortical level, but also, and equally importantly, at the subcortical level. Specifically, it is likely that the thalamus benefited from this more globular environment given its strategic, central position.

Sitting right at the center of the brain, the thalamus is in an ideal position to modulate the activity of distant cortical (and subcortical) structures and render them equidistant. In addition, by the logic of co-evolution and correlated growth, parietal expansion is likely to lead to a similar expansion of thalamic nuclei strongly associated with the parietal lobe (dorsal pulvinar). This strikes us as particularly relevant, as the pulvinar has been claimed to be crucially involved in cortico-cortical information management (Theyel et al. 2010, Saalmann et al. 2012). As a matter of fact, the most conspicuous cortical region linked to parietal bulging, the precuneus (Bruner et al. 2014a, b), is most strongly connected to the pulvinar (Cavanna and Trimble 2006 and references therein).

The notion of co-evolution strengthens our belief that the thalamus is key in the domain of human cognition. A recent study by Bohsali et al. (2015) has revealed significant structural connectivity between Broca's area and the thalamus, specifically the pulvinar. As is well known, Broca's area has undergone a spectacular enlargement in recent human evolution (Schenker et al. 2010), which must have been associated with a comparable recruitment and expansion of the thalamic nuclei to which Broca's area is connected. Interestingly, there is independent evidence for a significant expansion of the pulvinar in our species. The pulvinar nucleus of the thalamus is known to have expanded greatly in primate evolution, with its expansion linked to cortical areas associated with integration of visual information, attention, and movement planning. But in humans, higher-order association nuclei like the pulvinar and the medio-dorsal nucleus have been said to be disproportionately large (Striedter 2005: 331) (see also Armstrong 1981). Such an enlargement is often related to an evolutionary novelty: in human development alone, the pulvinar receives extra GABAergic cells migrating from a telencephalic structure known as the ganglionic eminence (Rakic and Sidman 1969, Letinic and Rakic 2001).

Interestingly as well, there is indirect evidence of a recruitment of thalamic nuclei like the pulvinar for linguistic purposes. Taken into account the reduction of the visual cortex in humans, especially in *Homo sapiens* (Pearce et al. 2013), one could hypothesize that the pulvinar, classically associated with visual attention, was recruited for purposes of linguistic cognition, as we wish to suggest in the present paper.

A final consideration leading to a reappraisal of the role of the thalamus is more pragmatic, as it pertains to technological considerations. There may indeed have been a technological reason why early biological concepts of language were predominantly corticocentric. The thalamus lies deep inside the brain, and is therefore far less easily accessible than the more surface cortical regions. However, as reviewed in Klostermann et al. (2013), the emergence of new technical possibilities is slowly correcting for the cortical bias. For instance, David et al. (2011) provide the first demonstration in cognitive neuroscience that subcortical-cortical loops can be empirically investigated using noninvasive electrophysiological recordings. Specifically, they show that "a hidden source, modeling magnetically silent deep nuclei, is required to explain the[ir] data best". Based on data from intracranial recordings with similar language material in Wahl et al. (2008), the so-called "deep source" in David et al. (2011) is assumed to correspond to the thalamus and the respective thalamocortical loops. While it remains true that subcortical structures like the basal ganglia and the cerebellum continue to figure more prominently in this new, less cortico-centric paradigm (see, e.g., Miller and Buschman 2007, Antzoulatos and Miller 2011, Kotz and Schwartze 2010, Ullman 2001, 2004, 2006), the thalamus cannot be ignored, as it provides the way for these other subcortical structures to interface with the cortex in a reciprocal fashion (for recent suggestive evidence of this fruitfulness of this perspective, see Teichmann et al. 2015). In fact, the data to be reviewed below pertaining to the involvement of the thalamus in mental health suggest that the thalamus may well be the answer to Friederici's (2006) question: 'What's in control of language?'

## 3 The thalamus: far more than a relay station

It is common to regard the thalamus as a mere relay station (relay bias). Indeed, the thalamus has traditionally been thought of as a necessary link in the flow of information from the periphery to the cortex. But there are good reasons to view the higher-order nuclei making up the dorsal thalamus as crucially implicated in higher cognition. This is what we wish to highlight in this section.

As we have already pointed out, the function most commonly attributed to the thalamus is that of 'channeling' information from the subcortical parts to the neocortex. Indeed, every sensory system (with the exception of the olfactory system) includes a thalamic nucleus that receives sensory signals and sends them to the associated primary cortical areas. If we were to stop here, we could say that the thalamus essentially determines the bandwidth capacity of the brain. But its role is far more complex. In fact, the thalamus evaluates all the incoming stimuli and segregates them, so that they can proceed to the neocortex. And this early segregation of input to modality-specific information is actually responsible for the localization of functional systems in the neocortex. That is, the thalamic performance justifies to a great extent the designations of the cortical systems as visual, auditory, somatosensory, motor etc. So, on the basis of this alone, it is impossible to inspect issues related to lateralization and the distribution of the neocortical loci without taking into serious consideration the role of the thalamus.

Building on considerations of the sort just mentioned, Buzsáki (2006: 177) regards the thalamus as the 'afferent and efferent expansion' of the neocortex. But this efferent-afferent function is far from random; rather, it is both 'economical' and 'sophisticated'. The 'economy' of the thalamus can be construed in two senses: firstly, as 'selectivity', in that its efferent function is expressed subcortically, relaying important inputs from the hippocampus, the cerebellum, the amygdala, the basal ganglia, the brainstem and the superior colliculus, and its afferent function towards the neocortex; secondly, the thalamus is 'economical' as far as its location in the brain is concerned, especially in the context of a globular brain. Being equidistant from all the cortical areas, it is the most adequate and flexible spatiotemporal gate of information, in as much as spatial wiring length entails temporal axonal communication.

But the thalamus is not a mere segregator-distributor of information. It acts as an oscillatory pacemaker, or better put, it is the part of the brain assigned to tune the oscillations of the other subcortical structures. From this perspective, its efferent role can be regarded as setting the dynamic oscillatory networks subcortically, and its afferent role as making them robust brain constellations. Thus, the 'tuning-role' can also shed light to the aforementioned 'selectivity', since the transthalamic network orchestration is echoed in the neocortex (we could speak here of an 'allocator role').

The orchestrator role of the thalamus makes it a key node in a variety of networks identified in the literature independently of language: the global workspace model formulated by Dehaene et al. (1998) and also the model of Tononi and Edelman (1998) in the domain of consciousness, the multiple-demand system of Duncan (2010, 2013), the 'connective core' model of Shanahan (2012), or the integrative architecture for general intelligence and executive function in Barbey et al. (2012). It is also likely implicated in the functioning of the top-down, frontoparietal attentional regulation network (Miller and Buschman 2012), the "frontoparietal control system" (Vincent et al. 2008), and the already mentioned default mode network. For any of these networks, the thalamus no doubt qualifies as a "high Hub Traffic" node in the "rich club", "high cost, high-capacity backbone for global brain communication" in van den Heuvel and Sporns (2011) and van den Heuvel et al. (2012).

Though not identical, all these networks have core properties that we wish to attribute to the combinatorial power of the language faculty. Thus, many of the networks just cited are said to be involved in mind-wandering and inner speech (Gruberger et al. 2011), to be uniquely positioned to integrate information coming from various systems and to "adjudicate between potentially competing inner-versus outer-directed processes" (Vincent et al. 2008), to be characterized by an "amplification, global propagation and integration of brain signals", "ignit[ing] a network of distributed areas", implicating in the " temporary maintenance, global sharing, and flexible routing" of information (passages from Dehaene et al. 2014 defining "conscious perception").

Much as Miller (2013) defines the "top-down attention control", the brain network envisaged here is meant to be responsible for "intelligent behavior", understood as "the human ability to adaptively implement a wide variety of tasks" (Cole et al. 2013). As Duncan (2013) writes in the context of his frontoparietal Multiple-Demand system, which he takes to provide "a core basis for the psychometric concept of fluid intelligence", "accompanying activity is commonly seen in subcortical regions including basal ganglia, thalamus, and cerebellum." In this context, Bohlken et al. (2013) write that "of all subcortical volumes measured, only thalamus volume is significantly correlated with intellectual functioning". Bohlken et al. remind us that the thalamus, with its widespread cortical connections, is likely to play a key role in human intelligence. Connectivity indeed appears to be the key, given the claims that more efficient or long-range connectivity is associated with higher intelligence test performance, as highlighted in Duncan 2013 and references therein, especially van den Heuvel et al. 2009. (Incidentally, more recent studies provide evidence of a strong correlation between thalamus volume and cognitive abilities; see, e.g., Scharinger et al. 2014.)

## 4 The thalamus, language, and mental health

If we are right in ascribing to the thalamus such a leading role in cognition, we ought to expect to find numerous associations between thalamic damage and cognitive disorders. This section shows that this expectation appears to be met.

There is indeed rapidly accumulating evidence that cognitive disorders that are routinely associated with language and the distinctive mode of thought it entails, such as schizophrenia, autism, dementia, major depression, verbal working memory impairments, etc., crucially involve thalamic disorders, especially as they affect the higher-order nuclei, such as the mediodorsal nucleus and the pulvinar.

Some of this evidence goes as far back as Stern (1939), where a case of severe dementia was associated with bilateral symmetrical degeneration of the thalamus, but likely due to the reigning cortico-centrism of the time, the significance of this finding appeared to have been lost on its contemporaries. Recently, thalamic damage has been reported for a variety of cognitive disorders. Let us list some salient examples here, before examining what this might tell us about the neurobiological nature of linguistic cognition.

In the context of schizophrenia, Agarwal et al. (2008) report "microstructural thalamic changes"; Kemether et al. (2003), Alelú-Paz and Giménez-Amaya (2008) and Byne et al. (2009), abnormal thalamic volume (reduction); Thong et al. (2013), thalamic shape abnormalities; Popken et al. (2000) and Harms et al. (2007), neuron loss; and Pinault (2011), "dysfunctional thalamus-related networks". Martinsde Souza et al. (2010) consider the thalamus a potential biomarker candidate for schizophrenia (see also Brucato et al. 2015). At a more general level, Andreasen (1997) discusses the role of the thalamus in the pathogenesis of schizophrenia, building on original insights from Jones (1997).

Given Goldman-Rakic's (1994) claim that a defect in working memory may be the fundamental impairment leading to schizophrenic thought disorder, it may not come as a surprise to find reports of thalamic damage in the context of working memory impairments, in the context of schizophrenia and beyond (Dagenbach et al. 2001, Vigren et al. 2013, Parnaudeau et al. 2013).

In the context of autism, a similar state of affairs obtains: Egawa et al. (2011) and Tsatsanis et al. (2003) report reduced thalamic volume; Nair et al. (2013), impaired thalami-cortical connectivity; and Shi et al. (2013), gray and white matter abnormalities in the thalamus.

Thalamic atrophy is also reported following severe brain injury, and is found to correlate with cognitive impairment (Lutkenhoff et al. 2013). Abnormal neuron numbers in thalamus nuclei are found in cases of major depression (Young et al. 2004), and finally, thalamic damage is also reported in leukomalacia (Ligam et al. 2009), attention deficit hyperactivity disorder (Mills et al. 2012), dementia (Kovacs et al. 2013), multiple sclerosis (Cifelli et al. 2002), major/treatment-resistant depression (Greicius et al. 2007), Cerebral Autosomal-Dominant Arteriopathy with Subcortical Infarcts and Leukoencephalopathy (O'Sullivan et al. 2004), and Alzheimer's disease (Binnewijzend et al. 2012, Zhou et al. 2013). In the context of Alzheimer's disease, it is worth mentioning that thalamic volume was found to be predictive of performance on tests of cognitive speed and decreases in healthy aging (Van Der Werf et al. 2001): "A significant decrease in volume of the thalamus with increasing age was found, relatively stronger than and independent of the decrease of total brain volume".

So frequent is the association between thalamic shape and cognitive disorders that Vila-Rodriguez et al. (2008) talks of thalamic shape as a potential "endophenotype".

Thalamic reductions have also been found to correlate with verbal fluency impairment in those born prematurely (Giménez et al. 2006). Interestingly, in this case, the aspects of verbal fluency most affected are not phonological, but syntactico-semantic, which is reminiscent of Wahl et al. (2008), where syntactosemantic violations were shown to activate the thalamus of normal adults more than phonological violations.

Perhaps the most frequently reported involvement of the thalamus in pathological contexts has to do with impaired consciousness (see, e.g., Crone et al. (2014), Fernández-Espejo et al. (2010), Maxwell et al. (2006), Zhou et al. (2011)), which is relevant in the context of this paper, given the way consciousness is understood by Dehaene et al. (2014), as discussed in the previous section.

Finally, given the strategic position occupied by the thalamus, one expects to find aberrant thalamic circuitry in impairments that are regularly associated with the basal ganglia, such as Parkinson's or Huntington's diseases (on thalamic damage in such cases, see McKeown et al. 2008, Henderson et al. 2000, Aron et al. 2003). Likewise, given the anatomical interface of the thalamus and the characterization of dyslexia as a disconnection syndrome (Boets et al. 2013), we should expect thalamic damage there too (Galaburda and Eidelberg 1982, Díaz et al. 2012). Reports of thalamic damage giving rise to certain types of acalculia (Jensen 2010, Mendez et al. 2003) should also be expected, if indeed at least some kinds of mathematical computations are offshoots of our language faculty, as is often suggested (Chomsky 1988, 2008).

To conclude this brief survey on the thalamus and mental health, we would like to mention that minor physical anomalies (MPAs) occur more frequently in neurodevelopmental disorders such as autism (Cheung et al. 2011). As Cheung et al. write, MPAs are neurodevelopmental markers which manifest as unusual morphological features of the face or physique. They occur in more than 14 per cent of healthy newborn but significantly more often (as often as 60 per cent of the cases) in neurodevelopmental disorders such as schizophrenia, autism, hyperactivity, epilepsy, or mental retardation. We mention this in the context of the present paper because, as Boeckx and Benítez-Burraco (2014) discuss extensively, the thalamus plays a crucial rule in proper brain development, so much so that MPAs could 'mirror' (even subtle) aberrant neurodevelopment involving the thalamus and could therefore serve as convenient biomarkers, and eventually provide robust targets for efficient therapies.

Bish et al.'s (2004) report of thalamic reductions in children with chromosome 22q11.2 deletion syndrome fits well in this context. Chromosome 22q11.2 deletion syndrome (22q) encompasses DiGeorge and velocardiofacial (VCFS) syndromes. It is a congenital condition resulting from a deletion at chromosome 22q11.2 and has a prevalence of at least one in 4,000 live births. According to Bish et al. (2004), the most systematically observed manifestations of 22q include cleft palate, heart defects, T-cell abnormalities, and neonatal hypocalcemia, as well as facial dysmorphisms and mild to moderate cognitive deficits. Along with an overall delay in early cognitive, psychomotor, and language development and an overall IQ typically in the range 70–85, a subset of deficits are evident in the areas of visuospatial and numerical performance. Children with 22q also show an extremely high incidence of psychopathology, especially schizophrenia, as they reach adulthood. That we find thalamic damage (especially affecting the pulvinar) in these children suggests that it would be a serious mistake to ignore the role of the thalamus in language and cognition.

The point of this section is not to provide a comprehensive catalog of cases of thalamic damage associated with cognitive/language disorders. Rather, we want to use the evidence just reviewed to shed light on what it is that the thalamus does to make human linguistic cognition possible in the first place. In other words, our intention is to follow a long tradition of studies exploiting atypical situations to understand the norm.

The majority of the disorders mentioned in this section are routinely characterized as disorders of information processing, 'dysregulation' syndromes, or, even more frequently, 'disconnection syndromes' (see, e.g., Schmahmann and Pandya 2008). They are viewed as dysfunctions in the coordination of distributed neural activity between and within functionally specialized regions of the brain. In the relevant literature, the guiding idea is that connectivity in some sense determines behavior, and that as soon as a node in the relevant network fails, the whole network is affected; the 'traffic' stops, or at the very least is severely affected. In this context, the thalamus appears like a central pacemaker (Buzsáki 1991) or "universal synchronizer" (Ghosh et al. 2014), a "mediator" among networks (Barron et al. 2015), pretty much like the conductor of an orchestra: if one of the players fails, the conductor's job is affected. If the conductor fails, the jobs of all the players are affected. This may be a good place to recall Patterson's (1987) remark: "if one were to single out a brain structure that displayed the possibility for central 'timing' functions in brain, it would most likely be the thalamus".

For disorders like schizophrenia and autism, seen as 'information-processing impairments', it is common to find talks of an imbalance between inhibition and excitation (Yizhar et al. 2011). The central idea here is that for the type of cognition we are interested in to emerge, there must be a suitable, perhaps even optimal, mix between local and long-distance connections, with the traffic (information flow) being regulated by a suitable balance of excitation and inhibition (with inhibition mostly coming from the thalamus, a point we return to in the next section). But likely due to its evolutionary novelty, the network's balance is fragile, with even subtle structural anomalies giving rise to cortico-thalamic misconnection (Agarwal et al. 2008, Blatt 2012, Sun et al. 2013). For instance, Parnaudeau et al. (2013) found that a subtle decrease in the activity of the mediodorsal thalamus is sufficient to trigger selective impairments in prefrontal-dependent cognitive tasks. In a similar vein, Zikopoulos and Barbas (2007) talk of an "imbalance in the communication between the thalamus and cortex at the core of a host of psychiatric and neurological conditions." (Numerous authors cited above talk of hypo- or hyper-connectivity syndromes.)

The idea of disrupted balance emerged in the context of the role of the thalamus in consciousness (see Llinás and Steriade 2006, Steriade 2004, Contreras et al. 1996, Steriade et al. 1993), but it has since then been extended to other cognitive impairments. "Aberrant rhythmic activity" giving rise to "excitation leaks" (Frantseva et al. 2014), or "abnormal thalamic modulation" of cortical or subcortical structures (Normand et al. 2013), now regularly feature as causal explanations for endophenotypes. As Normand et al. (2013) observe, the relevance of the thalamus lies in the fact that this is a brain structure that "feature complex feedback loops and widespread reciprocal connectivity that could amplify and spread the effects of a slight functional imbalance", an effect that we saw may be even reflected in minor physical anomalies, cf. Cheung et al. 2011.

As Kircher and Glendenning (2002) point out, an expanded brain that is out of control is not helpful. There must be modulation of this enhanced cortex. Kircher and Glendenning show that a primary source of this modulation comes from the enhanced inhibitory capabilities of the thalamus, and the increased number of neurons sensitive to the most common inhibitory neurotransmitter found (GABA). By its influence on our neocortex, the thalamus provides greater control of neural processing. Kircher and Glendenning propose that it may be our ability to inhibit our cortex that has resulted in our increased 'intelligence', or, to put it in terms that we prefer, our specific, linguistic mode of cognition. In this context, it is worth recalling that many studies reporting volume abnormalities in the thalamus are talking of the size of the whole thalamus relative to total brain size. Thus, Tsatsanis et al. (2003) report for autistics that the size of the whole thalamus was less strongly correlated to total brain size, and the sizes of the right and left thalamus were less strongly correlated in autistic versus control subjects. In addition, within the subgroup with larger brain sizes, thalamic volume was significantly reduced in individuals with autism compared with normal controls.

Due to its strategically central position, both in the course of brain development, and in the *sapiens*-specific, globular brain, the thalamus impacts the organization of neural pathways and functional connectivity in the developing brain (see Courchesne and Pierce (2005) and Tsatsanis et al. (2003) for clear articulations of this idea in the context of autism). Damage to the thalamus is bound to result in compromised functional integration between brain regions, reducing the efficiency of the whole network ("network fragmentation" as the phenomenon is called in the literature on the default-mode network cited above).

It is for this reason that many of the disorders discussed so far are called "neurodevelopmental disconnection syndromes" (Shi et al. 2013).

## 5 Thalamus-generated oscillations and oscillopathies

Perhaps the best way to understand the crucial connecting, regulatory function of the thalamus is in terms of brain oscillations. When it comes to brain functioning, anatomical connections are not enough. Rather, dynamic, functional connectivity between anatomically connected regions is key. Reinhart and Woodman (2013) put it this way: "Oscillatory synchronization has been proposed to dynamically establish large-scale networks among brain areas; the network relevant to each cognitive operation should form and dissolve through dynamically regulating the strength of functional connectivity between brain areas via oscillatory synchronization or coupling."

As reviewed in Akam and Kullmann (2010), task-dependent increases in strength and interregion coherence of network oscillations have been reported in numerous brain systems, with strong, coherent, synchronized oscillation between sending and receiving regions during communication. Uhlhaas and Singer (2006) review evidence that certain brain disorders, such as schizophrenia, epilepsy, autism, Alzheimer's disease, and Parkinson's are associated with abnormal neural synchronization. This is not surprising, considering that these deficits concern functions, such as working memory, attention, and perceptual organization, that have been argued to involve synchronization of oscillatory activity in specific frequency bands.

Interestingly, local and long-distance synchrony appear to take place at different frequencies, and so Uhlhaas and Singer (2006) suggest that selective deficiencies in the ability of cortico-thalamocortical loops to engage in precisely synchronized oscillations at particular frequencies may provide the right level of explanation for the disorders at issue. In particular, we think that this possibility may provide us with a deeper understanding of what the thalamus does, and how language is implemented in the brain, taking us from the connectome to the dynome. This is the topic of this section.

Uhlhaas and Singer (2006) point out that "cognitive dysfunctions were particularly pronounced for tasks requiring interactions between widely distributed brain areas, such as integration of polymodal stimulus attributes, dynamic perceptual grouping, working memory, and executive processes." If we are correct, the thalamus plays a key role establishing and maintaining large-scale integration of activity. In other words, we expect many of the disorders reviewed above to be accounted for in terms of thalamic dysrhythmia.

Daitch et al. (2013) pursue this very idea, starting with the claim that "different attention processes (holding vs. shifting attention)" are associated with synchrony at different frequencies, and crucially appeal to the role of the thalamus in this context. As they observe, "the pulvinar nucleus of the thalamus, for example, has broadly distributed cortical connections and exhibits both attention-modulated spike rate and synchrony with visual areas. The inhibitory thalamic reticular nucleus, which receives overlapping projections from prefrontal cortex and higher-order sensory cortical areas, may also play a role in modulating incoming sensory information with selective attention."

Daitch et al. (2013) go on to point out that "low-frequency oscillations [of the alpha type] have been shown to entrain to the rhythm of a task, when stimuli are presented in a rhythmic, predictable manner. On the other hand, [gamma] synchrony, which enhances the efficacy of communication between regions, but which is more metabolically costly than low-frequency activity, may operate under conditions of temporal uncertainty. Notably, both the pulvinar and thalamic reticular nucleus, mentioned above as a potential generators of cortical synchronous activity, operate in both phasic (i.e., transient) and tonic (i.e., sustained) modes. These two modes may underlie the low frequency versus sustained high-frequency coherence correlates of attention, respectively."

As Daitch et al. (2013) note, the oscillations modulated by a task are determined by many factors, including the distribution of brain regions recruited, and so it is not implausible to hypothesize that once we move away from vision, which is the focus of their study, and appeal to other cortical regions, the very same mechanisms and rhythms may be at work in generative linguistic cognition. These regions are plausibly those discussed above in the context of globularity, the default-mode network, and so on: most prominently, the frontal and parietal association cortices.

Parietal and frontal cortices are indeed key nodes of the working memory network, and they have been claimed to be modulated by the alpha oscillations generated by the thalamus during working memory tasks (Roux and Uhlhaas 2014). Roux and Uhlhaas (2014) is in line with Dipoppa and Gutkin (2013). Both studies seek to ascribe functional meaning to the various frequencies detected during specific cognitive tasks. As they write, "although initially thought to reflect cortical idling, a growing body of evidence critically implicates rhythmic activity in the alpha band in cortical communication and cognition. Theta activity occurs preferentially in tasks that involve sequential coding of multiple working memory items, whereas alpha oscillations tend to occur during tasks that require maintenance of simultaneously presented visual or spatial information."

Pursuing this idea, Roux and Uhlhaas (2014) note that "modulation of alpha activity may enable the gating of task-relevant working memory items, for example during suppression of distracting information during working memory encoding and maintenance. In this context, recruitment of alpha-band oscillations for inhibition of distracting information can also occur during tasks requiring the maintenance of sequential information, highlighting that theta/alpha frequencies are closely linked to the task demands during working memory and not limited to the format of working memory representations". For Roux and Uhlhaas (2014), "gamma-band oscillations represent a generic mechanism for the representation of individual working memory items, irrespective of working memory content and format."

As Uhlhaas et al. (2010) remark, "alpha activity has been associated not only with inhibitory functions, but also with the long-distance coordination of gamma oscillations, and theta activity has been proposed to support large-scale integration of subsystems serving the formation and recall of memories." That is to say, "short distance synchronization tends to preferentially occur at higher frequencies (gamma band) than long-distance synchronization, which often manifests itself in the beta- but also in the theta- and alpha-frequency range." The thalamic generation of alpha activity fits very well with proposals like those of Shipp (2003), Saalmann et al. (2012), Saalmann and Kastner (2011), according to which the thalamus (specifically, the pulvinar) synchronizes oscillations between interconnected cortical areas, thereby modulating the efficacy of cortico-cortical information transfer. (On thalamocortical phase synchronization entrains gamma oscillations during long-term memory retrieval, see also Staudigl et al. (2012).)

The thalamus, then, is key to providing oscillatory coherence, acting as the network's metronome, as it were; regulating the flow of neural 'traffic' via rhythmic synchrony, which Miller (2013) views as brain's way of achieving coherent, meaningful thought. Viewed in this light, a healthy brain amounts to a 'consonant' brain. When parts of the network are damaged, synchronization breaks down, leading to the fragmentation of thought. "Imprecise synchrony leads to imprecise connectivity that can only support imprecise temporal dynamics" (Uhlhaas et al. 2008). Singer (2011) goes as far as claiming that "abnormal communicaton is the central pathophysiological feature of neuropsychiatric disorders". Focusing on schizophrenia, Uhlhaas and Singer (2010) claim that in this disorder the relevant brain circuits are "characterized by imprecise temporal dynamics are unable to support the neural coding regime [required]". Yizhar et al. (2011) note that behavioral impairment in both autism and schizophrenia has been associated with elevated baseline (non-evoked) high-frequency activity in the 30–80 Hz range, which could account for descriptions of these disorders in terms of "over excitation", "excitation leaks" or "under-inhibition".

In light of this, we can only concur with Tsatsanis et al. (2003), who write that "an examination of the thalamus may provide insight into autism as an information-processing disorder, particularly with regard to shaping the neural architecture of the brain in a way that leads to less functional connectivity and thus a less synchronized network." Given the central role of the thalamus in the context of early brain development, as well as in the processing of sensory information from the environment, it is indeed the case that both genetic or epigenetic disturbances of the mechanisms responsible for the generation of temporally structured activity patterns coordinated by the thalamus may impede the activity-dependent specification of developing circuitry which in turn leads to abnormal temporal patterns. This latter point is stressed by Uhlhaas et al. (2010), who write: "Considering the important role of neural synchrony in the shaping of cortical circuits at different developmental periods, we hypothesise that [in autism] abnormal brain maturation during early prenatal and postnatal periods results in cortical circuits that are unable to support the expression of high-frequency oscillations during infancy. These impaired oscillations might in turn reduce the temporal precision of coordinated firing patterns and thereby disturb activity-dependent circuit selection during further development." Synchronization, then, is not only relevant during normal performance, it is equally important in the context of the initial setting up of the network that will eventually perform the relevant tasks. In both cases, the centrality of the thalamus cannot be ignored. (Although we will not have the opportunity to expand on this here, for lack of space, it may be worth mentioning that the centrality of the thalamus finds its clearest expression in its self-sustained oscillatory mechanism during sleep-mode and sleep-wake regulation. Alterations in sleep-wake patterns and the homeostatic function of the thalamus is known to impact learning, memory, and cognition generally (Brown et al. 2012).)

## 6 Conclusion

This article began with a plea to take subcortical structures into account in an attempt to offer an adequate neurobiological account of the human language faculty and its contribution to species-specific cognitive traits. In doing so, we have added ourselves to the list of authors who suggest that "the focus on the neocortex as the principle area of change in primate brain evolution might have been excessive, and that some attention should now be paid to cortico-[subcortical] circuits and the behavioural functions they may mediate" (Whiting and Barton 2003). As Finlay et al. (2001) write in an evolutionary context, "there is no reason to presume selection pressures for cortically based functions drove brain expansion at all. . . . [T]he brain grows as a covarying whole, increasing in size according to a fairly straightforward log function. It is just as likely, therefore, that pressures for enhanced archicortical, corticoid, or subcortical processing could have triggered the adjustment of global timing constraints that led, incidentally, to much bigger isocortices."

To correct for the classical view that cognition is the exclusive domain of the cortex, we have exploited the literature on the thalamus. Numerous authors (e.g., Carrera and Bogousslavsky (2006)) have already pointed out that thalamic damage can mimic all cortical syndromes, suggesting that a focus on the thalamus may be a productive alternative to the classical view. In addition, as Theyel et al. (2010) remark, "much of the information transfer between cortical areas involves cortico-thalamo-cortical circuits, . . . cortico-thalamo-cortical information transfer may [then] represent an important addition to, or even replacement of, the current dogma that corticocortical transfer of primary information exclusively involves direct corticocortical pathways". Our overall conclusion from our survey of the literature agrees with Saalmann and Kastner (2011), when they say: "Only with detailed knowledge of thalamic processing and thalamocortical interactions will it be possible to fully understand cognition."

In addition to correcting for the cortical bias, we have insisted on the need to go beyond inappropriately modular perspective on language and its neural implementation. As Treves (2009) nicely puts it, "understanding the neural basis of higher cognitive functions such as those involved in language requires in fact a shift from a localization approach to an analysis of network operation". This is slowly becoming the consensus view. One-to-one brain-behavior mappings of complex functions like "language processing" have largely been replaced by explanations of regional brain function in terms of more abstract computational properties and context-specific interactions with anatomically distributed networks. This perspective was already present in Lenneberg's mind, when he wrote (Lenneberg 1967: 54) that "We cannot expect to find any kind of new protuberance or morphological innovation which deals exclusively with a particular behavior. Any modification on the brain is a modification on the entire brain. Thus species-specific behavior never has a confined, unique, neuroanatomic correlate, but always and necessarily must involve reorganization of processes that affect most of the central nervous system."

For this emerging research program to bear its fruits, we must be clear about what language does. In our view, it is much closer to global cognitive roles usually discussed in the literature on working memory, consciousness, executive control, attentional processes, and the like. The picture of linguistic cognition that emerges is one where the appropriate integration and processing of information is key. For this, it is necessary to study the recruitment of large neuronal assemblies, mediated by the thalamus. This neuronal activity, made coherent thanks to the intrinsic oscillatory potential of thalamic nuclei, arguably provides the means to link various neuronal populations as a response to particular aspects of an experience with regard to its significance and to guiding behavior, a way to organize the world around us and to make sense of it.

Taking the thalamus as a central region of interest has the added benefit of shedding light on cognitive disorders. We hope to have provided enough references in support of the role of the thalamus in the pathobiology of a variety of cognitive impairments often taken to be human-specific. In this context, let us stress that the point here is not to argue that the thalamus is a primary causal mechanism in all these mental diseases. This is not how we wish to interpret the data reviewed in earlier sections of this paper. Rather, in agreement with Tsatsanis et al. (2003), we see "great value in investigating the role of the thalamus in connection with larger neural systems and clinical dimensions" relevant to understanding breakdowns of mental health. In particular, in line with a growing literature on this topic, we have suggested that studying the role of the thalamus in controlling brain rhythms is very promising.

Given the centrality of the thalamus above and beyond language, down to embryonic development, its involvement in linguistic cognition may explain why it appears impossible for the core combinatorial aspects of language to completely go missing (Moro 2008: 179). Far from being the puzzle that Moro makes it looks to be, it may be a natural consequence of how the essence of the human language faculty is implemented in the brain.

## Note

\* The present work was made possible through a Marie Curie International Reintegration Grant from the European Union (PIRG-GA-2009–256413), research funds from the Fundació Bosch i Gimpera, the Generalitat de Catalunya (2014-SGR-200), and the Spanish ministry of economy and competitiveness (FFI2013–43823-P and FFI2014-61888-EXP).

## References

Aboitiz, Francisco. 1995. Working memory networks and the origin of language areas in the human brain. *Medical Hypotheses* 44:504–506.

Aboitiz, Francisco, Ricardo R. García, Conrado Bosman, and Enzo Brunetti. 2006. Cortical memory mechanisms and language origins. *Brain and Language* 98:40–56.

Agarwal, Nivedita, Gianluca Rambaldelli, Cinzia Perlini, Nicola Dusi, Omer Kitis, Marcella Bellani, Roberto Cerini, Miriam Isola, Amelia Versace, Matteo Balestrieri, Anna Gasparini, Roberto Pozzi Mucelli, Michele Tansella, and Paolo Brambilla. 2008. Microstructural thalamic changes in schizophrenia: A combined anatomic and diffusion weighted magnetic resonance imaging study. *Journal of Psychiatry & Neuroscience: JPN* 33:440.

Akam, Thomas, and Dimitri M Kullmann. 2010. Oscillations and filtering networks support flexible routing of information. *Neuron* 67:308–320.

Alelú-Paz, Raúl, and José Manuel Giménez-Amaya. 2008. The mediodorsal thalamic nucleus and schizophrenia. *Journal of Psychiatry & Neuroscience: JPN* 33:489.

Andreasen, Nancy C. 1997. The role of the thalamus in schizophrenia. *Canadian Journal of Psychiatry* 42:27–33.

Antzoulatos, Evan G., and Earl K. Miller. 2011. Differences between neural activity in prefrontal cortex and striatum during learning of novel abstract categories. *Neuron* 71:243–249.

Armstrong, E. 1981. A quantitative comparison of the hominoid thalamus. iv. posterior association nuclei – The pulvinar and lateral posterior nucleus. *American Journal of Physical Anthropology* 55:369–383.

Aron, A. R., F. Schlaghecken, P. C. Fletcher, E. T. Bullmore, M. Eimer, R. Barker, B. J. Sahakian, and T. W. Robbins. 2003. Inhibition of subliminally primed responses is mediated by the caudate and thalamus: Evidence from functional MRI and Huntington's disease. *Brain* 126:713–723.

Barbey, Aron K., Roberto Colom, Jeffrey Solomon, Frank Krueger, Chad Forbes, and Jordan Grafman. 2012. An integrative architecture for general intelligence and executive function revealed by lesion mapping. *Brain* 135:1154–1164.

Barron, D. S., S. B. Eickhoff, M. Clos, and P. T. Fox. 2015. Human pulvinar functional organization and connectivity. *Human Brain Mapping* 2015 36(7):2417–2431.

Barton, R. A. 2012. Embodied cognitive evolution and the cerebellum. *Philosophical Transactions of the Royal Society B: Biological Sciences* 367:2097–2107.

Barton, R. A., and C. Venditti. 2014. Rapid evolution of the cerebellum in humans and other great apes. *Current Biology* 24:2440–2444.

Beaton, Alan, and Peter Mariën. 2010. Language, cognition and the cerebellum: Grappling with an enigma. *Cortex* 46:811–820.

Berwick, R. C., A. D. Friederici, N. Chomsky, and J. J. Bolhuis. 2013. Evolution, brain, and the nature of language. *Trends in Cognitive Sciences* 17(2):89–98.

Binnewijzend, Maja A. A., Menno M. Schoonheim, Ernesto Sanz-Arigita, Alle Meije Wink, Wiesje M. van der Flier, Nelleke Tolboom, Sofie M. Adriaanse, Jessica S. Damoiseaux, Philip Scheltens, Bart N. M. van Berckel, F. Barkhof. 2012. Resting-state FMRI changes in Alzheimer's disease and mild cognitive impairment. *Neurobiology of Aging* 33:2018–2028.

Bish, Joel P., Vy Nguyen, Lijun Ding, Samantha Ferrante, and Tony J. Simon. 2004. Thalamic reductions in children with chromosome 22q11. 2 deletion syndrome. *Neuroreport* 15:1413–1415.

Blatt, Gene J. 2012. The neuropathology of autism. *Scientifica* Volume 2012 (2012), Article ID 703675, 16 pages.
Boeckx, C. 2011a. The emergence of language, from a biolinguistic point of view. In *The Oxford handbook of language evolution*, ed. M. Tallerman and K. Gibson, 492–501. Oxford: Oxford University Press.
Boeckx, Cedric. 2011b. Some reflections on Darwin's Problem in the context of Cartesian Biolinguistics. In *The biolinguistic enterprise: New perspectives on the evolution and nature of the human language faculty*, ed. A.-M. Di Sciullo, and C. Boeckx, 42–64. Oxford: Oxford University Press.
Boeckx, C. 2013a. Biolinguistics: Forays into human cognitive biology. *Journal of Anthropological Sciences* 91:63–89.
Boeckx, C. 2013b. Merge: Biolinguistic considerations. *English Linguistics* 30:463–483.
Boeckx, Cedric, and Antonio Benítez-Burraco. 2014. The shape of the language-ready brain. *Frontiers in Psychology* 5:282.
Boets, Bart, Hans P. Op de Beeck, Maaike Vandermosten, Sophie K. Scott, C.éline R. Gillebert, Dante Mantini, Jessica Bulthé, Stefan Sunaert, Jan Wouters, and Pol Ghesqui'ere. 2013. Intact but less accessible phonetic representations in adults with dyslexia. *Science* 342:1251–1254.
Bohlken, Marc M., Rachel M. Brouwer, René C. W. Mandl, Neeltje E. M. Haren, Rachel G. H. Brans, G. Caroline M. Baal, Eco J. C. Geus, Dorret I. Boomsma, Renée S. Kahn, Hulshoff Pol. 2013. Genes contributing to subcortical volumes and intellectual ability implicate the thalamus. *Human Brain Mapping* 35(6):2632–2642.
Bohsali, A. A., W. Triplett, A. Sudhyadhom, J. M. Gullett, K. McGregor, D. B. FitzGerald, T. Mareci, K. White, and B. Crosson. 2015. Broca's area – Thalamic connectivity. *Brain and Language* 141:80–88.
Brown, R. E., R. Basheer, J. T. McKenna, R. E. Strecker, and R. W. McCarley. 2012. Control of sleep and wakefulness. *Physiological Reviews* 92:1087–1187.
Brucato, N., T. Guadalupe, B. Franke, S. E. Fisher, and C. Francks. 2015. A schizophrenia-associated HLA locus affects thalamus volume and asymmetry. *Brain, Behavior, and Immunity* 46:311–318. .
Bruner, E. 2004. Geometric morphometrics and paleoneurology: Brain shape evolution in the genus homo. *Journal of Human Evolution* 47:279–303.
Bruner, E. 2010. Morphological differences in the parietal lobes within the human genus. *Current Anthropology* 51:S77–S88.
Bruner, E., J. M. de la Cuétara, M. Masters, H. Amano, and N. Ogihara. 2014a. Functional craniology and brain evolution: From paleontology to biomedicine. *Frontiers in Neuroanatomy* 8:19.
Bruner, E., G. Manzi, and J. L. Arsuaga. 2003. Encephalization and allometric trajectories in the genus homo: Evidence from the neandertal and modern lineages. *Proceedings of the National Academy of Sciences* 100:15335–15340.
Bruner, E., G. Rangel de Lázaro, J. M. Cuétara, M. Martín-Loeches, R. Colom, and H. Jacobs. 2014b. Midsagittal brain variation and MRI shape analysis of the precuneus in adult individuals. *Journal of Anatomy* 224:367–376.
Buckner, Randy L, Jessica R Andrews-Hanna, and Daniel L Schacter. 2008. The brain's default network. *Annals of the New York Academy of Sciences* 1124:1–38.
Buzsáki, G. 1991. The thalamic clock: Emergent network properties. *Neuroscience* 41:351–364.
Buzsáki, György. 2006. *Rhythms of the brain*. Oxford: Oxford University Press.

Byne, William, Erin A Hazlett, Monte S Buchsbaum, and Eileen Kemether. 2009. The thalamus and schizophrenia: Current status of research. *Acta Neuropathologica* 117:347–368.

Campion, G., and G. Elliot-Smith. 1934. *The neural basis of thought*. New York: Harcourt Brace Jovanovich.

Carrera, Emmanuel, and Julien Bogousslavsky. 2006. The thalamus and behavior effects of anatomically distinct strokes. *Neurology* 66:1817–1823.

Cavanna, A. E., and M. R. Trimble. 2006. The precuneus: A review of its functional anatomy and behavioural correlates. *Brain* 129:564–583.

Cheung, Charlton, Grainne M. McAlonan, Yee Y. Fung, Germaine Fung, K. Yu Kevin, Kin-Shing Tai, Pak C. Sham, and Siew E. Chua. 2011. Mri study of minor physical anomaly in childhood autism implicates aberrant neurodevelopment in infancy. *PloS ONE* 6:e20246.

Chomsky, N. 1988. *Language and problems of knowledge: The Managua lectures*. Cambridge, MA: MIT Press.

Chomsky, N. 2008. On phases. In *Foundational issues in linguistics*, ed. C. Otero, R. Freidin, and M.-L. Zubizarreta, 133–166. Cambridge, MA: MIT Press.

Chomsky, N. 2012. *The science of language: Interviews with James McGilvray*. Cambridge, UK: Cambridge University Press.

Cifelli, Alberto, Marzena Arridge, Peter Jezzard, Margaret M Esiri, Jacqueline Palace, and Paul M Matthews. 2002. Thalamic neurodegeneration in multiple sclerosis. *Annals of Neurology* 52:650–653.

Cole, Michael W., Jeremy R. Reynolds, Jonathan D. Power, Grega Repovs, Alan Anticevic, and Todd S. Braver. 2013. Multi-task connectivity reveals flexible hubs for adaptive task control. *Nature Neuroscience* 16:1348–1355.

Contreras, Diego, Alain Destexhe, Terrence J. Sejnowski, and Mircea Steriade. 1996. Control of spatiotemporal coherence of a thalamic oscillation by corticothalamic feedback. *Science* 274:771–774.

Coolidge, Frederick L., and Thomas Wynn. 2005. Working memory, its executive functions, and the emergence of modern thinking. *Cambridge Archaeological Journal* 15:5–26.

Coolidge, Frederick L., and Thomas Wynn. 2007. The working memory account of neandertal cognition – How phonological storage capacity may be related to recursion and the pragmatics of modern speech. *Journal of Human Evolution* 52:707–710.

Courchesne, Eric, and Karen Pierce. 2005. Brain overgrowth in autism during a critical time in development: implications for frontal pyramidal neuron and in-terneuron development and connectivity. *International Journal of Developmental Neuroscience* 23:153–170.

Crone, Julia Sophia, Andrea Soddu, Yvonne Höller, Audrey Vanhaudenhuyse, Matthias Schurz, Jürgen Bergmann, Elisabeth Schmid, Eugen Trinka, Steven Laureys, and Martin Kronbichler. 2014. Altered network properties of the fronto-parietal network and the thalamus in impaired consciousness. *NeuroImage: Clinical* 4:240–248.

Dagenbach, Dale, Alycia K. Kubat-Silman, and John R. Absher. 2001. Human verbal working memory impairments associated with thalamic damage. *International Journal of Neuroscience* 111:67–87.

Daitch, Amy L., Mohit Sharma, Jarod L. Roland, Serguei V. Astafiev, David T. Bundy, Charles M. Gaona, Abraham Z. Snyder, Gordon L. Shulman, Eric C. Leuthardt, and Maurizio Corbetta. 2013. Frequency-specific mechanism links human brain networks for spatial attention. *Proceedings of the National Academy of Sciences* 110:19585–19590.

David, O., B. Maess, K. Eckstein, and A. D. Friederici. 2011. Dynamic causal modeling of subcortical connectivity of language. *The Journal of Neuroscience* 31:2712–2717.

Dehaene, Stanislas, Lucie Charles, Jean-Rémi King, and Sébastien Marti. 2014. Toward a computational theory of conscious processing. *Current Opinion in Neurobiology* 25:76–84.

Dehaene, Stanislas, Michel Kerszberg, and Jean-Pierre Changeux. 1998. A neuronal model of a global workspace in effortful cognitive tasks. *Proceedings of the National Academy of Sciences* 95:14529–14534.

Díaz, Begoña, Florian Hintz, Stefan J Kiebel, and Katharina von Kriegstein. 2012. Dysfunction of the auditory thalamus in developmental dyslexia. *Proceedings of the National Academy of Sciences* 109:13841–13846.

Dipoppa, Mario, and Boris S. Gutkin. 2013. Flexible frequency control of cortical oscillations enables computations required for working memory. *Proceedings of the National Academy of Sciences* 110:12828–12833.

Duff, Melissa C., and Sarah Brown-Schmidt. 2012. The hippocampus and the flexible use and processing of language. *Frontiers in Human Neuroscience* 6:69.

Duncan, John. 2010. The multiple-demand (md) system of the primate brain: Mental programs for intelligent behaviour. *Trends in Cognitive Sciences* 14:172–179.

Duncan, John. 2013. The structure of cognition: Attentional episodes in mind and brain. *Neuron* 80:35–50.

Egawa, Jun, Yuichiro Watanabe, Hideaki Kitamura, Taro Endo, Ryu Tamura, Naoya Hasegawa, and Toshiyuki Someya. 2011. Reduced thalamus volume in non-right-handed male patients with autism spectrum disorders. *Psychiatry and Clinical Neurosciences* 65:395–395.

Fernández-Espejo, Davinia, Carme Junque, Montserrat Bernabeu, Teresa Roig-Rovira, Pere Vendrell, and Jose M Mercader. 2010. Reductions of thalamic volume and regional shape changes in the vegetative and the minimally conscious states. *Journal of Neurotrauma* 27:1187–1193.

Ferstl, Evelyn C., Jane Neumann, Carsten Bogler, and D. Yves Von Cramon. 2008. The extended language network: A meta-analysis of neuroimaging studies on text comprehension. *Human Brain Mapping* 29:581–593.

Finlay, Barbara L., Richard B. Darlington, and Nicholas Nicastro. 2001. Developmental structure in brain evolution. *Behavioral and Brain Sciences* 24:263–278.

Frantseva, Marina, Jie Cui, Faranak Farzan, Lakshminarayan V. Chinta, Jose Luis Perez Velazquez, and Zafiris Jeffrey Daskalakis. 2014. Disrupted cortical conductivity in schizophrenia: Tms–eeg study. *Cerebral Cortex* 24:211–221.

Friederici, Angela D. 2006. What's in control of language? *Nature Neuroscience* 9:991–992.

Galaburda, Albert M., and David Eidelberg. 1982. Symmetry and asymmetry in the human posterior thalamus: Ii. thalamic lesions in a case of developmental dyslexia. *Archives of Neurology* 39:333.

Garofoli, D., and M.N. Haidle. 2013. Epistemological problems in cognitive archaeology: An anti-relativistic proposal towards methodological uniformity. *Journal of Anthropological Sciences* 92:7–41.

Ghosh, Subrata, Krishna Aswani, Surabhi Singh, Satyajit Sahu, Daisuke Fujita, and Anirban Bandyopadhyay. 2014. Design and construction of a brain-like computer: A new class of frequency-fractal computing using wireless communication in a supramolecular organic, inorganic system. *Information* 5:28–100.

Giménez, Mónica, Carme Junqué, Ana Narberhaus, Francesc Botet, Núria Bargalló, and Josep Maria Mercader. 2006. Correlations of thalamic reductions with verbal fluency impairment in those born prematurely. *Neuroreport* 17:463–466.

Goldman-Rakic, Patricia S. 1994. Working memory dysfunction in schizophrenia. *The Journal of Neuropsychiatry and Clinical Neurosciences*, 6(4):348–357.

Greicius, Michael D., Benjamin H. Flores, Vinod Menon, Gary H. Glover, Hugh B. Solvason, Heather Kenna, Allan L. Reiss, and Alan F. Schatzberg. 2007. Restingstate functional connectivity in major depression: Abnormally increased contributions from subgenual cingulate cortex and thalamus. *Biological Psychiatry* 62:429–437.

Gruberger, Michal, Eti Ben-Simon, Yechiel Levkovitz, Abraham Zangen, and Talma Hendler. 2011. Towards a neuroscience of mind-wandering. *Frontiers in Human Neuroscience* 5:56.

Gunz, P., S. Neubauer, L. Golovanova, V. Doronichev, B. Maureille, and J.-J. Hublin. 2012. A uniquely modern human pattern of endocranial development. Insights from a new cranial reconstruction of the neandertal newborn from mezmaiskaya. *Journal of Human Evolution* 62:300–313.

Gunz, P., S. Neubauer, B. Maureille, and J.-J. Hublin. 2010. Brain development after birth differs between neanderthals and modern humans. *Current Biology* 20:R921–R922.

Harms, Michael P., Lei Wang, Daniel Mamah, Deanna M. Barch, Paul A. Thompson, and John G. Csernansky. 2007. Thalamic shape abnormalities in individuals with schizophrenia and their nonpsychotic siblings. *The Journal of Neuroscience* 27:13835–13842.

Hauser, Marc D. 2009. The possibility of impossible cultures. *Nature* 460:190–196.

Henderson, J. M., K. Carpenter, H. Cartwright, and G. M. Halliday. 2000. Loss of thalamic intralaminar nuclei in progressive supranuclear palsy and parkinson's disease: Clinical and therapeutic implications. *Brain* 123:1410–1421.

Heuvel, Martijn P. van den, René S. Kahn, Joaquín Goñi, and Olaf Sporns. 2012. High-cost, high-capacity backbone for global brain communication. *Proceedings of the National Academy of Sciences* 109:11372–11377.

Heuvel, Martijn P. van den, and Olaf Sporns. 2011. Rich-club organization of the human connectome. *The Journal of neuroscience* 31:15775–15786.

Heuvel, Martijn P. van den, Cornelis J. Stam, René S. Kahn, and Hilleke E. Hulshoff Pol. 2009. Efficiency of functional brain networks and intellectual performance. *The Journal of Neuroscience* 29:7619–7624.

Jensen, Matthew B. 2010. The accountant who lost arithmetic: A case report of acalculia with a left thalamic lesion. *Journal of Medical Cases* 1:92.

Jones, Edward G. 1997. Cortical development and thalamic pathology in schizophrenia. *Schizophrenia Bulletin* 23:483–501.

Kemether, Eileen M, Monte S Buchsbaum, William Byne, Erin A Hazlett, Mehmet Haznedar, Adam M Brickman, Jimcy Platholi, and Rachel Bloom. 2003. Magnetic resonance imaging of mediodorsal, pulvinar, and centromedian nuclei of the thalamus in patients with schizophrenia. *Archives of General Psychiatry* 60:983–991.

Kircher, Adam, and Karen K Glendenning. 2002. The evolution of human intelligence and neural inhibition. *Universidade Estadual de Campinas Brain and Mind Magazine* 1–10.

Klostermann, Fabian, Lea K. Krugel, and Felicitas Ehlen. 2013. Functional roles of the thalamus for language capacities. *Frontiers in Systems Neuroscience* 7:32.

Kotz, Sonja A., and Michael Schwartze. 2010. Cortical speech processing unplugged: A timely subcortico-cortical framework. *Trends in Cognitive Sciences* 14:392–399.
Kovacs, Gabor G., Alexander Peden, Serge Weis, Romana Höftberger, Anna S. Berghoff, Helen Yull, Thomas Ströbel, Stefan Koppi, Regina Katzenschlager, Dieter Langenscheidt, Hamid Assar, Elisabeth Zaruba, Albrecht Gröner, Till Voigtländer, Gina Puska, Eva Hametner, Astrid Grams, Armin Muigg, Michael Knoflach, Lajos László, James W. Ironside, Mark W. Head, and Herbert Budka. 2013. Rapidly progressive dementia with thalamic de-generation and peculiar cortical prion protein immunoreactivity, but absence of proteinase k resistant prp: A new disease entity? *Acta Neuropathologica Communications* 1:72.
Lashley, Karl S. 1950. In search of the engram. *Symposia of the Society for Experimental Biology* 4:454–482.
Lenneberg, Eric H. 1967. *Biological foundations of language*. New York: Wiley.
Letinic, K., and P. Rakic. 2001. Telencephalic origin of human thalamic gabaergic neurons. *Nature neuroscience* 4:931–936.
Lieberman, Philip. 2002. On the nature and evolution of the neural bases of human language. *American Journal of Physical Anthropology* 119:36–62.
Lieberman, Philip. 2009. *Human language and our reptilian brain: The subcortical bases of speech, syntax, and thought*. Cambridge, MA: Harvard University Press.
Lieberman, Philip. 2013. Synapses, language, and being human. *Science* 342: 944–945.
Ligam, Poonam, Robin L. Haynes, Rebecca D. Folkerth, Lena Liu, May Yang, Joseph J. Volpe, and Hannah C Kinney. 2009. Thalamic damage in periventricular leukomalacia: Novel pathologic observations relevant to cognitive deficits in survivors of prematurity. *Pediatric Research* 65:524–529.
Llinás, Rodolfo R., and Mircea Steriade. 2006. Bursting of thalamic neurons and states of vigilance. *Journal of Neurophysiology* 95:3297–3308.
Lutkenhoff, Evan S., David L. McArthur, Xue Hua, Paul M. Thompson, Paul M. Vespa, and Martin M. Monti. 2013. Thalamic atrophy in antero-medial and dorsal nuclei correlates with six-month outcome after severe brain injury. *NeuroImage: Clinical* 3:396–404.
Martins-de Souza, Daniel, Giuseppina Maccarrone, Thomas Wobrock, Inga Zerr, Philipp Gormanns, Stefan Reckow, Peter Falkai, Andrea Schmitt, and Christoph W Turck. 2010. Proteome analysis of the thalamus and cerebrospinal fluid reveals glycolysis dysfunction and potential biomarkers candidates for schizophrenia. *Journal of Psychiatric Research* 44:1176–1189.
Maxwell, William L., Mary Anne MacKinnon, Douglas H. Smith, Tracy K. McIntosh, and David I. Graham. 2006. Thalamic nuclei after human blunt head injury. *Journal of Neuropathology & Experimental Neurology* 65:478–488.
McKeown, Martin J, Ashish Uthama, Rafeef Abugharbieh, Samantha Palmer, Mechelle Lewis, and Xuemei Huang. 2008. Shape (but not volume) changes in the thalami in parkinson disease. *BMC Neurology* 8:8.
Mendez, Mario F., Nora C. Papasian, and Gerald T. H. Lim. 2003. Thalamic acalculia. *The Journal of Neuropsychiatry and Clinical Neurosciences* 15:115–116.
Miller, Earl K. 2013. The "working" of working memory. *Dialogues in Clinical Neuroscience* 15:411–418.
Miller, Earl K., and Tim J. Buschman. 2012. Cortical circuits for the control of attention. *Current Opinion in Neurobiology* 23:1–7.

Miller, Earl K., and Timothy J. Buschman. 2007. Bootstrapping your brain: How interactions between the frontal cortex and basal ganglia may produce organized actions and lofty thoughts. In *Neurobiology of Learning and Memory (2nd Edition)*, ed. Raymond P. Kesner and Joe L. Martinez, 339–354. London: Elsevier.

Mills, Kathryn L., Deepti Bathula, Taciana G. Costa Dias, Swathi P. Iyer, Michelle C. Fenesy, Erica D. Musser, Corinne A. Stevens, Bria L. Thurlow, Samuel D. Carpenter, Bonnie J. Nagel, Joel T. Nigg and Damien A. Fair. 2012. Altered corticostriatal – Thalamic connectivity in relation to spatial working memory capacity in children with ADHD. *Magnetic Resonance Imaging of Disturbed Brain Connectivity in Psychiatric Illness* 3:2.

Moro, A. 2008. *The boundaries of babel: The brain and the enigma of impossible languages*. Cambridge, MA: MIT Press.

Murdoch, B. E. 2010. The cerebellum and language: Historical perspective and review. *Cortex* 46:858–868.

Nair, Aarti, Jeffrey M. Treiber, Dinesh K. Shukla, Patricia Shih, and Ralph-Axel Müller. 2013. Impaired thalamocortical connectivity in autism spectrum disorder: a study of functional and anatomical connectivity. *Brain* 136:1942–1955.

Neubauer, S., P. Gunz, and J.-J. Hublin. 2010. Endocranial shape changes during growth in chimpanzees and humans: A morphometric analysis of unique and shared aspects. *Journal of Human Evolution* 59:555–566.

Normand, Elizabeth A., Shane R. Crandall, Catherine A. Thorn, Emily M. Murphy, Bettina Voelcker, Catherine Browning, Jason T. Machan, Christopher I. Moore, Barry W. Connors, and Mark Zervas. 2013. Temporal and mosaic¡ i¿ tscl¡/i¿ deletion in the developing thalamus disrupts thalamocortical circuitry, neural function, and behavior. *Neuron* 78:895–909.

O'Sullivan, Michael, Sumeet Singhal, Rebecca Charlton, and Hugh S. Markus. 2004. Diffusion tensor imaging of thalamus correlates with cognition in cadasil without dementia. *Neurology* 62:702–707.

Parnaudeau, Sebastien, Pia-Kelsey O'Neill, Scott S. Bolkan, Ryan D. Ward, Atheir I. Abbas, Bryan L. Roth, Peter D. Balsam, Joshua A. Gordon, and Christoph Kellendonk. 2013. Inhibition of mediodorsal thalamus disrupts thalamofrontal connectivity and cognition. *Neuron* 77:1151–1162.

Patterson, Terry. 1987. Studies toward the subcortical pathogenesis of schizophrenia. *Schizophrenia Bulletin* 13:555–576.

Pearce, Eiluned, Chris Stringer, and Robin I. M. Dunbar. 2013. New insights into differences in brain organization between neanderthals and anatomically modern humans. *Proceedings of the Royal Society of London. Series B, Biological Sciences* 280:20130168.

Penfield, W., and L. Roberts. 1959. *Speech and brain mechanisms*. Princeton: Princeton University Press.

Pinault, Didier. 2011. Dysfunctional thalamus-related networks in schizophrenia. *Schizophrenia Bulletin* 37:238–243.

Popken, Gregory J., William E. Bunney, Steven G. Potkin, and Edward G. Jones. 2000. Subnucleus-specific loss of neurons in medial thalamus of schizophrenics. *Proceedings of the National Academy of Sciences* 97:9276–9280.

Rakic, P., and R. L. Sidman. 1969. Telencephalic origin of pulvinar neurons in the fetal human brain. *Zeitschrift für Anatomie und Entwicklungsgeschichte* 129:53–82.

Rakic, Pasko. 2009. Evolution of the neocortex: A perspective from developmental biology. *Nature Reviews Neuroscience* 10:724–735.

Reinhart, Robert M. G., and Geoffrey F. Woodman. 2013. Oscillatory coupling reveals the dynamic reorganization of large-scale neural networks as cognitive demands change. *Journal of Cognitive Neuroscience* 26(1):175–188.

Roux, Frédéric, and Peter J. Uhlhaas. 2014. Working memory and neural oscillations: Alpha – gamma versus theta – gamma codes for distinct wm information? *Trends in Cognitive Sciences* 18:16–25.

Rubin, Rachael D., Patrick D. Watson, Melissa C. Duff, and Neal J. Cohen. 2014. The role of the hippocampus in flexible cognition and social behavior. *Frontiers in Human Neuroscience* 8:742.

Saalmann, Yuri B., and Sabine Kastner. 2011. Cognitive and perceptual functions of the visual thalamus. *Neuron* 71:209–223.

Saalmann, Yuri B., Mark A. Pinsk, Liang Wang, Xin Li, and Sabine Kastner. 2012. The pulvinar regulates information transmission between cortical areas based on attention demands. *Science* 337:753–756.

Scharinger, M., M. J. Henry, J. Erb, L. Meyer, and J. Obleser. 2014. Thalamic and parietal brain morphology predicts auditory category learning. *Neuropsychologia* 53:75–83.

Schenker, N. M., W. D. Hopkins, M. A. Spocter, A. R. Garrison, C. D. Stimpson, J. M. Erwin, P. R. Hof, and C. C. Sherwood. 2010. Broca's area homologue in chimpanzees (pan troglodytes): probabilistic mapping, asymmetry, and comparison to humans. *Cerebral Cortex* 20:730–742.

Schmahmann, Jeremy D., and Deepak N. Pandya. 2008. Disconnection syndromes of basal ganglia, thalamus, and cerebrocerebellar systems. *Cortex* 44:1037–1066.

Shanahan, M. 2012. The brain's connective core and its role in animal cognition. *Philosophical Transactions of the Royal Society B: Biological Sciences* 367:2704–2714.

Shi, Feng, Li Wang, Ziwen Peng, Chong-Yaw Wee, and Dinggang Shen. 2013. Altered modular organization of structural cortical networks in children with autism. *PloS one* 8:e63131.

Shipp, S. 2003. The functional logic of cortico – Pulvinar connections. *Philosophical Transactions of the Royal Society of London. Series B: Biological Sciences* 358:1605–1624.

Singer, Wolf. 2011. Dynamic formation of functional networks by synchronization. *Neuron* 69:191–193.

Staudigl, Tobias, Tino Zaehle, Jürgen Voges, Simon Hanslmayr, Christine Esslinger, Hermann Hinrichs, Friedhelm C Schmitt, Hans-Jochen Heinze, and Alan Richardson-Klavehn. 2012. Memory signals from the thalamus: Early thalamocortical phase synchronization entrains gamma oscillations during long-term memory retrieval. *Neuropsychologia* 50(14):3519-27.

Steriade, Mircea. 2004. Local gating of information processing through the thalamus. *Neuron* 41:493–494.

Steriade, Mircea, David A. McCormick, and Terrence J. Sejnowski. 1993. Thalamocortical oscillations in the sleeping and aroused brain. *Science* 262:679–685.

Stern, K. 1939. Severe dementia associated with bilateral symmetrical degeneration of the thalamus. *Brain* 4:157–171.

Striedter, G. F. 2005. *Principles of brain evolution*. Sunderland, MA: Sinauer Associates.

Sun, Limin, Nazareth Castellanos, Christine Grützner, Dagmar Koethe, Davide Rivolta, Michael Wibral, Laura Kranaster, Wolf Singer, Markus F. Leweke, and Peter J. Uhlhaas. 2013. Evidence for dysregulated high-frequency oscillations during sensory processing in medication-naïve, first episode schizophrenia. *Schizophrenia Research* 150:519–525.

Teichmann, M., C. Rosso, J.-B. Martini, I. Bloch, P. Brugi'eres, H. Duffau, S. Lehéricy, and A.-C. Bachoud-Lévi. 2015. A cortical – subcortical syntax pathway linking broca's area and the striatum. *Human Brain Mapping* 36(6):2270–2283.

Theofanopoulou, C., and C. Boeckx. submitted. Neural Syntax.

Theyel, Brian B., Daniel A. Llano, and S. Murray Sherman. 2010. The corticothalamocortical circuit drives higher-order cortex in the mouse. *Nature Neuroscience* 13:84–88.

Thong, Jamie Yu Jin, Anqi Qiu, Min Yi Sum, Carissa Nadia Kuswanto, Ta Ahn Tuan, Gary Donohoe, Yih Yian Sitoh, and Kang Sim. 2013. Effects of the neuro-granin variant rs12807809 on thalamocortical morphology in schizophrenia. *PloS one* 8:e85603.

Tononi, G., and G. M. Edelman. 1998. Consciousness and complexity. *Science* 282:1846–1851.

Treves, Alessandro. 2009. Spatial cognition, memory capacity, and the evolution of mammalian hippocampal networks. *In Cognitive Biology: Evolutionary and Developmental Perspectives on Mind, Brain, and Behavior*, ed. Luca Tommasi, Mary A. Peterson, and Lynn Nadel, 41–59. Cambridge, MA: MIT Press.

Tsatsanis, Katherine D., Byron P. Rourke, Ami Klin, Fred R. Volkmar, Domenic Cicchetti, and Robert T. Schultz. 2003. Reduced thalamic volume in high-functioning individuals with autism. *Biological Psychiatry* 53:121–129.

Uhlhaas, Peter J., Corinna Haenschel, Danko Nikoli´C., and Wolf Singer. 2008. The role of oscillations and synchrony in cortical networks and their putative relevance for the pathophysiology of schizophrenia. *Schizophrenia Bulletin* 34:927–943.

Uhlhaas, Peter J., Frédéric Roux, Eugenio Rodriguez, Anna Rotarska-Jagiela, and Wolf Singer. 2010. Neural synchrony and the development of cortical networks. *Trends in Cognitive Sciences* 14:72–80.

Uhlhaas, Peter J., and Wolf Singer. 2006. Neural synchrony in brain disorders: Relevance for cognitive dysfunctions and pathophysiology. *Neuron* 52:155–168.

Uhlhaas, Peter J., and Wolf Singer. 2010. Abnormal neural oscillations and synchrony in schizophrenia. *Nature Reviews Neuroscience* 11:100–113.

Ullman, Michael T. 2001. A neurocognitive perspective on language: The declarative/procedural model. *Nature Reviews Neuroscience* 2:717–726.

Ullman, Michael T. 2004. Contributions of memory circuits to language: The declarative/procedural model. *Cognition* 92:231–270.

Ullman, Michael T. 2006. Is broca's area part of a basal ganglia thalamocortical circuit? *Cortex* 42:480–485.

Van Der Werf, Ysbrand D., Danielle J. Tisserand, Pieter Jelle Visser, Paul A. M. Hofman, Eric Vuurman, Harry Uylings, and Jelle Jolles. 2001. Thalamic volume predicts performance on tests of cognitive speed and decreases in healthy aging: A magnetic resonance imaging-based volumetric analysis. *Cognitive Brain Research* 11:377–385.

Vigren, Patrick, Anders Tisell, Maria Engström, Thomas Karlsson, Olof Leinhard Dahlqvist, Peter Lundberg, and Anne-Marie Landtblom. 2013. Low thalamic NAA-concentration corresponds to strong neural activation in working memory in kleinelevin syndrome. *PloS One* 8:e56279.

Vila-Rodriguez, Fidel, Leon French, Vilte Barakauskas, Carri-Lyn Mead, and Babak Khorram. 2008. Thalamic shape: A possible endophenotype. *The Journal of Neuroscience* 28:3533–3534.

Vincent, Justin L., Itamar Kahn, Abraham Z. Snyder, Marcus E Raichle, and Randy L. Buckner. 2008. Evidence for a frontoparietal control system revealed by intrinsic functional connectivity. *Journal of Neurophysiology* 100:3328–3342.

Wahl, Michael, Frank Marzinzik, Angela D. Friederici, Anja Hahne, Andreas Kupsch, Gerd-Helge Schneider, Douglas Saddy, Gabriel Curio, and Fabian Klostermann. 2008. The human thalamus processes syntactic and semantic language violations. *Neuron* 59:695–707.

Walker, Earl. 1938. *The primate thalamus*. Chicago: University of Chicago Press.

Whiting, B. A., and R. A. Barton. 2003. The evolution of the cortico-cerebellar complex in primates: anatomical connections predict patterns of correlated evolution. *Journal of Human Evolution* 44:3–10.

Wynn, Thomas, and Frederick L. Coolidge. 2007. Did a small but significant enhancement in working memory capacity power the evolution of modern thinking. *Rethinking the Human Revolution: New Behavioural and Biological Perspectives on the Origin and Dispersal of Modern Humans*, ed. Paul Mellars, Katie Boyle, Ofer Bar-Yosef, and Chris Stringer, 79–90. Cambridge, UK: McDonald Institute Monographs.

Yizhar, Ofer, Lief E. Fenno, Matthias Prigge, Franziska Schneider, Thomas J. Davidson, Daniel J. O'Shea, Vikaas S. Sohal, Inbal Goshen, Joel Finkelstein, Jeanne T. Paz, Katja Stehfest, Roman Fudim, Charu Ramakrishnan, John R. Huguenard, Peter Hegemann, and Karl Deisseroth. 2011. Neocortical excitation/inhibition balance in information processing and social dysfunction. *Nature* 477:171–178.

Young, Keith A., Leigh A. Holcomb, Umar Yazdani, Paul B. Hicks, and Dwight C. German. 2004. Elevated neuron number in the limbic thalamus in major depression. *American Journal of Psychiatry* 161:1270–1277.

Zhou, Bo, Yong Liu, Zengqiang Zhang, Ningyu An, Hongxiang Yao, Pan Wang, Luning Wang, Xi Zhang, and Tianzi Jiang. 2013. Impaired functional connectivity of the thalamus in alzheimer's disease and mild cognitive impairment: A resting-state fMRI study. *Current Alzheimer Research* 10:754–766.

Zhou, Jingsheng, Xiaolin Liu, Weiqun Song, Yanhui Yang, Zhilian Zhao, Feng Ling, Anthony G. Hudetz, and Shi-Jiang Li. 2011. Specific and nonspecific thalamo-cortical functional connectivity in normal and vegetative states. *Consciousness and Cognition* 20:257–268.

Zikopoulos, Basilis, and Helen Barbas. 2007. Parallel driving and modulatory pathways link the prefrontal cortex and thalamus. *PLoS One* 2:e848.

# 16 A biolinguistic approach to language disorders
## Towards a paradigm shift in clinical linguistics

*Antonio Benítez-Burraco*

## 1 Clinical linguistics: a messy scenario

On paper, clinical categories like dyslexia or specific language impairment (SLI) refer to cognitive disorders in which only language becomes impaired and that can be distinguished from other similar categories at all levels of analysis (phenotypic, cognitive, neurobiological, genetic, etc.). For example, people suffering from dyslexia have difficulties reading texts and spelling words (Lyon et al. 2003). These problems are thought to be caused by the dysfunction of the phonological component of the working memory (Shaywitz et al. 1998). Additionally, the brains of dyslexics show anomalies that are both structural (Galaburda et al. 1985, Deutsch et al. 2005) and functional (Shaywitz et al. 1998, Maisog et al. 2008) and which concern many of the brain areas involved in reading and spelling in the non-affected population (see Démonet et al. 2004 for review). Finally, most of the several candidate genes for dyslexia identified to date regulate axonal growth and neuronal migration in the cortex, plausibly accounting for the structural and functional anomalies attested in the brains of dyslexics (see Benítez-Burraco 2010 for review).

Nonetheless, for clinical linguists, things are usually less clear-cut and more difficult to handle. To begin with, patients commonly show symptoms that are compatible with more than one disorder (linguistic or not linguistic by nature), to the extent that comorbidity is a frequent outcome of clinical practice. Using again dyslexia as an example, reading difficulties are observed in many cognitive disorders. Actually, dyslexia is frequently comorbid with other language disorders, including SLI (Smith et al. 1996; Catts et al. 2005) and speech-sound disorder (SSD) (Smith et al. 1996; Shriberg et al. 1999; Stein et al. 2004), but also with attention deficit hyperactivity disorder (ADHD) (Purvis and Tannock 1997; Shaywitz 1998). Secondly, people affected by one disorder generally display linguistic abilities that are quite variable. In order to apprehend this variability, different subtypes of the same disorder need to be posited, in which one (among several) specific aspect(s) of language becomes more impaired. However, variation is also observed throughout development, to the extent that affected children can switch from one subtype to another of the same disorder as they grow (Botting and Conti-Ramsdem 2004). As

Karmiloff-Smith and Mills (2006: 585) put it "one cannot simply assume that deficits in the phenotypic outcome are the same as those apparent in the infant start state". Importantly, deficits in language performance may arise from cognitive impairment in a non-direct fashion. This circumstance substantially increases the observed variation at the symptomatic/clinical level. Obviously, it substantially complicates the categorization of disorders. As Karmiloff-Smith puts it (2008: 693), "to understand any developmental syndrome, it is essential to distinguish between the behavioral phenotype (based on scores from standardized tests of overt behavior) and the cognitive phenotype (based on in-depth analyses of the mental processes underlying the overt behavior)". In fact, "sometimes equivalent behavioral scores camouflage very different cognitive processes" (Donnai and Karmiloff-Smith 2000: 167).

On the whole, it seems that different disorders (or different subtypes of the same disorder) may result from the same (broad) cognitive deficit, which can manifest differentially in different populations and/or environmental conditions (hence the alleged heterogeneity and/or comorbidity). At the same time, different deficits (that may or may not be specifically linguistic) can contribute to the same disorder, this implying that clinical categories may be well construed as conglomerates of several cognitive deficits, yet characterised by substantially similar symptomatic profiles. Moreover, the diverse subtypes of a particular disorder may represent conditions in which one of such deficits prevails. Ultimately, in other different populations and/or environments any of these underlying deficits can manifest as a different disorder (hence the purported heterogeneity and/or comorbidity). For instance, the dysfunction of the phonological component of working memory gives rise not just to dyslexia, but also to SLI (Bishop 2002) and SSD (Shriberg et al. 1999). Conversely, several other deficits have claimed to contribute to dyslexia, including problems with categorical perception (Serniclaes et al. 2004), difficulties for correctly processing (and discriminating between) brief acoustic impulses (Temple et al. 2000), cerebellar dysfunctions (Nicolson and Fawcett 2006), problems with visual processing (Lovegrove et al. 1980), or a dysfunction of the magnocellular pathway (Livingstone et al. 1991; Stein and Walsh 1997).

Finally, it is frequently observed that problems with language in the affected people concern quite broad aspects of language, to the extent that the attested deficits do not normally match the units, levels, or operations that underlie linguistic theory (Newmeyer 1997). As a consequence, clinical typologies are not always acceptable under a linguistic lens. For instance, some speech therapists claim that three basic subtypes of SLI do exist: phonological, expressive and expressive-receptive (e.g. Rapin and Allen 1983; American Psychiatric Association 1994). Similarly, a syntactic-pragmatic subtype is also included in some classifications (Rapin and Allen 1983). However, these are separate levels in most usual accounts of language.

Comorbidity, heterogeneity, and variability are observed at the neurobiological level too. Hence, the affected regions (structurally or functionally) in one disorder may well be impaired in people suffering from other different condition.

Consider, for instance, the ventral portion of the occipito-temporal region. This area contains one of the two processing subsystems needed for reading that are located in the posterior region of the left hemisphere. Not surprisingly, this area is underactive in dyslexics during reading tasks (Horwitz et al. 1998; Shaywitz et al. 1998; Paulesu et al. 2001). However, this area seems to be involved as well in the recognition of faces and it has been linked to prosopagnosia too (Sorger et al. 2007; Dricot et al. 2008). At the same time, similar (abnormal) neurobiological profiles can be observed in different clinical conditions. For instance, an increase of the gray-matter density in the perisylvian cortex has been documented in ADHD, Williams syndrome, and fetal alcohol syndrome (Toga et al. 2006). All these conditions have diverse aetiologies and different neurocognitive profiles, but all of them encompass language deficits (Mervis and Becerra 2007; Rapport et al. 2008; Wyper and Rasmussen 2011) and may be comorbid (O'Malley and Nanson 2002; Rhodes et al. 2011). Overall, it is not clear whether the involved regions are multifunctional by nature or perform instead some basic process that is recruited for language and for other cognitive abilities. Moreover, it is frequently observed that affected regions may give rise to mixed symptoms. Lastly, it commonly occurs that their boundaries are located differently in different individuals.

Finally, things are not easier to interpret at the molecular level. Different candidate genes and risk factors for different language disorders have been identified to date. However, as we have seen in the case of dyslexia, it is not one but many genes that usually contribute to each disorder (*polymorphism*). Typically, several pathogenic variants of each candidate gene have been identified. At the same time, other polymorphisms may contribute to the language abilities of the non-affected population. Importantly, the same mutation in the same gene may cause the disorder in some individuals, but not in others (*variable penetrance*). Conversely, pathogenic variants of a gene may be well absent in people affected by the disorder (*phenocopy*). Moreover, the same mutation can give rise to different disorders in different populations, to the extent that candidate genes for a particular disorder may be found mutated in other conditions (*pleiotropy*). Ultimately, mutations in genes encoding proteins that are functionally related to one particular candidate (i.e. they belong to the same interactome) may give rise to different disorders in different subjects or environments. *FOXP2* (the famous "language gene") and some of its functional partners nicely illustrate this complex scenario. The linguistic (and the cognitive) profile of people bearing the well-known mutation R553H (KE family) is not homogeneous (Vargha-Khadem et al. 1995; Watkins et al. 2002). Moreover, several other pathogenic mutations of the gene, entailing diverse linguistic and cognitive deficits, have been identified thus far (Vargha-Khadem et al. 1995; Watkins et al. 2002; Vargha-Khadem et al. 2005; Shriberg et al. 2006; Roll et al. 2010). Additionally, unlike *FOXP2* itself, the mutation of *CNTNAP2* (one of its physiological targets) give rise to canonical SLI (Vernes et al. 2008), but also to stuttering (Petrin et al. 2010), language and mental delay (Sehested et al. 2010), and autism (Alarcón et al. 2008; Bakkaloglu et al. 2008). However, some polymorphisms of *CNTNAP2* affects language development of healthy children (Whitehouse et al. 2011) and language processing in adult healthy people (Whalley et al. 2011).Conversely, the mutation of *SRPX2* (another of FOXP2's targets) (Roll et al. 2010) gives rise to rolandic (or sylvian) epilepsy with speech dyspraxia (Roll et al. 2006) or to bilateral perisylvian polymicrogiria with dysartria and mild mental retardation (Roll et al. 2006).

## 2 Unsatisfactory efforts (though still worth trying)

To some extent, the problem reviewed above should benefit from the improvement of the diagnostic tools currently used in clinical linguistics. Specifically, it is important to maximize the linguistic nature of the experimental tasks used for the diagnosis in order to only evaluate specific components/operations of language that may be selectively impaired in language disorders. However, this is not easy to achieve. Actually, it may well be impossible if other cognitive devices besides language itself are involved in passing from competence to performance (Newmeyer 1997). At the same time, diagnostic tests should evaluate real neurolinguistic entities only. As we discuss below, in some cases the linguistic features, units, categories, rules, or computations under analysis (phonological features, agreement patterns and the like) may be not compatible with the sort of computations the brain is able to make in real time. A related concern is the reliability and relevance of the parameters under evaluation in the tests. For instance, a shortfall in repeating pseudowords or in generating inflected verbal forms have been proposed as core psycholinguistic markers for SLI (Bishop et al. 1996). Nonetheless, the former deficit is also a relevant hallmark of children with dyslexia (Mayringer and Wimmer 2000, Quaglino et al. 2008), or with Down syndrome (Jarrold et al. 2000). At the same time, pseudoword reading could actually be a misinforming measure if either phonological processing ability or phonological awareness are to be evaluated in children below 4 years (Thomson et al. 2006). Concerning inflection, because different computational processes are involved in agreement, lower scores in tests evaluating inflectional morphology can be actually due to diverse underlying deficits.

Moreover, several diagnostic tests may exist for the same disorder. If they follow different criteria, it can be (erroneously) concluded that the condition is caused by different underlying deficits, and/or is comorbid with other language impairments (and/or other cognitive disorders), or (quite typically) that several subtypes of the disorder actually exists. A related concern is the fact that disorders are commonly diagnosed categorically (you have it or you haven't). As a consequence, people having one disorder usually show symptoms that are not homogeneous (this ultimately explains why different subtypes of a disorder are frequently postulated; see above). In practice, clinical categories are cover terms for pathological groups that are diverse both symptomatically and aetiologically (see Parisse and Maillart 2009 on SLI). Because their definition commonly entails some sort of homogenization of the observed data, it is important to always rely on properly normalised statistical procedures when establishing them. Nowadays language disorders are usually characterised as continuous variables, that is, as specific intervals within a continuum also encompassing the linguistic competence of the non-affected population. However, we should always wonder about the biological reliability and significance of the clinical frontiers we may eventually draw (between the affected and the non-affected populations, across different disorders, or those delimiting different subtypes of the same disorder) (see Shaywitz et al. 2008 on dyslexia for a discussion).

Similarly, we need to improve the confidence and the resolution of the neuroimaging devices used for analysing the disordered brain. Current techniques do not allow us to always discern whether multifunctional areas are composed or not of different neuronal populations performing different kinds of computations. Moreover, functional neuroimaging just provides us with (low-resolution) images of the physiological changes (in terms of blood flux, electrical activity, and so on) elicited by the experimental tasks used for the diagnosis. However, these pictures cannot be equated with the representations and computations that are important for language (and for linguistic theory). As Poeppel (2012) puts it, mapping is not explaining. In order to explain what we observe, we first need to address two important shortcomings of current neurolinguistic studies. First, "[l]inguistic and neuroscientific studies of language operate with objects of different granularity" (Poeppel and Embick 2005: 105). Neurolinguistics makes broad conceptual distinctions (syntax vs. semantics, morphology vs. syntax, etc.), which usually involve the admixture of multiple components or processes of diverse nature. Second, "the fundamental elements of linguistic theory cannot be reduced or matched up with the fundamental biological units identified by neuroscience" (Poeppel and Embick 2005: 105). Overall, we first need to spell out language (and language deficits) "in computational terms that are at the appropriate level of abstraction (i.e. can be performed by specific neuronal populations)" (Poeppel and Embick 2005: 106) (we will return to this problem in section 3). Ultimately, if other cognitive systems besides language are compulsorily involved in passing from competence to performance, we should not expect that neuroimaging techniques provide us with 'sharp' images of language at the neural level.

Finally, it is necessary as well to optimise the tools employed for analysing the molecular underpinnings of language disorders. Of course, the concerns raised above (in particular, they way in which clinical subjects are diagnosed) are also important at this level. However, these tools have different caveats and shortcomings. For example, approaches based on quantitative trait loci cannot properly detect highly polymorphic loci. As a consequence, it may be wrongly concluded that the disorders are caused by the mutation of a few principal genes. Similarly, positional cloning just renders statistical correlations between specific phenotypes and genes. Nonetheless, this needs to be validated in other populations and environments. Finally, genome-wide analysis (GWAs) allows for identifying candidate genes across the whole genome, but strong statistical corrections need to be implemented.

On the other hand, we also need to optimize current typologies of the disorders, both those based on symptoms and those based on their aetiology.

Concerning typologies based on symptoms, because disorders usually show a continuous distribution, it may be worth taking into account the severity of the symptoms (see Monfort and Monfort 2012 for a discussion). However, this may not be enough. Actually, we should expect that clinical categories still have different aetiologies. Moreover, some of them (or some of their subtypes) may be unreal if they merge units, levels, or operations of language. With regard to aetiological classifications, it has been suggested that clinical approaches to disorders can be substantially improved if different kinds of data are considered: genetic, neurobiological, cognitive, and even evolutionary. However, as we reviewed above, the same dysfunctional pieces may be shared across disorders that have distinctive symptomatic profiles.

## 3 Exploring new avenues

The strategies reviewed above will surely contribute to a better understanding and handling of language disorders. Nonetheless, they may not be enough. (Some kinds of) clinical linguistics still rely on naïve approaches to the biological underpinnings of language and language disorders. Hence, as we highlighted in the first section of the paper, gene mutations are expected to affect brain areas involved in language processing only, and ultimately, to give rise to linguistic deficits only (e.g. Falcaro et al. 2008). Similarly, language disorders are expected to be homogeneous categories (at all levels of analysis) across populations and throughout development. And this is not the case. In our opinion, we need an improved approach to language disorders in the spirit of the Biolinguistic turn in language sciences (see Boeckx and Benítez-Burraco 2014a for a review). Eventually, a change of focus (or a paradigm shift) may also be needed (see section 4).

At the very least, it is urgent to take both linguistics and biology seriously when analysing language disorders. On the Linguistics side, language disorders should be construed (and examined) in terms of the primitives (units and computations) that are central in current linguistic theories (of course, only of those that can be computed by the brain in real time). On the biology side, some key lessons about the way in which living beings are organized and develop should be taken into account. To begin with, genes are not blueprints. Non-genetic factors also play a key role in controlling development. At the same time, development (and this is particularly true of the brain) is not fully predetermined before birth, since it also depends on environmental factors. As a consequence, the phenotype is always indirectly related to the genotype.

Let us examine this problem in some detail. Genes just codify biochemical products (either proteins or non-coding RNAs [ncRNAs]) that perform specific functions inside or outside the cell. However, genes are not able to do this by themselves (not to mention to give rise to phenotypic traits!). Genes are transcribed into RNA and (some of them) are subsequently translated into proteins by complex biological machinery. In conjunction with gene regulatory regions, this machinery determines when, where, and how much a gene is expressed, and which functional products are going to be synthesised (several functional products can be synthesised from the same gene, which will affect different traits). We are just about getting a rough idea about the intricacy of this regulatory machinery. For example, we have recently learnt that gene expression heavily relies on ncRNAs and not only on DNA sequences and regulatory proteins (Mattick et al. 2009; Mattick 2011). Similarly, we have found that DNA is widely epigenetically modified, that is, it is modified to modulate how regulatory factors will interact with it. Importantly, these modifications are inheritable too (Isles and Wilkinson 2000), and they have been linked to basic brain processes (such as neural proliferation and differentiation, and particularly, to neural plasticity) and, eventually, to key cognitive abilities for language acquisition and processing, such as learning and memory (Levenson and Sweatt 2006; Gräff and Mansuy 2008; Mehler 2008). Of course, many internal (e.g. proteins, hormones, chemiotactic factors, etc.) and external (i.e. environmental cues) may affect the transcriptional (and epigenetic) state of a gene. Overall, we now believe that development (and ultimately, the emergence of pathological traits) depends more on the transcriptional state of the cell than on genetic sequences themselves (Mattick et al. 2009). A corolary is that we cannot go on regarding DNA mutations as the only (or even the major) aetiological factor of inherited language disorders.

Nonetheless, even if a gene is expressed in the proper place, time window and amount, a direct link with a particular phenotype is not granted. Gene products usually undergo posttranscriptional and/or posttranslational modifications, rendering different transcripts and/or diverse proteins or ncRNAs. Very frequently these molecules need to be assembled in multimolecular complexes. Importantly, gene products usually interact in the form of intricate regulatory networks (Geschwind and Konopka 2009). These complex interactions make the phenotype linked to the mutation of a particular gene pretty variable and hardly predictable. This explains why the mutation of one of these genes can give rise to different language and/or cognitive deficits and disorders, as we pointed out in section 1. Likewise, other diverse factors influence (the variability of) the trajectories ultimately followed by developmental processes. For example, viscoelasticity or differential diffusion and oscillation (acting in combination with basic properties of the cell like polarity or differential adhesion) modulate the way in which all the involved elements (proteins, ncRNAs, hormones, etc.) behave, interact, and function. This ultimately affects basic dimensions of tissue development and organization, such as regionalization patterns, and eventually, to phenotypic traits (Newman and Comper 1990; Goodwin 1994; Newman et al. 2006). Lastly, developmental processes are, to some extent, stochastic phenomena. This is why "identical developmental processes [and consequently, identical gene sequences] in identical environments produce different outcomes" (Balaban 2006: 320).

When it comes to the brain, it is important to notice that this complex regulatory mechanism does not give rise to neural devices that are fully operative. On the contrary, additional changes in neural architecture are needed.

They usually result from feedback effects from other brain regions or from external stimuli. Consequently, a direct link between language and the brain should never be expected. Actually, as we pointed out in the previous section, the neural devices emerging from development cannot be directly equated to (the neural substrate of) linguistic features or operations. On the contrary, "differently structured cortical areas are specialized for performing different types of computations, and [ . . . ] some of these computations are necessary for language but also for other cognitive functions" (Poeppel and Embick 2005: 112). This is why the impairment of any of these areas may affect more than one cognitive functions and ultimately, give rise to symptoms that are suggestive of more than one (comorbid) disorders. Incidentally, this disqualifies language from being a module in the Fodorian sense. On the contrary, language is a cross-modular cognitive function, resulting from the interface of diverse neuronal devices performing basic functions (Hauser et al. 2002; Balari and Lorenzo 2013; Boeckx and Benítez-Burraco 2014b). Such *cognitive modules* (as Griffiths 2007 calls them) are always the outcome of major changes in the brain architecture and function occurred during development under environmental cues, although their basic wiring is achieved before birth under genetic instructions (see Karmiloff-Smith 2010 for discussion). Consequently, we cannot go on construing disorders as static entities. On the contrary, we should expect that the phenotypic profile of the affected people (and the biological and cognitive machinery supporting their linguistic abilities) is different at different stages of development. As pointed out by Karmiloff-Smith (2009: 58): "to understand developmental outcomes, it is vital to identify full developmental trajectories, to assess how progressive change occurs from infancy onwards, and how parts of the developing system may interact with other parts differently at different times across ontogenesis". Moreover, similar cognitive profiles can rely on different brain architectures. As Karmiloff-Smith (2010: 182) puts it: "the same behaviour may be subserved by different neural substrates at different ages during development". Because there may be more than one way of implementing a (more or less) functional faculty of language at the term of growth (see Hancock and Bever 2013 for discussion), we (urgently) need a good developmental account of language disorders.

## 4 A new paradigm

The improved biological (or biolinguistic) account of language and of language disorders outlined in the previous section is more in line with how biologists think about development and evolution, how neuroscientists think about the brain and how psychologists think about cognitive development (and even how most linguists outside Chomskyan circles think about language). Nonetheless, further evidence suggests that we may actually need a new theoretical framework in clinical linguistics if we want to properly understand (and deal with) language disorders.

We do believe that developmental processes are the key to understanding what we observe in the adult state. As we noted in the previous section, variation pervades language (and language disorders) at all levels, from genes to molecules to brain networks to psycholinguistic measures. However, it is crucial to note that variation is quite constrained too. And the same holds for developmental disturbances. In truth, the developing brain is able to compensate many kind of damage, to the extent that quite preserved linguistic abilities can be achieved in spite of many kind of mutations, brain anomalies and severe cognitive impairments (Sirois et al. 2008). Interestingly, while some aspects of language are nearly never disturbed or are always compensated (for example, basic phrasal rules), others are impaired in many (if not all) disorders (for example, verbal inflection). Ultimately, the number of language disorders is far smaller than the number of aetiological factors involved. Moreover, we observe that although disorders show specific symptomatic profiles, their prevalent symptoms usually result from the impairment of low-level, more generalized processes. Actually, we find in them diffuse effects on brain architecture and function, and on different cognitive capacities (this being compatible with a greater impairment of certain functions). This is why disorders are better construed as the outcome of anomalous associations across domains (instead of the juxtaposition of impaired and preserved modules). Ultimately, as we have already noted, the linguistic profile of affected people changes from one group to another, and from one developmental stage to the next. Overall, quite preserved linguistic capacities can be achieved in spite of deeper cognitive impairments relying on different (and changing) brain architectures and cognitive abilities. At the same time, there are not so many ways of implementing language at the brain level.

Our main point is that this messy scenario (as we called it in section 1) is easier to interpret if we move to a new theoretical paradigm, namely, an evo-devo account of disorders, which builds on the evo-devo theories that interpret the deep links between development and evolution in biology. Actually, what we observe in language disorders (to a greater degree than in the normal population) is that language is both sensitive to environmental changes (that is, *plastic*) and resistant to environmental perturbations (that is, *canalized*), both prompted to evolve (that is, *evolvable*) and resistant to modification (that is, *robust*). Whenever canalization fails to cope with developmental perturbations (deleterious gene mutations, brain damage, and the like), certain cognitive deficits arise, certain linguistic abilities are not properly achieved and/or certain developmental milestones are not reached or its acquisition becomes delayed because they are achieved via compensatory mechanisms. Plausibly, these anomalies concern only neural networks that are endowed with less robust compensatory mechanisms because of their evolutionary novelty (Toro et al. 2010). Conversely, the components of language (genetic, physiological, or cognitive) that are more resistant to damage (and that are not affected in disorders) have a long evolutionary history. According to Gibson (2009), de-canalization explains the high prevalence of complex diseases (including language disorders) among human populations. We believe that specific mutations, demographic bottlenecks and cultural changes caused a phase transition from ape cognition to human cognition that prompted the emergence of language as a result of the interface among basic cognitive blocks which are particularly robust after millions of years of stabilizing selection (Balari and Lorenzo 2013; Boeckx and Benítez-Burraco 2014b). However, this transition uncovered cryptic variation which increased the prevalence of language disorders. Moreover, because of its evolutionary novelty and the decreased resilience of the networks they rely on, these new interfaces are very sensitive to damage. Plausibly, this explains why the same deficits are found in nearly all disorders and why they usually concern morpho-phonology and the most demanding tasks in computational terms (e.g. agreement).

Importantly, we also believe that the limited set of pathological conditions characterised by clinical linguists may be the only possible set of phenotypes resulting from the combination of the diverse factors regulating the development of the brain. In evo-devo theories, these limited set of phenotypes are usually characterised as restricted areas within the morpho-space or adaptive landscape of one species (McGhee 2006). We think that this fresh account of disorders may be of great interest for clinical linguistics. Accordingly, we should expect that each disorder is located in a different place of the language morpho-space (which also includes the language faculty of the non-affected population). What we need, then, is to find the best parameters defining the language morpho-space. For example, we might rely on (aberrant) gene expression profiles in the cell to define the stable states attracted through development (remember that we expect pathological instances to be also stable ontogenetic states, but endowed with idiosyncratic, less functional properties). Another promising possibility is the kind of networks resulting from the measurement of the syntactic relationships between words (or morphemes) in the utterances produced by speakers in real conversations. This approach accurately characterises language growth in the child as phase transitions in the syntactic complexity of her discourse. Different disorders are expected to show different, disorder-specific profiles in terms of the topographical features of these networks and the timing of the transitions (if any) between different kinds of networks, to the extent that they emerge as robust endophenotypes (or early clinical hallmarks) of the disorders (for details, see Barceló-Coblijn et al. 2015).

Nonetheless, a better candidate for properly defining the morpho-space of language growth in the species (either pathological or not) is brain rhythms. Brain rhythms are primitive components of brain function, and we expect them to be connected to some computational primitives of language, allowing us to understand (and not just to localize) brain functions. For example, basic operations in the minimalist account of language, like 'Spell Out' or 'Unify' (that is, the regulation of Merge by means of its interfacing with, or its embedding inside, the cognitive systems responsible for interpretation and externalization) (Jackendoff 2002; Hagoort 2005), can be interpreted as the embedding of high frequency oscillations inside oscillations operating at a slower frequency (see Benítez-Burraco and Boeckx 2014 for details). Similarly, some rhythmic features of speech are related to specific brain oscillations (Giraud and Poeppel 2012). Importantly, the hierarchy of brain oscillations has remained remarkably preserved within mammals during evolution (Buzsáki et al. 2013). Consequently, we should expect that the human-specific pattern of brain activity is a slight variation of the pattern observed in other primates. Interestingly, different cognitive disorders have probed to correlate with specific profiles of abnormal brain activity (Buzsáki and Watson 2012). We believe that these anomalous patterns may correspond to different points within the adaptive landscape of the language faculty. If we succeed in this translation, we may be able to diagnose language disorders earlier and in a more accurate way, because each disorder is expected to result from a selective, disorder-specific alteration of the same brain oscillation grammar. Importantly, also, these brain rhythms are expected to be highly quantifiable and heritable traits and thus, confident endophenotypes of the disorders.

## 5 Future prospects

The paradigm shift in clinical linguistics we advocate is not easy to achieve. If we really want to gain a better characterisation (and understanding) of language disorders and also to optimize our therapeutic tools, we need to improve our current understanding of the biological underpinnings of the language faculty (disordered or intact). In this we can rely on recent achievements of biolinguistics, which is progressively moving from a naïve account of the biology of language to more biologically grounded views of language facts (see Boeckx and Benítez Burraco 2014a for review). Concerning language disorders and the new account of these complex conditions we have argued for in this paper, we should persevere in several lines of research: (i) disentangle the molecular mechanisms that channel (and fail to channel) variation at all levels, (ii) improve evo-devo-friendly depictions of the modularization of the disordered brain; (iii) optimize current models of the linguistic ontogeny in people with disorders; and (iv) pay attention to emergent properties (and to properties that fail to emerge), since language is surely a complex system (Deacon 2005).

## References

Alarcón, M., Abrahams, B. S., Stone, J. L., Duvall, J. A., Perederiy, J. V., Bomar, J. M., Sebat, J., Wigler, M., Martin, C. L., Ledbetter, D. H., Nelson, S. F., Cantor, R. M., and Geschwind, D. H. (2008) Linkage, association, and gene-expression analyses identify *CNTNAP2* as an autism-susceptibility gene. *Am. J. Hum. Genet.* 82:150–159.

American Psychiatric Association (1994) *Diagnostic and Statistical Manual of Mental Disorders (DSM-IV)*. Washington, DC: American Psychiatric Association.

Bakkaloglu, B., O'Roak, B. J., Louvi, A., Gupta, A. R., Abelson, J. F., Morgan, T. M., Chawarska, K., Klin, A., Ercan-Sencicek, A. G., Stillman, A. A., Tanriover, G., Abrahams, B. S., Duvall, J. A., Robbins, E. M., Geschwind, D. H., Biederer, T., Gunel, M., Lifton, R. P., and State, M. W. (2008) Molecular cytogenetic analysis and resequencing of contactin associated protein-like 2 in autism spectrum disorders. *Am. J. Hum. Genet.* 82:165–173.

Balaban, E. (2006) Cognitive developmental biology: History, process and fortune's wheel. *Cognition* 101:298-332.

Balari, S., and Lorenzo, G. (2013) *Computational Phenotypes*. Oxford: Oxford University Press.

Barceló-Coblijn, L., Benítez-Burraco, A., and Irutzun, A. (2015) Syntactic networks as an endophenotype of developmental language disorders: an evo-devo approach to clinical linguistics. *Biolinguistics* 9:43-49.

Benítez-Burraco, A. (2010) Neurobiology and neurogenetics of dyslexia. *Neurología* 25:563-581.

Benítez-Burraco, A., and Boeckx, C. (2014) Universal Grammar and biological variation: An EvoDevo agenda for comparative biolinguistics. *Biol. Theor.* 9:122-134.

Bishop, D. V. M. (2002). Motor immaturity and specific speech and language impairment: Evidence for a common genetic basis. *Am. J. Med. Genet.* 114:56-63.

Bishop, D. V. M., North, D., and Donlan, C. (1996) Nonword repetition as a behavioural marker for inherited language impairment: Evidence from a twin study. *J. Child Psychol. Psychiatry* 36:1-13.

Boeckx, C., and Benítez-Burraco, A. (2014a) Biolinguistics 2.0. In *The Design, Development and Evolution of Human Language: Biolinguistic Explorations* (K. Fujita, N. Fukui, N. Yusa & M. Ike-Uchi, eds). Tokyo: Kaitakusha, pp. 8-30.

Boeckx C., and Benítez-Burraco A. (2014b) The shape of the human language-ready brain. *Front. Psychol.* 5:282.

Botting, N., and Conti-Ramsden, G. (2004) Characteristics of children with specific language impairment. In *Classification of Developmental Language Disorders* (L. Verhoeven and H. Van Balkom, eds.). Mahwah: Erlbaum, pp. 23-38.

Buzsáki, G., Logothetis, N., and Singer, W. (2013) Scaling brain size, keeping timing: Evolutionary preservation of brain rhythms. *Neuron* 80:751-764.

Buzsáki, G., and Watson, B. O. (2012) Brain rhythms and neural syntax: Implications for efficient coding of cognitive content and neuropsychiatric disease. *Dialogues Clin. Neurosci.* 14:345-367.

Catts, H. W., Adlof, S. M., Hogan, T. P., and Weismer, S. E. (2005) Are specific language impairment and dyslexia distinct disorders? *J. Speech Lang. Hear. Res.* 48:1378-1396.

Deacon, T. W. (2005) Language as an emergent function: Some radical neurological and evolutionary implications. *Theoria* 54:269-286.

Démonet, J.-F., Taylor, M. J., and Chaix, Y. (2004) Developmental dyslexia. *Lancet* 363:1451-1460.

Deutsch, G. K., Dougherty, R. F., Bammer, R., Siok, W. T., Gabrieli, J. D., and Wandell, B. (2005) Children's reading performance is correlated with white matter structure measured by diffusion tensor imaging. *Cortex* 41:354-363.

Donnai, D., and Karmiloff-Smith, A. (2000) Williams syndrome: From genotype through to the cognitive phenotype. *Am. J. Med. Genet.* 97:164-171.

Dricot, L., Sorger, B., Schiltz, C., Goebel, R., and Rossion, B. (2008) Evidence for individual face discrimination in non-face selective areas of the visual cortex in acquired prosopagnosia. *Behav. Neurol.* 19:75-79.

Falcaro, M., Pickles, A., Newbury, D. F., Addis, L., Banfield, E., Fisher, S. E., Monaco, A. P., Simkin, Z., Conti-Ramsden, G., and the SLI Consortium (2008) Genetic and phenotypic effects of phonological short-term memory and grammatical morphology in specific language impairment. *Genes Brain Behav.* 7:393-402.

Galaburda, A. M., Sherman, G. F., Rosen, G. D., Aboitiz, F., and Geschwind, N. (1985). Developmental dyslexia: Four consecutive patients with cortical anomalies. *Ann. Neurol.* 18:222–233.

Geschwind, D. H., and Konopka, G. (2009) Neuroscience in the era of functional genomics and systems biology. *Nature* 461:908–915.

Gibson, G. (2009) Decanalization and the origin of complex disease. *Nature Rev. Genet.* 10:134–140.

Giraud, A. L., and Poeppel, D. (2012) Cortical oscillations and speech processing: Emerging computational principles and operations. *Nat. Neurosci.* 15:511–517.

Goodwin, B. (1994) *How the Leopard Changed its Spots. The Evolution of Complexity*. New York: Charles Scribner's Sons.

Gräff, J., and Mansuy. I. M. (2008) Epigenetic codes in cognition and behaviour. *Behav. Brain Res.* 192:70–87.

Griffiths, P. E. (2007) Evo-Devo meets the mind: Towards a developmental evolutionary psychology. In *Integrating Evolution and Development: Form Theory to Practice* (R. Brandon and R. Sansom, eds.). Cambridge, MA: MIT Press, pp. 195–225.

Hagoort, P. (2005). On Broca, brain, and binding: A new framework. *Trends Cogn. Sci.* 9, 416–423.

Hancock, R., and Bever, T. G. (2013) Genetic factors and normal variation in the organization of language. *Biolinguistics* 7:75–95.

Hauser, M. D., Chomsky, N., and Fitch, W. T. (2002) The faculty of language: What is it, who has it, and how did it evolve? *Science* 298:1569–1579.

Horwitz, B., Rumsey, J. M., and Donohue, B. C. (1998) Functional connectivity of the angular gyrus in normal reading and dyslexia. *Proc. Natl. Acad. Sci. U.S.A* 95:8939–8944.

Isles, A. R., and Wilkinson, L. S. (2000) Imprinted genes, cognition and behaviour. *Trends Cogn. Sci.* 4:309–318.

Jackendoff, R. (2002) *Foundations of Language: Brain, Meaning, Grammar, Evolution*. Oxford: Oxford University Press.

Jarrold, C., Baddeley, A. D., and Hewes, A. K. (2000) Verbal short-term memory deficits in Down syndrome: A consequence of problems in rehearsal? *J. Child Psychol. Psychiatry* 41:233–244.

Karmiloff-Smith, A. (2008) Research into Williams syndrome: The state of the art. In *Handbook of Developmental Cognitive Neuroscience* (C. A. Nelson and M. Luciana, eds.). Cambridge, MA: MIT Press, pp. 691–700.

Karmiloff-Smith, A. (2009) Nativism versus neuroconstructivism: Rethinking the study of developmental disorders. *Dev. Psychol.* 45:56–63.

Karmiloff-Smith, A. (2010) A developmental perspective on modularity. In *Towards a Theory of Thinking* (B. M. Glatzeder, V. Goel, and A. von Müller, eds.). Berlin: Springer-Verlag, pp. 179–187.

Karmiloff-Smith, A., and Mills, D. L. (2006) Williams Syndrome. In *Encyclopedia of Language and Linguistics* (K. Brown, ed.). Vol. 13. Oxford: Elsevier, pp. 585–589.

Levenson, J. M., and Sweatt, J. D. (2006) Epigenetic mechanisms: A common theme in vertebrate and invertebrate memory formation. *Cell. Mol. Life Sci.* 63:1009–1016.

Livingstone, M. S., Rosen, G. D., Drislane, F. W., and Galaburda, A. M. (1991) Physiological and anatomical evidence for a magnocellular defect in developmental dyslexia. *Proc. Natl. Acad. Sci. U.S.A.* 88:7943–7947.

Lovegrove, W. J., Bowling, A., Badcock, D., and Blackwood, M. (1980) Specific reading disability: Differences in contrast sensitivity as a function of spatial frequency. *Science* 210:439–440.

Lyon, G., Shaywitz, S. y Shaywitz, B. (2003) A definition of dyslexia. *Ann. Dyslexia* 53:1–14.

Maisog, J. M., Einbinder, E. R., Flowers, D. L., Turkeltaub, P. E., and Eden, G. F. (2008) A meta-analysis of functional neuroimaging studies of dyslexia. *Ann. N. Y. Acad. Sci.* 1145:237–259.

Mattick, J. S. (2011) The central role of RNA in human development and cognition. *FEBS Lett.* 585:1600–1616.

Mattick, J. S., Taft, R. J., and Faulkner, G. J. (2009) A global view of genomic information: Moving beyond the gene and the master regulator. *Trends Genet.* 26:21–28.

Mayringer, H., and Wimmer, H. (2000) Pseudoname learning by German-speaking children with dyslexia: Evidence for a phonological learning deficit. *J. Exp. Child Psychol.* 75:116–133.

McGhee, G. R. (2006) *The Geometry of Evolution, Adaptive Landscapes and Theoretical Morphospaces*. Cambridge, UK: Cambridge University Press.

Mehler, M. F. (2008) Epigenetic principles and mechanisms underlying nervous system functions in health and disease. *Prog. Neurobiol.* 86:305–341.

Mervis, C. B., and Becerra, A. M. (2007) Language and communicative development in Williams syndrome. *Ment. Retard. Dev. Disabil. Res. Rev.* 13:3–15.

Monfort, I., and Monfort, M. (2012) Utilidad clínica de las clasificaciones de los trastornos del desarrollo del lenguaje. *Rev. Neurol.* 54: S147–S154.

Newman, S. A., and Comper, W. D. (1990) 'Generic' physical mechanisms of morphogenesis and pattern formation. *Development* 110:1–18.

Newman, S. A., Forgacs, G., and Müller, G. B. (2006) Before programs: the physical origination of multicellular forms. *Int. J. Dev. Biol.* 50:289–299.

Newmeyer, F. J. (1997) Genetic dysphasia and linguistic theory. *J. Neurolinguistics* 10:47–73.

Nicolson, R. I., and Fawcett, A. J. (2006) Do cerebellar deficits underlie phonological problems in dyslexia? *Dev. Sci.* 9:259–262.

O'Malley, K. D., and Nanson, J. (2002) Clinical implications of a link between fetal alcohol spectrum disorder and attention-deficit hyperactivity disorder. *Can. J. Psychiatry* 47:349–354.

Parisse, C., and Maillart, C. (2009) Specific language impairment as systemic developmental disorders. *J. Neurolinguistics* 22:109–122.

Paulesu, E., Demonet, J.-F., Fazio, F., McCrory, E., Chanoine, V., Brunswick, N., Cappa, S. F., Cossu, G., Habib, M., Frith, C. D., and Frith, U. (2001) Dyslexia: Cultural diversity and biological unity. *Science* 291:2165–2167.

Petrin, A. L., Giacheti, C. M., Maximino, L. P., Abramides, D. V., Zanchetta, S., Rossi, N. F., Richieri-Costa, A., and Murray, J. C. (2010) Identification of a microdeletion at the 7q33-q35 disrupting the CNTNAP2 gene in a Brazilian stuttering case. *Am. J. Med. Genet. A* 152A:3164–3172.

Poeppel, D. (2012) The maps problem and the mapping problem: Two challenges for a cognitive neuroscience of speech and language. *Cogn. Neuropsychol.* 29:34–55.

Poeppel, D., and Embick, D. (2005) Defining the relation between linguistics and neuroscience. In *Twenty-first Century Psycholinguistics: Four Cornerstones* (A. Cutler, ed.). Hillsdale: Lawrence Erlbaum, pp. 103–120.

Purvis, K. L., and Tannock, R. (1997) Language abilities in children with attention deficit hyperactivity disorder, reading disabilities, and normal controls. *J. Abnorm. Child Psychol.* 25:133–144.

Quaglino, V., Bourdin, B., Czternasty, G., Vrignaud, P., Fall, S., Meyer, M. E., Berquin, P., Devauchelle, B., and de Marco, G. (2008) Differences in effective connectivity between dyslexic children and normal readers during a pseudoword reading task: An fMRI study. *Neurophysiol. Clin.* 38:73–82.

Rapin, I., and Allen, D. A. (1983) Developmental language disorders: Nosologic considerations. In *Neuropsychology of Language, Reading and Spelling* (U. Kirk, ed.). New York: Academic Press, pp. 155–184.

Rapport, M. D., Alderson, R. M., Kofler, M. J., Sarver, D. E., Bolden, J., and Sims, V. (2008) Working memory deficits in boys with attention-deficit/hyperactivity disorder (ADHD): the contribution of central executive and subsystem processes. *J. Abnorm. Child Psychol.* 36:825–837.

Rhodes, S. M., Riby, D. M., Matthews, K., and Coghill, D. R. (2011) Attention-deficit/hyperactivity disorder and Williams syndrome: Shared behavioral and neuropsychological profiles. *J. Clin. Exp. Neuropsychol.* 33:147–156.

Roll, P., Rudolf, G., Pereira, S., Royer, B., Scheffer, I. E., Massacrier, A., Valenti, M. P., Roeckel-Trevisiol, N., Jamali, S., Beclin, C., Seegmuller, C., Metz-Lutz, M. N., Lemainque, A., Delepine, M., Caloustian, C., de Saint Martin, A., Bruneau, N., Depétris, D., Mattéi, M. G., Flori, E., Robaglia-Schlupp, A., Lévy, N., Neubauer, B. A., Ravid, R., Marescaux, C., Berkovic, S. F., Hirsch, E., Lathrop, M., Cau, P., and Szepetowski, P. (2006) *SRPX2* mutations in disorders of language cortex and cognition. *Hum. Mol. Genet.* 15:1195–1207.

Roll, P., Vernes, S. C., Bruneau, N., Cillario, J., Ponsole-Lenfant, M., Massacrier, A., Rudolf, G., Khalife, M., Hirsch, E., Fisher, S. E., and Szepetowski, P. (2010) Molecular networks implicated in speech-related disorders: FOXP2 regulates the SRPX2/uPAR complex. *Hum. Mol. Genet.* 19:4848–4860.

Sehested, L. T., Møller, R. S., Bache, I., Andersen, N. B., Ullmann, R., Tommerup, N., and Tümer, Z. (2010) Deletion of 7q34-q36.2 in two siblings with mental retardation, language delay, primary amenorrhea, and dysmorphic features. *Am. J. Med. Genet. A* 152A:3115–3119.

Serniclaes, W., Van Heghe, S., Mousty, P., Carré, R., and Sprenger-Charolles, L. (2004) Allophonic mode of speech perception in dyslexia. *Journal of Experimental Child Psychology* 87: 336–361.

Shaywitz, S. E. (1998) Dyslexia. *N. Engl. J. Med.* 338:307–312.

Shaywitz, S., Morris, R., and Shaywitz, B. (2008) The education of dyslexic children from childhood to young adulthood. *Annu. Rev. Psychol.* 59, 451–475.

Shaywitz, S. E., Shaywitz, B. A., Pugh, K. R., Fulbright, R. K., Constable, R. T., Mencl, W. E., Shankweiler, D. P., Liberman, A. M., Skudlarski, P., Fletcher, J. M., Katz, L., Marchione, K. E., Lacadie, C., Gatenby, C., and Gore, J. C. (1998) Functional disruption in the organization of the brain for reading in dyslexia. *Proc. Nat. Acad. Sci. U.S.A.* 95:2636–2641.

Shriberg, L. D., Ballard, K. J., Tomblin, J. B., Duffy, J. R., Odell, K. H., and Williams, C. A. (2006) Speech, prosody, and voice characteristics of a mother and daughter with a 7,13 translocation affecting *FOXP2*. *J. Speech Lang. Hear. Res.* 49:500–525.

Shriberg, L. D., Tomblin, J. B., and McSweeny, J. L. (1999) Prevalence of speech delay in 6-year-old children and comorbidity with language impairment. *J. Speech Lang. Hear. Res.* 42:1461–1481.

Sirois, S., Spratling, M., Thomas, M. S., Westermann, G., Mareschal, D., and Johnson, M. H. (2008) Précis of neuroconstructivism: How the brain constructs cognition. *Behav. Brain Sci.* 31:321–331.

Smith, S. D., Gilger, J. W., and Pennington, B. F. (1996) Dyslexia and other specific learning disorders. In *Principles and practice of medical genetics* (D. L. Rimoin, J. M. Connor and R. E. Pyeritz, eds.). New York: Churchill Livingstone, pp. 1767–1789.

Sorger, B., Goebel, R., Schiltz, C., and Rossion, B. (2007) Understanding the functional neuroanatomy of acquired prosopagnosia. *Neuroimage* 35:836–852.

Stein, J., and Walsh, V. (1997) To see but not to read: The magnocellular theory of dyslexia. *Trends Neurosci.* 20:147–152.

Stein, C. M., Schick, J. H., Taylor, H. G., Shriberg, L. D., Millard, C., Kundtz-Kluge, A., Russo, K., Minich, N., Hansen, A., Freebairn, L. A., Elston, R. C., Lewis, B. A., and Iyengar, S. K. (2004) Pleiotropic effects of a chromosome 3 locus on speech-sound disorder and reading. *American Journal of Human Genetics* 74: 283–297.

Temple, E., Poldrack, R. A., Protopapas, A., Nagarajan, S., Salz, T., Tallal, P., Merzenich, M. M., and Gabrieli, J. D. E. (2000) Disruption of the neural response to rapid acoustic stimuli in dyslexia: Evidence from functional MRI. *Proc. Nat. Acad. Sci. U.S.A.* 97:13907–13912.

Thomson, B., Crewther, D. P., and Crewther, S. G. (2006) Wots that werd? Pseudo-words (non-words) may be a misleading measure of phonological skills in young learner readers. *Dyslexia* 12:289–299.

Toga, A. W., Thompson, P. M., and Sowell, E. R. (2006) Mapping brain maturation. *Trends Neurosci.* 29:148–159.

Toro, R., Konyukh, M., Delorme, R., Leblond, C., Chaste, P., Fauchereau, F., Coleman, M., Leboyer, M., Gillberg, C., and Bourgeron, T. (2010) Key role for gene dosage and synaptic homeostasis in autism spectrum disorders. *Trends Genet.* 26:363–372.

Vargha-Khadem, F., Gadian, D. G., Copp, A., and Mishkin, M. (2005) *FOXP2* and the neuroanatomy of speech and language. *Nat. Rev. Neurosci.* 6:131–138.

Vargha-Khadem, F., Watkins, K. E., Alcock, K. J., Fletcher, P., and Passingham, R. E. (1995) Praxic and nonverbal cognitive deficits in a large family with a genetically transmitted speech and language disorder. *Proc Natl Acad Sci USA* 92:930–933.

Vernes, S. C., Newbury, D. F., Abrahams, B. S., Winchester, L., Nicod, J., Groszer, M., Alarcón, M., Oliver, P. L., Davies, K. E., Geschwind, D. H., Monaco, A. P., and Fisher, S. E. (2008) A functional genetic link between distinct developmental language disorders. *N. Engl. J. Med.* 359:2337–2345.

Watkins, K. E., Dronkers, N. F., and Vargha-Khadem, F. (2002) Behavioural analysis of an inherited speech and language disorder: Comparison with acquired aphasia. *Brain* 125:452–464.

Whalley, H. C., O'Connell, G., Sussmann, J. E., Peel, A., Stanfield, A. C., Hayiou-Thomas, M. E., Johnstone, E. C., Lawrie, S. M., McIntosh, A. M., and Hall, J. (2011) Genetic variation in *CNTNAP2* alters brain function during linguistic processing in healthy individuals. *Am. J. Med. Genet. B Neuropsychiatr. Genet.* 156:941–948.

Whitehouse, A. J., Bishop, D. V., Ang, Q. W., Pennell, C. E., and Fisher, S. E. (2011) *CNTNAP2* variants affect early language development in the general population. *Genes Brain Behav.* 10:451–456.

Wyper, K. R., and Rasmussen, C. R. (2011) Language impairments in children with fetal alcohol spectrum disorders. *J. Popul. Ther. Clin. Pharmacol.* 18: e364–376.

# Index

action grammar 148, 225
adaptive landscape, language disorders as areas within 265
agent 204–5
Agree 30–1, 42n7, 42n12, 43n26, 47–51, 53–5, 189; see also agreement
agreement 12, 16, 20, 22–4, 31–2, 36–7; attraction 218–20
Allen, C. L. 203–4
Aristotle 1
attention 104, 107–8, 114; focus and fringe 116–20

BA 44 224–6; see also Broca's area
BA 45 224–6; see also Broca's area
Baars, B. J. 106
Baddeley, A. 103–5
basal ganglia 231
Bind 39; see also binding
binding 30, 39, 41n3, 43n26
biolinguistics 1, 153–4
Bloom, P. 198
Boeckx, C. 118–19
Borer-Chomsky conjecture 129–31, 136
brain 163–4; and language disorders 263
broad syntax 209n5; see also narrow syntax
Broca's aphasics 143
Broca's area 109, 121–2, 223–5

case: abstract 199; alternation 59; feature 51–4
Case Filter 198; see also case
causation 174–5, 177–8, 181–2
central executive 104–5, 108
cerebellum 231
chain 39–40
chain-formation 30–2, 39; see also chain

CHILDES 74–5
Chomsky, N. 1, 9–15, 17–18, 21, 25n2, 25n3, 26n6, 33, 36, 69–70, 74, 115, 119, 129, 132–3, 144, 147, 153, 158–9, 165, 198, 201–2, 207, 209
chunk 105, 112–14, 119–22
CI Interface 200
classical ethology 180
cognitive: control 107–9, 112–14, 120–1; cycle 107–9, 116–20
comparative method 157
complexity 161
computational phenotype 164
conjunction reduction 203
convergent evolution 190
Corballis, M. 145
covariance relation 48–50, 52, 54
Crain, S. 220
cross-modularity 163

Darwin, C. 198
dependent-marking language 52–3
Depth 17–18, 38–41
descent with modification 2–3, 149
development 154–5, 159–61
Developmental Systems Theory (DST) 161
displacement 201, 208
domain specificity 154
Döring, P. 85–91
D-to-C movement 133–4, 136, 137
dual causation 181; see also causation
duality of semantics 202
Dynamic Symmetrization Condition (DSC) 19–21, 26

edge 54–6, 59–60
E-language 159–61
EM see External Merge

encoding (working memory operation) 110–12
endocentricity 9–13, 20, 24–5
endophenotypes of language disorders 265–6; syntactic networks as 265
English 53–5
environment 160, 162–4
episodic buffer 105
EPP 199
Equilibrium Intactness Condition (EIC) 20–1, 23–4, 26n6, 26n7, 26n8
evidence: archaeological/paleoanthropological 190–1; genetic 191–2
Evo-Devo 154–5, 159, 161; and language disorders 264
evolutionary biology 154–5
evolutionary linguistics 141, 146
evolvability 145
exaptation 142, 148
exocentricity 11, 20, 24
exoskeletality 118
expectation 85–91
experiencer 204–5
extended synthesis 150, 160; see also modern synthesis
externalization 130–1, 134, 136, 137, 142, 164, 200
External Merge 11–13, 15, 17, 19–21, 24, 33–5, 208, 209n5; see also Merge; Internal Merge

faculty of language 69; in the broad sense (FLB) 145, 154, 156–9; development of 261–4; in the narrow sense (FLN) 145, 154–9
fallacy: of communication 141; of continuity 146; of a single origin 145
featural symmetry 16–17, 19
feature-chain 36–7; see also agreement
feature-equilibrium 16, 18–19
feature Merge 201; see also Merge
feature selection 201
Fitch, W. T. 157–8
fMRI 223
formal feature 51–2
FP see functional projection
Frank, S. 217
Friederici, A. 224
Fujita, K. 226n5
functional projection 201

Galilean style 178
Galileo's science of motion 173–5, 179–80

generative linguistics 153
genes 159–63, 165
genetic blueprint 154
geneticism 162
genocentrism 161, 166
genotype 162–3, 165
German 85, 87–9
global broadcast 108, 116–18
global workspace 106–8, 114, 116–18, 121
globularity 164
goal 204–5
grammar 119–22
grammaticalization 161
Greenfield, P. M. 148

Haeberli, E. 199
Hauser, M. D. 158
head-marking languages 52–3
head-movement 13–15, 18–19, 25n4
Hindi 89–90
hippocampus 231
Huanglong 191
Husain, S. 89–90

I-language 159–61
IM see Internal Merge
inheritance 189
inhibition (working memory operation) 104, 108
internalization 142
Internal Merge 11–13, 15, 18–21, 24, 33–4, 40–1, 208, 209n5; see also Merge; External Merge
inverted-Y model 115–16

Jackendoff, R. 143, 158
Japanese 22–4, 49, 52–4, 56–9, 61–2
Jebel Faya 191
Jonides, J. 110–11

Keller, F. 87–91
Kisongo Maasai 133, 135–7
Konieczny, L. 85–91
Kuroda, S.-Y. 22

Label 38, 115, 117–20; see also labeling
labeling 10, 32–3, 37–9, 43n18, 51, 53, 60, 62n5, 189
language disorders: comorbidity between 256–7, 259, 263; de-canalization and 264–5; developmental trajectories of 256–7, 263; diagnostic tests for 259–60;

genotype-phenotype correlations in 261–2; heterogeneity of 256–7, 259; molecular analysis of 260; neuroimaging techniques applied to 260; phenocopy in 258; prevalence of 265; typologies of 260–1; variability of 256–7, 259, 264
language of communication 198, 200, 201, 208; see also language of thought
language of thought 198, 200, 201, 208
Lenneberg, E. 153–4, 158–9, 163, 165, 229, 245
Levy, R. 87–91
lexicon 34–5, 44n28
Linguistic Niche Hypothesis 165
linguistic phenotype 154, 164
LoC see language of communication
locality 86–91
lone mutant 142
LoT see language of thought
Lunadong 191

magical number four 105, 110
Marantz, A. 199
Mayr, E. 155
McElree, B. 110, 120–1
mechanical philosophy 174–6, 179, 181
mechanism 174, 179–80
memory: declarative 112–14; long-term 104–5, 110–14, 116–19; procedural 112–14, 121; short-term 103–5, 108–12, 120; see also working memory
Merge 2–3, 10–15, 19–21, 23, 25, 30–1, 33–5, 40–1, 43–4n28, 47–52, 54–5, 69, 71, 115–22, 142–3; domain-general 225; domain-specific 225; motor control origin of 149; see also External Merge; Internal Merge
Merge-only: evolution 149; hypothesis 150, 189–90
$\text{Merge}_0$ see 0-Merge
$\text{Merge}_0 \circ \text{Search}_0$ 31–41
mildly context-sensitive grammar 119
minimalist program 69, 189
minimality 38–41
minimal search 70–1, 78, 132–5
modern synthesis 150, 155, 162
modularity of language 263
Mohawk 52–3
Morgan, C. L. 146; see also Morgan's Canon

Morgan's Canon 147
Moro, A. 25, 25n5, 221
mother 36–7
Multiple Demand network 108–9, 112–14, 121
multiple occurrences of identical case 53, 55
Musso, M. 221
mutation rate 192
$M_0 \circ S_0$ see $\text{Merge}_0 \circ \text{Search}_0$

narrow syntax 209n5; see also broad syntax
nativism 162
negative inversion 222
Newtonian dynamics 175–6
nominative-genitive conversion (NGC) 56–7, 59–61, 63n11
non-local dependency 119–20
No-tampering Condition (NTC) 14
novelty 154–6, 159, 164

Oberauer, K. 116
Occam's razor 129
occurrence 35–7
Oman 190
oscillation 232, 241; see also cognitive cycle
oscillopathy 241; language disorders as 265–6
Out-of-Africa dispersal 190

parameter 128–9, 131
patient 204–5
penetrance of mutations in language genes 258
Pesetsky, D. 199
phase 48–9, 54–7
phenotype 155, 159–60, 162, 164
phenotypic: plasticity 162–3; space 164
$\phi$-feature 12, 17, 19, 21–4, 26n6, 26n7, 47–52
Phillips, C. 119
philosophical grammar 176–7
phonological loop 103, 105, 109
Pinker, S. 158, 198
Poeppel, D. 114–15, 163
polymorphisms of language genes 258
predicate-internal subject hypothesis 70–1, 78–9
prefrontal cortex 108–9, 112–13, 121
prefrontal-thalamic-basal ganglia network 112, 121
principle of gradation 224

protolanguage 143, 149
pulvinar 234

quirky subject 199

recipient 204–5
recursion 145, 156–7
refreshing (working memory operation) 105, 108
rule-ordering 132, 136

scenario of the emergence of Merge/UG 192
Search 30–3, 50–1
Search0 see 0-Search
self-organization 202
self-paced reading 91–9
SEM 10, 20, 30–2, 201, 208
sign language 160
Sigurðsson, H. Á. 203
Single-Cycle Hypothesis 209n3
single origin hypothesis 189
sister 40–1
SM interface 200
Smith, N. 220
spell-out 200
structural linguistics 177
structural prominence 17–18, 26n5, 38, 40; see also Depth
structure dependence 70–1, 73–4, 220–2
subject-auxiliary inversion 70–2
subject prominent language 206; see also topic prominent language
sustainment (working memory operation) 108, 110–12

thalamus 231–45
thematic hierarchy 204

theme 204–5
third factor 69, 71, 129–32, 134–7
Tinbergen, N. 180–2
topic prominent language 206
Torrego, E. 199
Transfer 59–61, 189, 201
Typing Mismatch Effect 94–5

UG see Universal Grammar
Ullman, M. T. 112
Unbounded Merge see Merge
underspecification 132–6
Universal Grammar 3, 69, 115, 221
updating (working memory operation) 105, 110

variation 129–32, 134–5
Vasishth, S. 89–90
verbal rehearsal 103, 105
V-final order 207
virtual conceptual necessity 209n3
Visibility: Condition 199; Requirement 205
visuospatial sketchpad 103–4, 109

working memory: the multicomponent model 103–6; the global workspace model (see global workspace)
workspace 35–41; see also global workspace
WS see workspace

Yusa, N. 222

0-Merge 31–5, 50–2
0-Search 31–41, 50–2
Zhirendong 191